Enzymes in Fruit and Vegetable Processing

Chemistry and Engineering Applications

Enzymes in Fruit and Vegetable Processing

Chemistry and Engineering Applications

edited by
Alev Bayındırlı

CRC Press
Taylor & Francis Group
Boca Raton   London   New York

CRC Press is an imprint of the
Taylor & Francis Group, an **informa** business

CRC Press
Taylor & Francis Group
6000 Broken Sound Parkway NW, Suite 300
Boca Raton, FL 33487-2742

First issued in paperback 2019

ISBN-13: 978-1-4200-9433-6 (hbk)
ISBN-13: 978-0-367-38412-8 (pbk)

Library of Congress Cataloging-in-Publication Data

Enzymes in fruit and vegetable processing : chemistry and engineering applications /
editor, Alev Bayindirli.
p. cm.
Includes bibliographical references and index.
ISBN 978-1-4200-9433-6 (hardcover : alk. paper)
1. Fruit--Processing. 2. Vegetables--Processing. 3. Enzymes--Biotechnology. I. Bayindirli, Alev. II. Title.

TP440.E69 2010
664'.8--dc22

2010009982

Visit the Taylor & Francis Web site at
http://www.taylorandfrancis.com

and the CRC Press Web site at
http://www.crcpress.com

Contents

Preface

Fruits and vegetables are consumed as fresh or processed into different type of products. Some of the naturally occurring enzymes in fruits and vegetables have undesirable effects on product quality, and therefore enzyme inactivation is required during fruit and vegetable processing in order to increase the product shelf-life. Commercial enzyme preparations are also used as processing aids in fruit and vegetable processing to improve the process efficiency and product quality, because enzymes show activity on specific substrates under mild processing conditions. Therefore, there has been a striking growth in the enzyme market for the fruit and vegetable industry.

While fruit and vegetable processing is the subject of many books and other publications, the purpose of this book is to give detailed information about enzymes in fruit and vegetable processing from chemistry to engineering applications. Chapters are well written by an authoritative author(s) and follow a consistent style. There are 12 chapters in this book, and the chapters provide a comprehensive review of the chapter title important to the field of enzymes and fruit and vegetable processing by focusing on the most promising new international research developments and their current and potential industry applications. Fundamental aspects of enzymes are given in Chapter 1. Color, flavor, and texture are important post-harvest quality parameters of fruits and vegetables. There are a number of product-specific details, dependent on the morphology, composition, and character of the individual produce. Chapters 2, 3, and 4 describe in detail the effect of enzymes on color, flavor, and texture of selected fruits and vegetables. Selection of the indicator enzyme for blanching of vegetables is summarized in Chapter 5. For enzymes as processing aids, Chapter 6 describes in detail the enzymatic peeling of citrus fruits and Chapter 7 presents the importance of enzymes for juice production from pome, stone, and berry fruits. Inactivation of enzymes is required to obtain cloudy juice from citrus fruits. Chapter 8 is related to citrus juices; orange juice receives particular attention. Enzymes also play an important role in winemaking. The application of industrial enzyme preparations in the wine industry is a common practice. The use of enzymes for wine production is the focus of Chapter 9. Chapter 10 provides serious review of the inactivation effect of novel technologies on fruit and vegetable enzymes to maximize product quality. Chapter 11 presents both chemical and technological information on enzyme-based biosensors for fruit and vegetable processing. The literature reported in each chapter highlights the current status of knowledge in the related area. Future trends for industrial use of enzymes are discussed in Chapter 12. The conclusion part of each chapter also presents the reader with potential research possibilities and applications.

This book is a reference book to search or learn more about fruit and vegetable enzymes and enzyme-based processing of fruit and vegetables according to the latest enzyme-assisted technologies and potential applications of new approaches obtained from university and other research centers and laboratories. Such knowledge is important for the companies dealing with fruit and vegetable processing to be competitive and also for the collaboration among industry, university, and research centers.

This book is also for the graduate students and young researchers who will play an important role for future perspectives of enzymes in fruit and vegetable processing.

Alev Bayındırlı

The Editor

Alev Bayındırlı is a professor in the Department of Food Engineering, Middle East Technical University, Ankara, Turkey. She has authored or co-authored 30 journal articles. She received a BS degree (1982) from the Department of Chemical Engineering, Middle East Technical University. MS (1985) and PhD (1989) degrees are from the Department of Food Engineering, Middle East Technical University. She is working on food chemistry and technology, especially enzymes in fruit and vegetable processing.

List of Contributors

J. Brian Adams
Formerly of Campden &
 Chorleywood Food Research
 Association (Campden BRI)
Chipping Campden
Gloucestershire, UK

Domingos P.F. Almeida
Faculdade de Ciências
Universidade do Porto
Porto, Portugal
Escola Superior de
 Biotecnologia Universidade
 Católica Portuguesa
Porto, Portugal

Ana Belén Bautista-Ortín
Departamento de Tecnología
 de Alimentos, Nutrición y
 Bromatología
Facultad de Veterinaria,
 Universidad de Murcia
Campus de Espinardo
Murcia, Spain

Alev Bayındırlı
Middle East Technical University
Department of Food Engineering
Ankara, Turkey

Inmaculada Romero-Cascales
Departamento de Tecnología
 de Alimentos, Nutrición y
 Bromatología
Facultad de Veterinaria,
 Universidad de Murcia
Campus de Espinardo
Murcia, Spain

Domenico Castaldo
Stazione Sperimentale per le
 Industrie delle Essenze e dei
 Derivati dagli Agrumi (SSEA)
Reggio Calabria, Italy

Domenico Cautela
Stazione Sperimentale per le
 Industrie delle Essenze e dei
 Derivati dagli Agrumi (SSEA)
Reggio Calabria, Italy

Liliana N. Ceci
Planta Piloto de Ingeniería Química
UNS-CONICET
Bahía Blanca, Argentina

Isabel Egea
Departamento Biología del Estrés
 y Patología Vegetal
Centro de Edafología y Biología
 Aplicada del Segura-CSIC
Espinardo, Murcia, Spain

Danielle Cristhina Melo Ferreira
Institute of Chemistry
Unicamp
Campinas, São Paulo, Brazil

Alfonso Giovane
Dipartimento di Biochimica e
  Biofisica
Seconda Università degli Studi di
  Napoli
Napoli, Italy

Vural Gökmen
Department of Food Engineering
Hacettepe University
Ankara, Turkey

Encarna Gómez-Plaza
Departamento de Tecnología
  de Alimentos, Nutrición y
  Bromatología
Facultad de Veterinaria,
  Universidad de Murcia
Campus de Espinardo
Murcia, Spain

Luis F. Goulao
Secção de Horticultura
Instituto Superior de Agronomia
Lisbon, Portugal
Centro de Ecofisiologia, Bioquimica
  e Biotecnologia Vegetal
Instituto de Investigação
  Cientifica Tropical
Oeiras, Portugal

Lauro Tatsuo Kubota
Institute of Chemistry
Unicamp
Campinas, São Paulo, Brazil

Jorge E. Lozano
Planta Piloto de Ingeniería Química
UNS-CONICET
Bahía Blanca, Argentina

Lucilene Dornelles Mello
UNIPAMPA
Campus Bagé
Bagé, RS, Brazil

Renata Kelly Mendes
Institute of Chemistry
Unicamp
Campinas, São Paulo, Brazil

Indrawati Oey
Department of Food Science
University of Otago
Dunedin, New Zealand

Cristina M. Oliveira
Secção de Horticultura
Instituto Superior de Agronomia
Lisbon, Portugal

Lucie Pařenicová
DSM Biotechnology Centre, DSM
Delft, The Netherlands

Maria Teresa Pretel
Escuela Politécnica Superior de
  Orihuela
Universidad Miguel Hernández
Alicante, Spain

Felix Romojaro
Departamento Biología del Estrés
  y Patología Vegetal
Centro de Edafología y Biología
  Aplicada del Segura-CSIC
Espinardo, Murcia, Spain

Johannes A. Roubos
DSM Biotechnology Centre, DSM
Delft, The Netherlands

Paloma Sánchez-Bel
Departamento Biología del Estrés
 y Patología Vegetal
Centro de Edafología y Biología
 Aplicada del Segura-CSIC
Espinardo, Murcia, Spain

Luigi Servillo
Dipartimento di Biochimica e
 Biofisica
Seconda Università degli Studi di
 Napoli
Napoli, Italy

Jun Song
Agriculture and Agri-Food Canada
Atlantic Food and Horticulture
 Research Centre
Nova Scotia, Canada

Marco A. van den Berg
DSM Biotechnology Centre, DSM
Delft, The Netherlands

chapter one

Introduction to enzymes

Alev Bayındırlı

Contents

1.1 Nature of enzymes

Enzymes are effective protein catalysts for biochemical reactions. The structural components of proteins are L-α-amino acids with the exception of glycine, which is not chiral. The four levels of protein structure are primary, secondary, tertiary, and quaternary structures. Primary structure is related to the amino acid sequence. The amino group of one amino acid is joined to the carboxyl group of the next amino acid by covalent bonding, known as a peptide bond. The amino acid side-chain groups vary in terms of their properties such as polarity, charge, and size. The polar amino acid side groups tend to be on the outside of the protein where they interact with water, whereas the hydrophobic groups tend to be in the interior part of the protein. Secondary structure (α-helix, β-pleated sheet, and turns) is important for protein conformation. Right-handed α-helix is a regular arrangement of the polypeptide backbone by hydrogen bonding between the carbonyl oxygen of one residue (i) and the nitrogenous proton of the other residue (i+4). β-pleated sheet is a pleated structure composed of polypeptide chains linked together through interamide hydrogen bonding between adjacent strands of the sheet. Tertiary structure refers to the three-dimensional structure of folded protein. Presence of disulfide bridges, hydrogen bonding, ionic bonding, and hydrophobic and van der Waals interactions maintain the protein conformation. Folding the protein brings together amino acid side groups from different parts of the amino acid sequence of the polypeptide chain to form the enzyme active site that consists of a few amino acid residues and occupies a relatively small portion of the total enzyme volume. The rest of the enzyme is important for the three-dimensional integrity. The quaternary structure of a protein results from the association of two or more polypeptide chains (subunits).

Specificity and catalytic power are two characteristics of an enzyme. Most enzymes can be extremely specific for their substrates and catalyze reactions under mild conditions by lowering the free energy requirement of the transition state without altering the equilibrium condition. The enzyme specificity depends on the conformation of the active site. The enzyme-substrate binding is generally explained by lock-and-key model (conformational perfect fit) or induced fit model (enzyme conformation change such as closing around the substrate). The lock-and-key model has been modified due to the flexibility of enzymes in solution. The binding of the substrate to the enzyme results in a distortion of the substrate into the conformation of the transition state, and the enzyme itself also undergoes a change in conformation to fit the substrate. Many enzymes exhibit stereochemical specificity in that they catalyze the reactions of one conformation but not the other. Catalytic power is increased by use of binding energy, induced-fit, proximity effect, and stabilization of charges in hydrophobic environment. The catalytic activity of many enzymes depends on the presence of cofactor for catalytic activity. If the organic compound as cofactor is loosely attached to enzyme, it is called a coenzyme. It is called a prosthetic group when the organic compound attaches firmly to the enzyme by covalent bond. Metal ion activators such as Ca^{++}, Cu^{++}, Co^{++}, Fe^{++}, Fe^{+++}, Mn^{++}, Mg^{++}, Mo^{+++}, and Zn^{++} can be cofactors. An enzyme without its cofactor is called an apoenzyme. An enzyme with a cofactor is referred as a haloenzyme. Enzymes catalyze the reactions by covalent catalysis or general acid/base catalysis.

1.2   Enzyme classification and nomenclature

Enzymes are classified into six groups (Table 1.1) according to the reaction catalyzed and denoted by an EC (Enzyme Commission) number. The first, second, and third–fourth digits of these numbers show class of the enzyme, type of the bond involved in the reaction, and specificity of the bond, respectively. Systematic nomenclature is the addition of the suffix -*ase* to the enzyme-catalyzed reaction with the name of the substrate. For example, naringinase, and α-L-Rhamnoside rhamnohydrolase are trivial and systematic names of the enzyme numbered as EC 3.2.1.40, respectively. Some of the enzyme-related databases are IUBMB, International Union of Biochemistry and Molecular Biology enzyme no menclature (www.chem.qmul. ac.uk/ iubmb/enzyme/); BRENDA, comprehensive enzyme information system

Table 1.1 Classification of Enzymes

EC Class	Reaction Catalyzed	Examples
EC1: Oxidoreductases	$A^- + B \rightleftarrows A + B^-$	Peroxidase Catalase Polyphenol oxidase Lipoxygenase Ascorbic acid oxidase Glucose oxidase Alcohol dehydrogenase
EC2: Transferases	$AB + C \rightleftarrows A + BC$	Amylosucrase Dextransucrase Transglutaminase
EC3: Hydrolases	$AB + H_2O \rightleftarrows AH + BOH$	Invertase Chlorophyllase Amylase Cellulose Polygalacturonase Lipase Galactosidase Thermolysin
EC4: Lyases	$\begin{matrix} X & Y \\ \vert & \vert \\ A - B \end{matrix} \rightleftarrows A = B + X - Y$	Pectin lyase Phenylalanine ammonia lyase Cysteine sulfoxide lyase Hydroperoxide lyase
EC5: Isomerase	$\begin{matrix} X & Y \\ \vert & \vert \\ A - B \end{matrix} \rightleftarrows \begin{matrix} Y & X \\ \vert & \vert \\ A - B \end{matrix}$	Glucose isomerase Carotenoid isomerase
EC6: Ligases	$A + B \rightleftarrows AB$	Hydroxycinnamate CoA ligase

(www.brenda-enzymes.org); the ExPASy, Expert Protein Analysis System enzyme nomenclature (www.expasy.org/enzyme/); and EBIPDB, European Bioinformatics Institute–Protein Data Bank enzyme structures database (www.ebi.ac.uk/thornton-srv/databases/enzymes).

1.3 Enzyme kinetics

Besides the quasi-steady-state kinetics (Briggs and Haldane approach), the rate of enzyme catalyzed reactions is generally modeled by the Michaelis-Menten approach. For a simple enzymatic reaction, binding of substrate (S) with free enzyme (E) is followed by an irreversible breakdown of enzyme-substrate complex (ES) to free enzyme and product (P). The substrate binding with E is assumed to be very fast relative to the breakdown of ES complex to E and P. Therefore, the substrate binding is assumed to be at equilibrium as shown in the following reaction scheme:

$$E+S \underset{k_{-1}}{\overset{k_1}{\rightleftarrows}} ES \xrightarrow{k_2} E+P \tag{1.1}$$

The Michaelis-Menten approach concerns the initial reaction rate where there is very little product formation. It is impossible to know the enzyme concentration in enzyme preparations. Therefore, the amount of the enzyme is given as units of activity per amount of sample. One international enzyme unit is the amount of enzyme that produces 1 micromole of product per minute. According to the applied enzyme activity assay, the enzyme unit definition must be clearly stated in research or application.

The total enzyme amount (E_o) equals the sum of the amount of E and ES complex. In terms of amounts, it can be represented as follows:

$$C_{E_o} = C_E + C_{ES} \tag{1.2}$$

The dissociation constant (K_m), which is also called the Michaelis-Menten constant, is a measure of the affinity of enzyme for substrate:

$$K_m = \frac{k_{-1}}{k_1} = \frac{C_E C_S}{C_{ES}} \tag{1.3}$$

If the enzyme has high affinity for the substrate, then the reaction will occur faster and K_m has a lower value. High K_m value means less affinity. K_m varies considerably from one enzyme to another and also with different substrates for the same enzyme.

For these elementary reactions 1.1, the initial reaction rate or reaction velocity (v) is expressed as

$$v = k_2 C_{ES} \tag{1.4}$$

If the enzyme is stable during the reaction, the maximum initial reaction rate (v_{max}) corresponds to

$$v_{max} = k_2 C_{E_o} \tag{1.5}$$

If the initial concentration of substrate (S_o) is very high during the reaction ($C_{S_o} \gg C_{E_o}$), the concentration of substrate remains constant during the initial period of reaction ($C_{S_o} \approx C_S$). Combining Equations 1.2–1.5, the Michaelis-Menten equation is obtained as

$$v = \frac{v_{max}C_S}{K_m + C_S} \qquad (1.6)$$

As an example, kinetic properties of polygalacturonase assayed in different commercial enzyme preparations were studied and the reactions in all samples followed Michaelis–Menten kinetics (Ortega et al., 2004)

Michaelis-Menten expression can be simplified as follows:

$$\text{zero order expression}: v = v_{max} \quad \text{for } C_s \gg K_m \qquad (1.7)$$

$$\text{first order expression}: v = \frac{v_{max}}{K_m}C_S \quad \text{for } C_s \ll K_m \qquad (1.8)$$

The Michaelis-Menten plot (Figure 1.1a) and Lineweaver-Burk plot (double-reciprocal plot, Figure 1.1b) are used for kinetic analyzes of data. While a plot of v as a function of C_S yields a hyperbolic curve, the double-reciprocal plot provides a straight line that is suitable for the estimation of the kinetic constants by linear regression.

An integrated form of the Michaelis-Menten equation is also used for the analysis of enzymatic reactions as follows:

$$-\frac{dC_S}{dt} = \frac{v_{max}C_S}{K_m + C_s} \qquad (1.9)$$

$$(C_{S_o} - C_S) + K_m \ln \frac{C_{S_o}}{C_S} = V_{max}t \qquad (1.10)$$

Another more realistic reaction scheme for one substrate reaction is the following:

$$E + S \rightleftarrows ES \rightleftarrows EP \rightleftarrows E + P \qquad (1.11)$$

where EP = enzyme product complex.

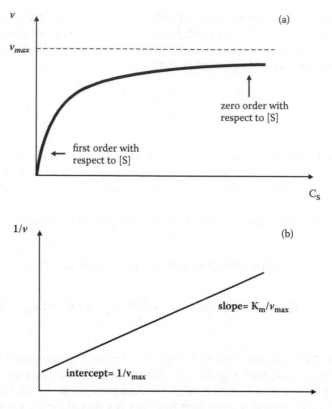

Figure 1.1 Michaelis–Menten plot (a) and Lineweaver–Burk plot for an enzyme-catalyzed reaction obeying Michaelis–Menten kinetics (b).

Following *ES* formation, an enzyme product complex (*EP*) is produced. This transformation may be reversible or irreversible. The velocity equation for this more realistic reaction is easily derived by considering rapid equilibrium conditions, and it will be a function of *S* and *P* concentrations.

The Michaelis-Menten approximation is for enzyme catalysis involving only a single substrate, but in many cases the reaction involves two or more substrates. The same approximation can be extended to two-substrate systems. There are two types of bisubstrate reactions: sequential and ping-pong reactions. Sequential reactions can be further classified as ordered sequential or random sequential mechanism. For the pingpong mechanism, there are two states of the enzyme: *E* and *F*. *F* is a modified state of *E* and often carries a fragment of S_1. The general pathways are as follows:

Ordered sequential mechanism:

$$E \underset{\longleftarrow}{\overset{S_1}{\longrightarrow}} ES_1 \underset{\longleftarrow}{\overset{S_2}{\longrightarrow}} ES_1S_2 \longrightarrow E + P_1 + P_2 \qquad (1.12)$$

Random sequential mechanism:

$$E \underset{\longleftarrow}{\overset{S_1}{\rightleftharpoons}} ES_1 \underset{\longleftarrow}{\overset{S_2}{\rightleftharpoons}} ES_1S_2 \longrightarrow E + P_1 + P_2 \qquad (1.13)$$

$$E \underset{\longleftarrow}{\overset{S_2}{\rightleftharpoons}} ES_2 \underset{\longleftarrow}{\overset{S_1}{\rightleftharpoons}} ES_1S_2 \longrightarrow E + P_1 + P_2 \qquad (1.14)$$

Ping-pong mechanism:

$$E \overset{S_1}{\rightleftharpoons} ES_1 \rightleftharpoons F + P_1 \qquad (1.15)$$

$$F \overset{S_2}{\rightleftharpoons} FS_2 \longrightarrow E + P_2 \qquad (1.16)$$

1.4 Factors affecting enzyme activity

Besides the presence of enzyme and substrate, pH, temperature, and the presence of inhibitors and activators are important factors for the rate of the enzymatic reactions. Table 1.2 shows the Michaelis-Menten kinetic approach for some simple enzyme inhibition types.

pH is also an important parameter for enzyme activity, since most of the enzyme catalysis is general acid–base catalysis. The activities of many enzymes vary with pH in a manner that can often be explained in terms of the dissociation of acids and bases. A simple approach to pH effect is the assumption of an enzyme with two dissociable protons and the zwitterion as the active form. The enzyme–substrate complex also may exist in three states of dissociation, such as

$$
\begin{array}{ccc}
E^- & & ES^- \\
{\scriptstyle H^+}\downarrow\uparrow & & {\scriptstyle H^+}\downarrow\uparrow \\
S+ \quad EH & \rightleftharpoons & EHS \longrightarrow EH + P \\
{\scriptstyle H^+}\downarrow\uparrow & & {\scriptstyle H^+}\downarrow\uparrow \\
EH_2^+ & & EH_2S^+
\end{array}
\qquad (1.17)
$$

For this case, the following velocity equation is obtained by using the Michaelis-Menten approach

$$v = \cfrac{v_{max}C_S}{K_m\left(1 + \cfrac{K_{a2}}{C_{H^+}} + \cfrac{C_{H^+}}{K_{a1}}\right) + [S]\left(1 + \cfrac{K'_{a2}}{C_{H^+}} + \cfrac{C_{H^+}}{K'_{a1}}\right)} \qquad (1.18)$$

where K_{a1}, K_{a2}, K'_{a1}, and K'_{a2} are related ionization constants.

Table 1.2 Simple Enzyme Inhibition Types and Substrate Inhibition

Type	Reaction	Maximum velocity and dissociation constants	Rate expression
Competitive inhibition	$E + S \underset{k_2}{\overset{k_1}{\rightleftharpoons}} ES \xrightarrow{k_2} E + P$ $E + I \underset{k_{-11}}{\overset{k_{11}}{\rightleftharpoons}} EI$	$v_{max} = k_2 C_{E_o}$ $C_{E_o} = C_E + C_{ES} + C_{EI}$ $K_m = \dfrac{C_E C_{ES}}{C_{ES}}$ $K_I = \dfrac{C_E C_I}{C_{EI}}$	$v = \dfrac{v_{max} C_S}{K_m\left(1 + \frac{C_I}{K_I}\right) + C_S}$

Examples for competitive inhibition:

Anacardic acid (C15:1) inhibition of the soybean lipoxygenase-1 (EC 1.13.11.12, Type 1) catalyzed the oxidation of linoleic acid (Kubo et al., 2006)

Kiwi fruit proteinaceous pectin methylesterase inhibitor for carrot pectin methyl esterase (Ly-Nguyen et al., 2004)

Tyrosinase (E.C. 1.14.18.1) inhibition by benzoic acid (Morales et al., 2002)

Type	Reaction	Maximum velocity and dissociation constants	Rate expression
Uncompetitive inhibition	$E + S \underset{k_2}{\overset{k_1}{\rightleftharpoons}} ES \xrightarrow{k_2} E + P$ $ES + I \underset{k_{-11}}{\overset{k_{11}}{\rightleftharpoons}} ESI$	$v_{max} = k_2 C_{E_o}$ $C_{E_o} = C_E + C_{ES} + C_{ESI}$ $K_m = \dfrac{C_E C_{ES}}{C_{ES}}$ $K_I = \dfrac{C_{ES} C_I}{C_{ESI}}$	$v = \dfrac{v_{max} C_S}{K_m + C_S\left(1 + \frac{C_I}{K_I}\right)}$

Examples for uncompetitive inhibition:

Effect of a nonproteinaceous pectin methylesterase inhibitor in potato tuber which is a heavy side-branched uronic chain, on pectin methylesterase from different plant species (McMillan and Perombelon, 1995)

p-Aminobenzenesulfonamide and sulfosalicilic acid for the catecholase activity of purified mulberry polyphenol oxidase (Arslan et al., 2004)

Sucrose inhibition of purified papaya pectinesterase (Fayyaz et al., 1995)

Noncompetitive inhibition

$$E + S \underset{k_2}{\overset{k1}{\rightleftharpoons}} ES \overset{k_2}{\longrightarrow} E + P$$

$$E + I \underset{k_{-11}}{\overset{k_{11}}{\rightleftharpoons}} EI$$

$$ES + I \underset{k_{-22}}{\overset{k_{22}}{\rightleftharpoons}} ESI$$

$$v_{max} = k_2 C_{E_o}$$

$$C_{E_o} = C_E + C_{ES} + C_{EI} + C_{ESI}$$

$$K_m = \frac{C_E C_S}{C_{ES}}$$

$$K_I = \frac{C_E C_I}{C_{EI}} = \frac{C_{ES} C_I}{C_{ESI}}$$

$$v = \frac{v_{max} C_S}{K_m \left(1 + \dfrac{C_I}{K_I}\right) + C_S \left(1 + \dfrac{C_I}{K_I}\right)}$$

Examples for noncompetitive inhibition:

Kiwi fruit proteinaceous pectin methylesterase inhibitor for banana and strawberry pectin methyl esterase (Ly-Nguyen et al., 2004)

Citric acid and oxalic acid inhibition of lettuce polyphenol oxidase (Altunkaya and Gökmen, 2008)

Sodium sulfate, citric acid and ascorbic acid inhibition of polyphenol oxidase from broccoli florets (Gawlik-Dzike et al., 2007)

Substrate inhibition

$$E + S \underset{k_2}{\overset{k_1}{\rightleftharpoons}} ES \overset{k_2}{\longrightarrow} E + P$$

$$ES + S \underset{k_{-11}}{\overset{k_{11}}{\rightleftharpoons}} ESS$$

$$v_{max} = k_2 C_{E_o}$$

$$C_{E_o} = C_E + C_{ES} + C_{ESS}$$

$$K_m = \frac{C_E C_S}{C_{ES}}$$

$$K_I = \frac{C_{ES} C_S}{C_{ESS}}$$

$$v = \frac{v_{max}}{\dfrac{K_m}{C_S} + 1 + \dfrac{C_S}{K_I}}$$

The simplification of this kinetic expression is possible by using low or high [S] and low or high [H+] for the simple analysis of the enzymatic system.

The enzymatic reaction rate increases with increasing temperature to a maximum level and then decreases with further increase of temperature due to enzyme denaturation. Temperature effect on reaction rate constant (k) can be expressed by Arrhenius' equation:

$$k = k_o e^{(-Ea/RT)}$$

(1.19)

where k_o and E_a are pre-exponential factor and activation energy, respectively. T is the absolute temperature in this equation.

Q_{10} value is also used to show the effect of temperature and is defined as the factor by which the rate constant is increased by increasing the temperature 10°C. By considering the reaction rate constants as k_1 and k_2 at temperatures of T_1 and T_2 ($T_1 + 10$), respectively, the ratio of k_2 to k_1 will be

$$\ln \frac{k_2}{k_1} = \frac{E_a}{R} \left(\frac{T_2 - T_1}{T_1 T_2} \right)$$

(1.20)

According to the definition of Q_{10} value, it equals to k_2/k_1 and the following equality is obtained:

$$\ln Q_{10} = \frac{E_a}{R} \left(\frac{10}{T_1 T_2} \right)$$

(1.21)

1.5   *Enzyme inactivation*

For each enzyme, there is an optimum temperature for activity, and increasing the temperature causes enzyme denaturation. Deamidation of asparagine and glutamine residues, hydrolysis of peptide bonds at aspartic acid residues, oxidation of cysteine residues, thiol-disulphide interchange, destruction of disulphide bonds, and chemical reaction between enzyme and other compounds such as polyphenolics can all cause irreversible enzyme inactivation at high temperatures (Volkin & Klibanov, 1989; Adams, 1991).

Thermal enzyme inactivation is generally modeled by first-order kinetics:

$$-\frac{dC_E}{dt} = kC_E$$

(1.22)

$$C_E = C_{E_0} e^{-kt}$$

(1.23)

For a non-zero enzyme activity upon prolonged heating time ($C_E = C_{E\infty}$ at $t = t_\infty$), the equation will be

$$C_E = C_{E_\infty} + (C_{E_0} - C_{E_\infty})e^{-kt} \qquad (1.24)$$

The other scheme for enzyme inactivation is the partial enzyme unfolding, followed by an irreversible reaction step:

$$E \rightleftharpoons E_U \longrightarrow E_I \qquad (1.25)$$

where E_U and E_I represent unfolded enzyme and inactive enzyme, respectively.

Native enzyme may consist of isoenzymes with different thermal stabilities. For this case, if the individual rate constants are sufficiently different from one another, a plot of the natural logarithm of the enzyme activity versus thermal treatment time shows a number of linear lines. Therefore the inactivation kinetic analysis of each isoenzyme ($E_1 \dots E_n$) may be more suitable than the kinetic analysis of total enzyme inactivation:

$$-\frac{dC_{E_1}}{dt} = k_1 C_{E_1} \dots -\frac{dC_{E_n}}{dt} = k_n C_{E_n} \qquad (1.26)$$

$$C_E = C_{E_1} + \dots + C_{E_n} \qquad (1.27)$$

Thermal stability of vegetable peroxidase isoenzymes was widely investigated to identify the mechanisms and corresponding kinetic models of inactivation (Güneş and Bayındırlı, 1993; Anthon and Barrett, 2002; Thongsook et al., 2007; Polata et al., 2009).

The dependence of enzyme inactivation on temperature can be expressed by the Arrhenius expression or in the same way that has been used for microbial inactivation with the use of D-value and z-value. By considering first-order enzyme inactivation, D-value is 1-log (90% of initial activity) reduction time at constant temperature. Modification of Equation 1.23 according to the definition of D-value provides the following expression:

$$D = \frac{2.303}{k} \qquad (1.28)$$

The rate of inactivation of enzymes increases in a logarithmic manner with increasing the temperature. Temperature change required for 1-log reduction in enzyme inactivation rate is z-value (thermal resistance constant). The following expression shows the relationship between activation energy and z-value:

$$\ln 10 = \frac{E_a}{R}\left(\frac{z}{T_1 T_2}\right) \qquad (1.29)$$

Alternatively, the plot of log C_E versus time can be used to obtain the D-value at each temperature, and the log D versus temperature plot provides the estimation of the z-value. From the slopes, the following equation can be easily obtained to show the relationship between D- and z-values by considering a reference temperature (T_{ref}).

$$\frac{D_T}{D_{ref}} = 10^{\left(T_{ref}-T\right)/Z} \tag{1.30}$$

From Equations 1.22, 1.28, and 1.30, the following equation can be obtained:

$$\log\left(\frac{C_E}{C_{Eo}}\right) = -\frac{t_T}{D_{ref}\,10^{\left(T_{ref}-T\right)/Z}} \tag{1.31}$$

pH of the product, water activity, and product composition also affect the rate of inactivation. Process-dependent factors such as processing equipment design, type of heating media, food size, and shape are the other important parameters. Thermal enzyme inactivation kinetics can also be analyzed by unsteady state approximation. For this case, the heat transfer equation must be used together with kinetic expressions to obtain the relationships between the rate at which a food product is heated and the inactivation of the enzyme under consideration. The study of Martens et al. (2001) is an example of the numerical modeling for the combined simulation of heat transfer and enzyme inactivation kinetics in the broccoli stem and asparagus. Mathematical model of heat transfer and enzyme inactivation in an integrated blancher cooler was developed (Arroqui et al., 2003). The model was validated by using different potato shapes (sphere, cube, and cylinder). Thermal inactivation of peroxidase and lipoxygenase during blanching of a solid food with a finite cylinder shape has been mathematically studied to evaluate the end point of the blanching process by Garrote et al. (2004). Heat transfer conditions, the total initial indicator enzyme activity, and the proportion of resistant and labile isoenzymes are considered during modeling.

Novel technologies, such as the application of high-pressure processing, ultrasound, or pulsed electric field, are being used increasingly for controlling enzymatic reactions in food products. The enzyme inactivation by non-thermal treatments can also be modeled by similar approximations. They may require other processing parameters to establish the inactivation reaction rate expression. For example, for high-pressure processing, the parameters are time, temperature, and pressure. Pressure dependence of the reaction rate constants was described by the Eyring equation,

$$k_p = k_{P_{ref}}\,e^{(-Va(P-P_{ref}/RT)} \tag{1.32}$$

where V_a is activation volume and P is pressure.

Ludikhuyze et al. (2003) presented a review from a quantitative point of view of the combined effects of pressure and temperature on enzymes affecting the quality of fruits and vegetables with kinetic characterization of the enzyme inactivation. Rauh et al. (2009) studied the inactivation of different enzymes in a short-time high-pressure process by considering different thermal boundary and initial conditions. Velocity, temperature, and enzyme activity fields in the treatment chamber were determined with numerical simulation.

Besides kinetic modeling, response surface methodology can be used and a polynomial model can be written to estimate the effects of different independent variables $(X_i \ldots X_n)$ and their interaction on enzyme activity:

$$C_E = \beta_0 + \sum_{i=1}^{n}\beta_i x_i + \sum_{i=1}^{n}\beta_{ii} x_i^2 + \sum_{i=1}^{n}\beta_{iii} x_i^3 + \sum_{i=1}^{n}\sum_{j=1 j\neq i}^{n}\beta_{ij} x_i x_j + \beta_n x_1 \ldots x_n \quad (1.33)$$

During the optimization of short-time blanching (steaming) of fresh-cut lettuce, the activity changes of peroxidase and polyphenol oxidase were modeled by using response surface methodology (Rico et al., 2008). Baron et al. (2006) used a factorial design with four factors (pressure, holding time, temperature, and waiting period between crushing of apples and high-pressure treatment) to study the high-pressure inactivation of pectin methylesterase in cloudy apple juice.

1.6   Enzymes in fruit and vegetable processing

Fruits and vegetables are a major source of fiber, minerals and vitamins, and different phytochemicals. Fruits and vegetables are consumed as fresh or processed to different types of safe products with high quality and high potential health benefits. Naturally present enzymes in fresh fruits and vegetables are degraded and metabolized after ingestion. They have not been associated with toxicity and are considered intrinsically safe. But these enzymes play a major role in the quality of fresh fruits and vegetables. Enzymes are very important for growth and ripening of fruits and vegetables and are active after harvesting and during storage. While most of the enzymes present in plant tissues are important for the maintenance of metabolism, some have also undesirable effects on color, texture, flavor, odor, and nutritional value. Activity of lipoxygenase affects flavor and odor development of some vegetables. Lipase and peroxidase are the other enzymes causing flavor and odor changes. Phenol oxidases are important for the discoloration of fruits and vegetables with adverse effects on taste and nutritional quality. Fruits and vegetables contain pectic substances that have major effects on the texture. The activity of pectinases causes fruit softening. α-amylases degrade starch to shorter polymeric fragments known as dextrins and affect the textural integrity. Fruits and vegetables contain ascorbic acid oxidase, which affects the vitamin availability.

During fruit and vegetable processing, nutritional availability, bioactive compound effectiveness, flavor, color, and texture or cloudiness according to the type of the product are important quality parameters. The inactivation of the quality deteriorative enzymes is very important during fruits and vegetables processing. The amino acid sequence, three-dimensional structure, and pH and temperature optima of the enzymes depend on the origin of the enzyme, and these properties of the enzymes influence processing conditions. For example, heat treatment of citrus juices is necessary for inactivating pectin-degrading enzymes to prevent cloud loss in juices. Generally, thermal treatment is used for industrial inactivation of enzymes. Novel technologies such as high pressure, pulsed electric fields, and ultrasound or a combination of these technologies with mild heat treatment can also be used as alternatives to thermal treatment at high temperatures to inactivate enzymes. Enzyme inactivation can be selected as the indicator to show the effectiveness of the process. For example, peroxidase inactivation is an indicator for the adequacy of heat treatment during blanching of vegetables.

Enzymes are also used as processing aids to improve the product quality or to increase the efficiency of operation such as peeling, juicing, clarification, and extraction of value-added products. Enzyme infusion techniques can be used for peeling and segmentation of fruits. Cellulases, amylases, and pectinases facilitate maceration, liquefaction, and clarification during fruit juice processing with the benefit of reducing processing costs and improving yields. The quality of juices manufactured and their stability have been enhanced through the use of enzymes. Amylases may be required after pressing to degrade starch for clear apple juice production. Naringinase improves the quality attributes of citrus juice by catalyzing the hydrolysis of bitter components. The activities of endogenous enzymes, or enzymes of microorganisms, are important in the development of the characteristic taste and flavor of the beverages. For example, enzyme activity is essential for the development of high-quality wines.

Enzymes as safe processing aids can be obtained mainly from pure cultures of selected strains of food-grade microorganisms. A soluble enzyme is not recovered after use; therefore, inexpensive extracellular microbial enzymes are used in soluble form. Enzymes isolated from microorganisms are not suitable for the conditions used in food processing. The enzyme production with higher yield is very important with the elimination of the production of toxic metabolites. Therefore, the use of recombinant DNA technology is required to manufacture novel enzymes suitable for food-processing conditions. Enzymes may be discovered by screening microorganisms sampled from different environments or developed by modification of known enzymes using modern methods of protein engineering or molecular evolution. Pariza and Johnson (2001) and Olempska-Beer et al. (2006) extensively discussed food-processing enzyme preparations obtained from recombinant microorganisms while evaluating the safety of microbial enzyme preparations.

An enzyme preparation typically contains the enzyme of interest and several added substances such as diluents, preservatives, and stabilizers. Enzyme preparations may also contain other enzymes and metabolites from the production organism and the residues of raw materials used in fermentation media. All these materials are expected to be of appropriate purity consistent with current good manufacturing practice. A list of commercial enzymes used in food processing can be found at the Web sites of the Enzyme Technical Association (http://www.enzymetechnicalassoc. org) and the Association of Manufacturers and Formulators of Enzyme Products (http://www.amfep.org).

Microbial enzymes are produced by batch fermentation under strictly controlled fermentation parameters such as temperature, pH, and aeration. The culture is periodically tested to ensure the absence of microbial contaminants. Fermentation media commonly contain nutrients and compounds such as dextrose, corn steep liquor, starch, soybean meal, yeast extract, ammonia, urea, minerals in the form of phosphates, chlorides or carbonates, antifoaming agents, and acid or alkali for pH adjustment. Most recombinant enzymes are secreted to the fermentation medium. After the fermentation, cells are separated by flocculation and filtration. Then the enzyme is subsequently concentrated by ultrafiltration or a combination of ultrafiltration and evaporation. Enzymes that accumulate within cells are isolated from the cellular mass, solubilized, and concentrated. The enzyme concentrate is then sterilized and formulated with compounds such as sucrose, maltose, maltodextrin, potassium sorbate, or sodium benzoate. For certain food applications, enzymes may also be formulated as granulates, tablets, or immobilized form. Enzymes are used in food processing at very low levels. Often, they are either not carried over to food as consumed or are inactivated. The final formulated enzyme product is assessed for compliance with specifications established for enzyme preparations by the Food Chemicals Codex (FCC, 2004) and Joint FAO/WHO Expert Committee on Food Additives (JECFA, 2001).

The number and variety of fruit and vegetable products have increased substantially. Investigations related to enzymes have been carried out with the objectives of (1) designing a new process or improving the available process, (2) development of new enzyme preparations with improved functionality, and (3) inactivation or inhibition of undesirable enzymatic activities in fruit and vegetable products. Biochemical and kinetic knowledge of the related enzymes and processing and preservation technologies will be important for these investigations.

Enzymes have important advantages for industrial application due to stereo- and regioselectivity; low temperature requirement for activity, which means low energy requirement for the process; less by-product formation; improvement of the quality of the product; non-toxic when correctly used; can be degraded biologically and can also be immobilized to reuse; increase stability; and easy separation from the environment.

Abbreviations

C_E: enzyme activity
C_S: substrate concentration
D: decimal reduction time
E: free enzyme
E_a: activation energy
E_f: free enzyme
E_I: inactive enzyme
E_o: total or initial activity of enzyme
E_U: unfolded enzyme
EP: enzyme-product complex
ES: enzyme-substrate complex
I: inhibitor
EI: enzyme-inhibitor complex
k: reaction rate constant
k_o: pre-exponential factor
K: dissociation constant
P: pressure
R: universal gas constant
S: substrate
V_a: activation volume
v: initial reaction rate
v_{max}:   maximum initial reaction rate
z: thermal resistance constant

References

Adams, J.B. 1991. Review: Enzyme inactivation during heat processing of foodstuffs *International Journal of Food Science and Technology* 26:1–20.

Altunkaya, A. and V. Gökmen. 2008. Effect of various inhibitors on enzymatic browning, antioxidant activity and total phenol content of fresh lettuce (*Lactuca sativa*). *Food Chemistry* 107:1173–1179.

Anthon, G.E. and D.M. Barrett. 2002. Kinetic parameters for the thermal inactivation of quality related enzymes in carrots and potatoes. *Journal of Agricultural and Food Chemistry* 50: 4119–4125.

Arroqui, C., A. Lopez, A. Esnoz, and P. Virseda. 2003. Mathematical model of heat transfer and enzyme inactivation in an integrated blancher cooler. *Journal of Food Engineering* 58 (2003) 215–225.

Arslan, O., M. Erzengin, S. Sinan and O. Ozensoy. 2004. Purification of mulberry (*Morus alba* L.) polyphenol oxidase by affinity chromatography and investigation of its kinetic and electrophoretic properties. *Food Chemistry* 88:479–484.

Baron, A., J.M. Denes, and C. Durier. 2006. High pressure treatment of cloudy apple juice *LWT-Food Science and Technology* 39:1005–1013.

Fayyaz, A., B.A. Asbi, H.M. Ghazali, Y.B. Che Man and S. Jinap. 1995. Kinetics of papaya pectinesterase. *Food Chemistry* 53:129–135.

Food Chemicals Codex (FCC), 2004. *Institute of Medicine of the National Academies.* 5th ed. The National Academies Press, Washington, DC.

Garrote, R.L., E.R. Silva, A. Ricardo, R.A. Bertone, and R.D. Roa. 2004. Predicting the end point of a blanching process. *LWT-Food Science and Technology.* 37:309–315.

Gawlik-Dziki, U., U. Szymanowska, and B. Baraniak. 2007. Characterization of polyphenol oxidase from broccoli (Brassica oleracea var. *Botrytis italic*) florets. *Food Chemistry.* 105: 1047–1053.

Güneş, B. and A. Bayındırlı. 1993. Peroxidase and Lipoxygenase inactivation during blanching of green beans, green peas and carrots. *LWT-Food Science and Technology* 26:406–410.

Joint FAO/WHO Expert Committee on Food Additives (JECFA), 2001. General specifications and considerations for enzyme preparations used in food processing. 57th session. *Compendium of Food Additive Specifications. FAO Food and Nutrition Paper 52* (Addendum 9).

Kubo, I., N. Masuoka, T.J. Ha, and K. Tsujimoto. 2006. Antioxidant activity of anacardic acids. *Food Chemistry* 99:555–562.

Ludikhuyze, L., A. Van Loey, Indrawati, C. Smout, and M. Hendrickx. 2003 Effects of combined pressure and temperature on enzymes related to quality of fruits and vegetables: From kinetic information to process engineering aspects. *Critical Reviews in Food Science and Nutrition,* 43:527–586.

Ly-Nguyen, B., A.M. Vann Loey, C. Smout, I. Verlent, T. Duvetter, and M.E. Hendrickx. 2009. Effect of intrinsic and extrinsic factors on the interaction of plant pectin methylesterase and its proteinaceous inhibitor from kiwi fruit. *Journal of Agricultural and Food Chemistry* 52: 8144–8150.

Martens, M., N. Scheerlinck, N. De Belie, and J. De Baerdemaeker. 2001. Numerical model for the combined simulation of heat transfer and enzyme inactivation kinetics in cylindrical vegetables. *Journal of Food Engineering* 47: 185-193

McMillan, G.P. and M.C.M. Perombelon. 1995. Purification and characterization of a high pI pectin methyl esterase isoenzymes and its inhibitor from tubers of *Solanum tuberosum* subsp. *Tuberosum* cv. Katahdin. *Physiological and Molecular Plant Pathology.* 46: 413–427.

Morales, M.D., S. Morante, A. Escarba, M.C. Gonzales, A.J. Reviejo, and J.M. Pingarron. 2002. Design of a composite amperometric enzyme electrode for the control of the benzoic acid content in food. *Talanta* 57: 1189–1198.

Olempska-Beer, Z.S., R.I. Merker, M.D. Ditto, and M.J. DiNovi. 2006. Food-processing enzymes from recombinant microorganisms: A review. *Regulatory Toxicology and Pharmacology* 45:144–158.

Ortega, N., S. de Diego, M. Perez-Mateos, and M.D. Busto. 2004. Kinetic properties and thermal behaviour of polygalacturonase used in fruit juice clarification. *Food Chemistry* 88: 209–217.

Pariza, M.W. and E.A. Johnson. 2001. Evaluating the safety of microbial enzyme preparations used in food processing: update for a new century. *Regulatory Toxicology and Pharmacology* 33, 173–186.

Polata, H., A. Wilinska, J. Bryjak, and M. Polakovic. 2009. Thermal inactivation kinetics of vegetable peroxidases *Journal of Food Engineering* 91: 387–391.

Rauh, C., A. Baars, and A. Delgado. 2009. Uniformity of enzyme inactivation in a short time high pressure process. *Journal of Food Engineering* 91:154–163.

Rico, D., A. B. Martin-Diana, C. Barry-Ryan, J. Frias, G. T.M. Henehan, and J. M. Barat. 2008. Optimisation of steamer jet-injection to extend the shelflife of fresh-cut lettuce. *Postharvest Biology and Technology* 48: 431–442.

Thongsook, T., J.R. Whitaker, G.M. Smith, and D.M. Barrett. 2007. Reactivation of broccoli peroxidases: Structural changes of partially denatured isoenzymes. *Journal of Agricultural and Food Chemistry* 55:1009–1018.

Volkin, D.B. and A.M. Klibanov. 1989. Minimising protein inactivation. In: *Protein Function: A Practical Approach*, T.E. Creighton, ed., pp. 1–24. Oxford: IRL Press at Oxford University Press.

chapter two

Effect of enzymatic reactions on color of fruits and vegetables

J. Brian Adams

Contents

2.1   Introduction

Enzymatic reactions can cause color changes in fruits and vegetables that significantly diminish consumer visual appeal and simultaneously reduce the levels of available vitamins and antioxidants. The enzymes of interest are those that lead to discolorations as a result of the formation of new pigments, or are involved in degradation of the naturally occurring pigments, and identifying these enzymes and the reactions they catalyze *in situ* is of prime importance. This is often not a straightforward task, as many enzymatically formed compounds are highly reactive and take part in a cascade of chemical reactions before the final pigments are formed (Adams and Brown, 2007). Additionally, interactions occur between enzymes and naturally occurring constituents that have an influence on enzyme stability and activity. In the case of raw material in an unprepared state, an understanding is needed of the enzymatic pathways that become active due to disruption of normal physiological processes. For prepared and processed material, knowledge is required of the enzymatic pathways that are active during peeling, dicing, storage, and processing stages. Methods can then be developed to control the activity of specific enzymes or to inhibit or inactivate them.

This chapter discusses the enzymatic reactions that can affect the color of raw and minimally processed fruits and vegetables. It covers the formation of discoloring pigments that can arise due to phenolic oxidation, sulfur-compound reactions, and starch breakdown, and the discolorations that occur on degradation of naturally occurring pigments with reference to anthocyanins, betacyanins, carotenoids, and chlorophylls. Specific fruits and vegetables are considered in each case.

2.2   Formation of discoloring pigments

2.2.1   Phenolic oxidation

The enzymatic oxidation of phenolic compounds can cause blackening or browning of fruits and vegetables either pre-harvest or during post-harvest storage. On abiotic wounding or biotic stress of fruits and vegetables, chemical signals originate at the site of injury that propagate into adjacent tissue where a number of physiological responses are induced including *de novo* synthesis of phenylalanine ammonia lyase (PAL), the initial rate-controlling enzyme in phenolic synthesis. This leads to the accumulation of phenolic compounds that can undergo enzymatic oxidation catalyzed by polyphenol oxidase (PPO) or by peroxidase (POD) leading to tissue discoloration. PPO utilizes oxygen to oxidize phenolic compounds to quinones that are highly reactive and can combine together and with other compounds to

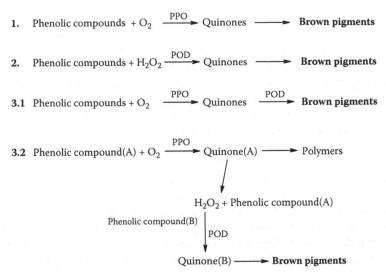

Figure 2.1 Alternative pathways for enzymatic browning of raw fruits and vegetables.

form brown pigments (Mayer, 2006) (Figure 2.1, reaction 1). The presence of PPO in fruits and vegetables means that the enzyme could be involved in the browning reactions if it is in an active form, and not inhibited by the phenolic oxidation products, and if oxygen and phenolic substrate concentrations are not limiting. These conditions probably exist initially on severe bruising, or cutting up fruits and vegetables, when the damage to the tissue allows the enzyme to come into contact with atmospheric oxygen and phenolic compounds at the cut surfaces. On internal browning of fruits and vegetables, however, this type of phenolic oxidation may not occur to the same extent, feasibly because of the compartmentalization of PPO in bound or particulate form, its existence as a latent enzyme, and because of low oxygen levels in the cellular environment. Alternatively or additionally, POD may be involved (Takahama, 2004), acting as an antioxidant enzyme to eliminate hydrogen peroxide (H_2O_2) present in excess as a result of the stress conditions imposed (Figure 2.1, reaction 2). H_2O_2 can be generated during the enzymatic degradation of ascorbic acid, and by the action of amine oxidases, oxalate oxidases, superoxide dismutase (SOD), and certain POD enzymes in chloroplasts, mitochondria, and peroxisomes. Autoxidation of the brown pigments can also lead to superoxide ion and H_2O_2. Elimination of excess H_2O_2 in higher plants involves oxidation of ascorbate to dehydroascorbate either by ascorbate-POD or by phenoxyl radicals formed by POD-dependent reactions. The dehydroascorbate is reduced back to ascorbate by NADH-dependent glutathione reductase in the ascorbate–glutathione cycle. When ascorbate and glutathione have been consumed, other POD enzymes, such as guaiacol-POD, could catalyze the oxidation of phenolic compounds by H_2O_2 and thereby cause browning. For some phenolic compounds, it has been proposed that PPO-derived quinone oxidation products may act directly as substrates for POD (Richard-Forget and Gauillard, 1997) (Figure 2.1, reaction 3.1), whereas for other phenolics the quinones spontaneously generate H_2O_2 that can be utilized by POD to oxidize a second phenolic compound (Murata et al., 2002) (Figure 2.1, reaction 3.2). This suggests that for some fruits and vegetables both PPO and POD could be involved in forming brown pigments and some evidence for this is presented in the examples given below.

Internal browning in fruits and vegetables is frequently correlated with low calcium levels. Calcium deficiency disorders arise when insufficient calcium is available to developing tissues possibly due to restricted transpiration. Alternatively, calcium deficiency may be due to hormonal mechanisms developed for the restriction of calcium transport to maintain rapid growth. Cytosolic calcium concentration is known to stabilize cell membranes and is a key regulator of plant defenses to such challenges as mechanical perturbation, cooling, heat shock, acute salt stress, hyperosmotic stress, anoxia, and exposure to oxidative stress. Thus, under low calcium conditions, cellular defense regulation may be disrupted and this may lead to accumulation of reactive oxygen species that cause oxidation of phenolic compounds via enzymatic activity.

2.2.1.1 *Phenolic oxidation in fruits*

2.2.1.1.1 *Apple browning*

Browning of raw apples and apple products is generally undesirable and is associated with (1) physiological disorders, such as bitter pit, superficial and senescent scalds, internal breakdown, watercore, and core flush; (2) bruising, a flattened area with brown flesh underneath, which is the most common defect of apples seen on the market; and (3) improper preparation of slices, purées, and juices leading to undesirable browning of the final products.

Superficial scald is one of several postharvest physiological disorders of apples. It appears as a diffuse browning on the skin, varying from light to dark, generally without the flesh being affected except in severe cases. Symptoms are not apparent at harvest time but after several months in chill storage, transfer to ambient temperature can lead to scald being expressed within a few days. Superficial scald tends to develop mainly on green-skinned apples and on the un-blushed areas on red cultivars. The severity of the disorder is influenced by many factors including apple cultivar, growing temperatures, cultural practice, harvest maturity, and postharvest chilling conditions. The disorder is more likely to occur in fruits with high nitrogen and potassium, and low calcium content. The oxidation of alpha-farnesene to toxic trienols has been correlated with superficial scald and may be involved in the disruption of skin cells that culminates in the enzymatic oxidation of phenols leading to formation of brown pigments. Following a nitrogen-induced anaerobiosis treatment of Granny Smith apples, scald was found to develop within a few minutes of transfer to air and its severity was positively correlated with the duration of the treatment (Bauchot et al., 1999). Expression of PPO was low while the fruit was held in nitrogen, suggesting that the regulation of PPO gene expression was dependent on oxygen. Once the fruit was transferred to air, browning occurred almost immediately, too rapidly for the initial development of browning symptoms in scald to be attributed to increased PPO gene expression. It was concluded, therefore, that PPO gene expression was not associated with the initial development of symptoms. PPO activity has been associated with superficial scald in the Granny Smith cultivar where quercetin glycosides were found to be the main polyphenol constituents in the apple skin (Piretti et al., 1996). It was hypothesized that, after glycosidase action, quercetin could be reduced to flavan-3,4-diol and then to proanthocyanidins. The latter could then be oxidized by PPO to quinone derivatives that react to form brown products covalently attached to skin proteins. In the case of onion scales described later, it has been shown that POD can oxidize quercetin to brown pigments suggesting that POD may have a role to play in apple scald. However, evidence presented for the involvement of POD isoenzymes in superficial scald was tenuous (Fernandez-Trujillo et al., 2003), and in a study of skin tissues of scald-resistant and scald-susceptible apple cultivars, no link was found between superficial scald susceptibility and POD or SOD enzymes (Ahn, Paliyath, and Murr, 2007).

Internal browning in apples can be related to chilling injury, senescent breakdown, or CO_2 injury in controlled atmosphere (CA) storage. It appears that only a certain proportion of apples are susceptible to browning in CA storage and this can range from a small spot of brown flesh to nearly the entire flesh being affected in severe cases. However, even in badly affected fruit, a margin of healthy, white flesh usually remains just below the skin. The browning shows well-defined margins and may include dry cavities resulting from desiccation. Browning develops early in CA storage and may increase in severity with extended storage time. The disorder is associated with high internal CO_2 levels in later-harvested, large, and overmature fruit. The enzymes involved in the CO_2-induced internal browning in apples are largely unidentified. Sensitivity to CO_2 can depend on cultivar, an effect that may be related to increased NADH oxidase and lower superoxide dismutase activities (Gong and Mattheis, 2003a). Chill storage of Braeburn apples in a low-oxygen CA caused internal browning that was correlated with superoxide ion accumulation caused by enhanced activity of xanthine oxidase and NAD(P)H oxidase, and reduced superoxide dismutase activity (Gong and Mattheis, 2003b). Using Pink Lady apple, it has been suggested that there is a closer association between internal browning and oxidant–antioxidants such as ascorbic acid and H_2O_2, than to the activity of PPO (de Castro et al., 2008). PPO activity increased on storage but was similar for apples kept in air or in CA storage and between undamaged and damaged fruit. The stem end was shown to have a significantly higher incidence of internal browning than the blossom end, and the cells in brown tissue were found to be dead while all healthy tissue in the same fruit contained living cells. Both the brown and the surrounding healthy tissues showed a decrease in ascorbic acid and an increase in dehydroascorbic acid during the first months of CA storage at low O_2 /high CO_2 levels, whereas undamaged fruit retained a higher concentration of ascorbic acid after the same time in storage. The level of H_2O_2 increased more in the flesh of CA-stored apples than in air-stored fruit, an indication of tissue stress. In addition, diphenylamine (DPA)-treatment significantly lowered H_2O_2 concentrations, and completely inhibited internal browning.

The *bruise- and preparation-related browning* in apples in the presence of atmospheric oxygen is generally accepted to be caused by PPO oxidation of apple phenolics (Nicolas et al., 1994). In cider apple juices, it has been found that the rate of consumption of dissolved oxygen did not correlate with PPO activity in the fruits and decreased faster than could be explained by the decrease of its phenolic substrates (Le Bourvellec et al., 2004). The evidence suggested that this was due to oxidized procyanidins having a higher inhibitory effect on PPO than the native procyanidins. Oxidation products of caffeoylquinic acid and (–)-epicatechin also inhibited PPO.

2.2.1.1.2   *Avocado blackening and browning*
Grey/black and brown discolorations can occur on the skin and in the flesh of avocados during chill storage probably due to enzymatic oxidation of phenolic compounds. The fruit is very susceptible during the climacteric rise, and the presence of ethylene and low calcium content increase sensitivity to chilling injury. Increases in PPO and guaiacol-POD activities have been observed both during chill storage and during shelf life at 20°C and, along with membrane permeability values, have been correlated with brown mesocarp discoloration (Hershkovitz et al., 2005).

2.2.1.1.3   *Olive browning*
It has been proposed that the browning reaction in bruised olives occurs in two stages (Segovia-Bravo et al., 2009). First, there is an enzymatic release of the phenolic compound hydroxytyrosol, due to the action of beta-glucosidases on hydroxytyrosol glucoside and esterases on oleuropein. In the second stage, hydroxytyrosol and verbascoside are oxidized by PPO, and by a chemical reaction that only occurs to a limited extent.

2.2.1.1.4   Peach browning

Raw peach and nectarine can undergo chilling injury manifested by browning of the flesh and pit cavity (internal browning). In general, peach is more susceptible than nectarine, and late season cultivars are most susceptible. Browning has been associated with restoring the fruit to room temperature while some ripening is still occurring (Luza et al., 1992). A study on changes in the PPO activity and phenolic content of peaches has shown that PPO increased up to the ripening stage and this was coincident with the maximum degree of browning as evaluated by absorbance measurements (Brandelli and Lopez, 2005). The browning potential closely correlated with the enzyme activity, but not with the phenolic content. Both PPO and POD have been extracted from peach fruit mesocarp (Jimenez-Atienzar et al., 2007). PPO was mainly located in the membrane fraction and was in a latent state, whereas POD activity was found in the soluble fraction. The roles of PPO and POD in peach internal browning have yet to be determined.

A higher level of cell membrane lipid unsaturation has been found to be beneficial in maintaining membrane fluidity and enhancing tolerance to low temperature stress in chill-stored peach fruit, the linolenic acid level feasibly being regulated by omega-3 fatty acid desaturase (Zhang and Tian, 2009).

2.2.1.1.5   Pear browning

As in the case of the apple, the pear is susceptible to superficial scald correlated with increased levels of alpha-farnesene though, as with the apple, further studies are required on the role of enzymes.

Two types of *internal brown* discolorations have been identified in pear that may be part of a continuum or possibly linked to different metabolic pathways. *Core browning* (or *core breakdown*) is mainly associated with wet tissue and structural collapse of the flesh, and often with a skin discoloration that resembles *senescent scald*, whereas *brown heart* is linked with the appearance of dry cavities and may show no symptoms externally. However, both discolorations can occur during the storage of pears under hypoxic conditions in the presence of increased CO_2 partial pressures. Thus, internal browning may be associated with a change from aerobic to anaerobic metabolism and, for core-browned pears, metabolic profiling has confirmed this to be the case (Pedreschi et al., 2009). The enzymes involved in the browning reactions have yet to be ascertained, though pears subject to CA storage showed a decrease in total ascorbic acid and an increase in oxidized ascorbate that corresponded with a sharp burst in ascorbate-POD and glutathione reductase activities (Larrigaudiere et al., 2001). A significant increase in SOD activity, higher amounts of H_2O_2, and a late decrease in catalase were also found. Increasing maturity at harvest has also been linked to internal browning in pears, feasibly due to reduced levels of antioxidants, an accumulation of superoxide ion and H_2O_2, and increased ascorbate-POD and POD activities (Franck et al., 2007).

Browning on bruising, peeling, or dicing of raw pears is generally considered to be due to the enzymatic oxidation of the main phenolic substrates 5-caffeoylquinic acid (chlorogenic acid) and (-)-epicatechin (Amiot et al., 1995). The phenolic content and susceptibility to browning were high in the peel, and appeared to depend on cultivar and to a lesser extent on maturity. It has been demonstrated using 4-methylcatechol, chlorogenic acid, and (–)-epicatechin as substrates in a model system that pear POD had no oxygen-dependent activity (Richard-Forget and Gauillard, 1997). However, in the presence of pear PPO, POD enhanced the phenol degradation. Moreover, when PPO was entirely inhibited by NaCl after different oxidation times, addition of POD led to a further consumption of the phenolic compound. It was concluded that pear PPO oxidation generated H_2O_2, the amount of which varied with the phenolic structure, and that either the H_2O_2 or the quinonic forms were used by pear POD as substrate.

2.2.1.1.6   Pineapple blackening
Internal blackening, or *blackheart*, is an important physiological disorder of pineapples that arises as a result of exposure to chilling temperatures. It has been found at harvest as a result of chilling in the field though more usually the early symptoms appear as dark spots at the base of fruitlets near the core after a few days at ambient temperature following chill storage. Longer storage causes the spots to coalesce and the tissue eventually becomes a dark mass, although considerable variability in intensity and incidence has been observed due to variations in growing conditions. It has been linked with oxygen levels in chill storage, with the presence of ethylene, and with calcium concentration and its distribution in the fruit. PPO activity, an enhanced PAL activity, and a reduction in the rate of increase of ascorbate-POD activity have also been correlated with the black discoloration (Zhou et al., 2003).

2.2.1.1.7   Tomato browning
Enzymatic browning occurs in tomato at a relatively slow rate partly due to the presence of antioxidants such as ascorbic acid and lycopene. PPO activity levels can vary with cultivar and this can lead to a greater susceptibility to enzymatic browning in some varieties (Spagna et al., 2005). Higher watering levels can also lead to a more marked expression of PPO (Barbagallo et al., 2008).

A physiological disorder of growing tomatoes linked to enzymatic browning is *blossom end rot*. Symptoms may occur at any stage in the development of the fruit but, most commonly, are first seen when the fruit is one-third to one-half full size. Initially a small, light brown spot appears at the blossom end, which enlarges and darkens as the fruit develops. The spot may cover up to half of the entire fruit surface, or it may remain small and superficial. Large lesions dry out and become flattened, black, and leathery in appearance. Blossom end rot is dependent upon a number of environmental conditions, especially those that affect the supply of water and calcium in the developing fruits. Calcium deficiency has been found to lead to an increase in caffeic and chlorogenic acids, and in PPO and POD activities (Dekock et al., 1980), and a strong correlation between blossom end rot and PPO activity has been observed (Casado-Vela et al., 2005).

2.2.1.2 Phenolic oxidation in vegetables

2.2.1.2.1 Carrot whitening

The white discoloration that can occur on the surface of abrasion-peeled carrots in storage has been associated with the formation of lignin (Howard and Griffin, 1993). PAL and POD activities were found to be stimulated and both enzyme activities remained elevated during storage. Soluble phenolics also increased during storage. Ethylene absorbents did not affect surface discoloration or lignification.

2.2.1.2.2 Jicama (yam bean) browning

Storage of jicama roots for a few days at 10°C or below causes external decay and a chill-induced brown discoloration of the flesh that is independent of cultivar. At lower temperatures, the flesh can take on a translucent appearance but not necessarily develop brown discoloration. The discoloration is associated with increased levels of soluble phenolic compounds and PAL activity and has generally been attributed to enzyme-catalyzed phenolic oxidation. Damage has been found to induce higher levels of both PPO and POD activities in external tissues (Aquino-Bolaños and Mercado-Silva, 2004). Lignin values were correlated with color changes and the lignin precursors, coumaric, caffeic and ferulic acids, coniferaldehyde, and coniferyl alcohol were good substrates for POD. It was therefore suggested that browning of cut jicama is related to the process of lignification in which POD plays an important role.

2.2.1.2.3 Lettuce browning

Physiological changes or physical damage can lead to enzymatic browning in raw lettuce. The physiological disorders include *tipburn* (browning of leaf margins in the field linked to low levels of calcium and other mineral distribution within the plant), *russet spotting* (brown spots on the midrib associated with exposure to ethylene in chill storage), *brown stain* (brown spots with darker borders on or near the midrib due to exposure to raised levels of CO_2), and *heart-leaf injury* (internal browning also due to CO_2 exposure). Wound-related browning has been correlated with induction of the PAL enzyme (Saltveit, 2000) (Figure 2.2). This leads to accumulation

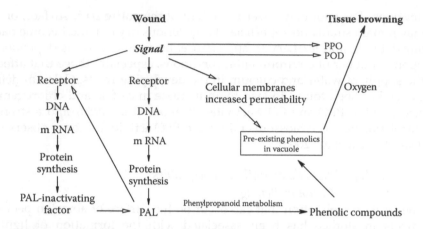

Figure 2.2 Proposed interrelationships between wounding and induced changes in lettuce leaf phenolic metabolism (Saltveit, M.E. 2000. Wound induced changes in phenolic metabolism and tissue browning are altered by heat shock. *Postharvest Biology and Technology* 21: 61–69; with permission).

of phenolic compounds such as chlorogenic acid, isochlorogenic acid and dicaffeoyl tartaric acid, whose levels vary with the type of stress (wounding, or exposure to ethylene), as well as the type of lettuce, and the temperature (Tomás-Barberán et al., 1997; Saltveit, 2000). The wound signal responsible for the increase in PAL activity is unclear although some preliminary evidence has suggested involvement of phospholipase D in the oxylipin pathway (Choi et al., 2005). A PAL-inactivating factor is induced a few hours after the induction of the enzyme in harvested lettuce, allowing time for enough phenolic compounds to be accumulated to cause browning. The enzymes involved in phenolic oxidation in both the physiological and the wound-related browning of lettuce are unknown, though POD activity on chlorogenic acid was found to increase significantly during chill storage of a browning susceptible cultivar (Degl'Innocenti et al., 2005).

2.2.1.2.4   Onion browning

The outer scales of onion bulbs dry out and turn brown on ageing. During this process, the death of scale cells could cause loss of cellular compartmentalization and release of enzymes that cause the major phenolic compounds, quercetin glucosides, to become deglucosidated (Takahama, 2004). Autoxidation of the quercetin formed led to superoxide ion that was then transformed to H_2O_2 and utilized by POD to oxidize quercetin to brown pigments. Catalase inhibited the oxidation, confirming the involvement of H_2O_2. Although PPO can also participate in quercetin oxidation to brown pigments, the enzyme was found to be below the level of detection in the assay employed. It was therefore concluded that PPO did not contribute to the oxidation of quercetin during the browning of onion scales.

2.2.1.2.5 Potato blackening/browning

Internal bruising or *blackspot* in healthy potatoes is a blue-grey zone that forms sub-epidermally in the vascular region of the potato tuber, the stem end of the tuber being most sensitive. It occurs over a period of 1–3 days after relatively minor impacts with the skin often showing no visible signs of damage.

The enzymes and substrates that lead to blackspot pigment formation are not well understood, though it is generally assumed that PPO is involved (McGarry et al., 1996). Partially purifying blackspot pigments has indicated that PPO-derived quinones reacted with nucleophilic amino acid residues in proteins, and that both tyrosine and cysteine were incorporated into the pigments (Stevens et al., 1998). Other investigations have shown that although there was no increase in the level of PPO on impact, subcellular redistribution of the enzyme occurred and this was found to coincide with a loss of membrane integrity (Partington et al., 1999). Evidence for superoxide ion formation at impact sites led to the hypothesis that superoxide is the preferred cosubstrate for PPO rather than any other form of molecular oxygen (Johnson et al., 2003). It was proposed that the transient shock-wave induced an initial burst of superoxide ion, possibly catalyzed by the plasma membrane enzyme NADPH-dependent oxidase, and this was followed by radical oxidative scission of cell wall pectin that initiated a signaling cascade causing a second, larger burst of superoxide synthesis. However, in earlier work, no evidence was found for any relationship with PPO activity (Schaller and Amberger, 1974) and POD involvement was suggested (Weaver and Hautala, 1970). More recently, PPO levels have been found to be higher in a resistant potato cultivar compared with a susceptible variety (Lærke et al., 2000). Also, the intensity of the dark color of the blackspot bruise was shown to be uncorrelated with the PPO activity or the concentration of phenolic compounds (Lærke et al., 2002). Thus, in the present state of knowledge, it is only possible to speculate that low energy impacts cause oxygen free radicals to be formed whose effects lead to a loss of membrane integrity and subsequent PPO or POD oxidation of phenolic compounds.

The *brown discoloration* in raw potatoes that occurs during handling and cutting prior to processing is widely accepted as being caused by PPO activity. Genetically engineered potato varieties with less PPO activity had less tendency to brown (Bachem et al., 1994), and the wild species *Solanum hjertingii*, containing low levels of certain PPO isoenzymes, did not exhibit enzymatic browning (Sim et al., 1997). Genotypes with two copies of a specific allele have been found to have the highest degree of discoloration as well as the highest level of PPO gene expression (Werij et al., 2007). Tyrosine may be an important substrate though levels could depend on the activity of a protease (Sabba and Dean, 1994). Factors correlated with lower ascorbate levels, such as chill storage, have been shown to lead to an increased browning potential in potatoes (Munshi, 1994).

On steam-peeling of potatoes, a dark brown ring can occur at the interface between heat-damaged and undamaged tissue, and on cooking unpeeled potatoes, an irregular brown ring may appear in the tissue adjacent to where the potatoes touch the bottom of the pan (Muneta and Kalbfleisch, 1987). The ring occurs where the dominant thermal effect is activation of the enzyme-catalyzed phenolic oxidation reaction.

2.2.1.2.6   Yam browning
In yam tubers, enzymatic brown discoloration is most intense at the stem end where there is a high concentration of phenolic compounds. The most economically important yam, the white yam, tends to show less browning because of its low substrate concentration, particularly (+)-catechin. Storage leads to accumulation of phenolic compounds, although this appears to be counterbalanced by loss of PPO activity. However, some cultivars of yams show browning that is poorly related to PPO activity (Omidiji and Okpusor, 1996). POD activity has been implicated, along with total phenolics, in the browning of yam flour and paste derived from it, the enzyme tending to be more stable after long times at lower blanching temperatures although initially less stable than PPO under all blanching conditions (Akissoé et al., 2005).

2.2.2   Reactions of sulfur-containing compounds

2.2.2.1   Garlic greening
Garlic products, such as purees and juices, and bottled garlic in vinegar or oil, can develop a blue/green discoloration unless a blanching step is interposed during production. In the traditional, homemade Chinese product ("Laba" garlic) where development of green color is desirable, the garlic requires several months of storage under chill conditions to terminate dormancy before it can turn green during pickling in vinegar.

Evidence has been presented that greening involves gamma-glutamyl transpeptidase (GGT) activity (Li et al., 2008) followed by alliinase degradation of S-(1- propenyl)- and S-(2-propenyl)-cysteine sulfoxides into the corresponding di(propenyl)thiosulfinates and pyruvate (Wang et al., 2008) (Figure 2.3). It was proposed that the di(1-propenyl)thiosulfinate reacts with amino acids to form pyrrolyl carboxylic acids. These are able to form blue pigments with the di(2-propenyl)thiosulfinate, and react with pyruvate to form yellow pigments. The mixture of blue and yellow pigments could then give rise to the green garlic color. Other pathways may coexist, however, as indicated by the isolation of a single species of green pigment with absorption maxima in the yellow and blue spectral regions (Lee et al., 2007). It was suggested that this green pigment contained sulfur, though elucidation of its structure proved problematic due to pigment instability.

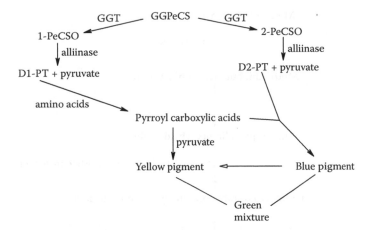

Figure 2.3 Proposed pathway responsible for garlic greening; GGPeCS: gamma-glutamyl propenyl cysteine sulfoxide; GGT: gamma-glutamyl transpeptidase; 1-PeCSO: S-(1-propenyl) cysteine sulfoxide; 2-PeCSO: S-(2-propenyl) cysteine sulfoxide; D1-PT: di(1-propenyl)thiosulfinate; D2-PT: di(2-propenyl) thiosulfinate (Wang, D. et al. 2008. 2-(1H- pyrrolyl)carboxylic acids as pigment precursors in garlic greening. *Journal of Agricultural and Food Chemistry* 56: 1495–1500; with permission).

2.2.2.2 *Onion pinking*

The pink discoloration that can occur in dehydrated and in some other forms of processed onion is a long-standing problem, though the pink pigments have yet to be isolated and chemically characterized. Model system studies have indicated a reaction mechanism analogous to the mechanism suggested for garlic greening as it involves the action of endogenous alliinase on propenyl cysteine sulfoxides. S-(1-propenyl)-L-cysteine sulfoxide (isoalliin) has been shown to be the primary precursor, and the importance of the 1-propenyl thiosulfinate degradation products has been confirmed (Kubec et al., 2004). Further model system studies have suggested that the thiosulfinates can react with amino acids to give substituted pyrrole precursors (Imai et al., 2006). Heating the pigment precursors with the thiosulfinate allicin yielded reddish-purple dipyrrole pigments.

2.2.2.3 *Radish pickle yellowing*

Salted winter radish root, Takuan-zuke, is a traditional Japanese food that naturally acquires a bright yellow color during salting and fermentation. However, the yellow dye is unstable under fluorescent lighting and this leads to a partial loss of color on retail display. This non-uniformity, coupled with the consumer perception that the color is artificial, has led to a desire by manufacturers to inhibit yellowing. Using model systems, it was found that the conditions for yellow pigment formation were the same

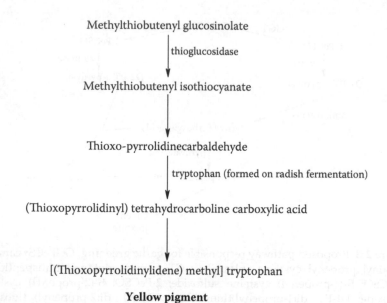

Methylthiobutenyl glucosinolate

| thioglucosidase

Methylthiobutenyl isothiocyanate

Thioxo-pyrrolidinecarbaldehyde

| tryptophan (formed on radish fermentation)

(Thioxopyrrolidinyl) tetrahydrocarboline carboxylic acid

[(Thioxopyrrolidinylidene) methyl] tryptophan

Yellow pigment

Figure 2.4 Formation of a yellow pigment in Japanese salted radish initiated by thioglucosidase (Data from Matsuoka et al., 2002).

as for thioglucosidase action, with ascorbic acid stimulating the production of the yellow pigment due to activation of the enzyme (Ozawa et al., 1993). Evidence has been presented that the yellow pigment is produced as a result of the reaction between methylthiobutenyl isothiocyanate from the thioglucosidase-catalyzed hydrolysis of the major radish glucosinolate, and tryptophan formed during the fermentation (Matsuoka et al., 2002) (Figure 2.4). The pigment was identified as [(thioxopyrrolidinylidene) methyl]-tryptophan.

2.2.3   Starch breakdown reactions

2.2.3.1   Potato browning

Occasionally, high levels of sucrose, glucose, and fructose can be formed on storage of potato tubers at chill temperatures. This can lead to an excessive level of Maillard reaction between the reducing sugars and amino acids that, in turn, can cause brown discoloration and non-uniform browning on manufacture of fried potato products. Considerable variation between different cultivars is due to the chilling susceptibility being largely inherited, although the accumulated sugar levels can vary with growing conditions within a single variety.

The sugars appear to be formed as a result of phosphorylytic starch degradation in the amyloplast catalyzed by phosphorylases, rather than hydrolytic degradation catalyzed by amylases. The hexose phosphates formed are converted by sucrose phosphate synthase to sucrose phosphate that can then be hydrolyzed by a phosphatase to sucrose and inorganic phosphate. Regulation of the enzymatic factors involved is not completely understood although the glycolytic breakdown of hexose phosphates is apparently restricted at low temperatures due to the cold-lability of the key enzymes phosphofructokinase and fructose-6-phosphate phospho-transferase, and this may be an important effect leading to the diversion of the hexose phosphates to sucrose. Hydrolysis of sucrose to glucose and fructose occurs probably due to acid invertase activity. Exposure to low temperatures also leads to enhanced synthesis of invertase and destruction of an invertase inhibitor. Other factors that may be involved in starch degradation include the resistance of the amyloplast membrane to enzymatic breakdown, and the starch granule composition as determined by the relative amounts of amylose and amylopectin.

2.3   Degradation of naturally occurring pigments

2.3.1   Anthocyanin degradation

The red anthocyanin pigments in raw fruits and unpasteurized juices can degrade as a result of anthocyanase (glycosidase) action, and due to oxidation to brown pigments caused by PPO or POD activities.

2.3.1.1   Blueberry browning

When fresh blueberries are blended, a rapid surface browning takes place that has been explained by the PPO catalyzed oxidation of chlorogenic acid to chlorogenoquinone and consequent anthocyanin degradation via a coupled oxidation or condensation (Kader et al., 1997). Any contribution of POD to this oxidation was excluded by addition of catalase to destroy H_2O_2. In contrast, POD activity appears to be the cause of anthocyanin degradation and browning in unpasteurized blueberry juice (Kader et al., 2002). Addition of H_2O_2 to the juice changed the red coloration to brown whereas addition of ascorbic acid prevented the formation of brown polymers. In model systems containing purified anthocyanins, blueberry POD, and H_2O_2, a low level of breakdown of anthocyanins occurred, suggesting that POD was able to act directly on the pigments. On addition of chlorogenic acid, a marked increase in the rate of anthocyanin degradation occurred, and it was therefore proposed that in the fruit juice, chlorogenoquinone degraded the pigment. However, it is unknown whether PPO was active in the blueberry juice and therefore its involvement in the anthocyanin degradation in juice is uncertain.

2.3.1.2   Grape browning

Browning of the pulp and the skin can be a problem in exported table grapes. It usually occurs during or after cold storage, and is often only detected once grape consignments have reached their destinations. Browning in raw grapes has been associated with phenolic compounds and PPO activity, and with low calcium and other mineral levels (Macheix et al., 1991). Degradation of flavan-3-ols in oxidizing musts has been reported to induce browning both of the must and of the resulting wine (Cheynier et al., 1995). This has also been suggested using grape must model systems, where enzymatically generated caffeoyltartaric acid quinones were found to oxidize catechin to brown polymers, and to degrade the grape anthocyanin, malvidin-3-glucoside.

2.3.1.3   Litchi browning

Postharvest browning of litchi fruit pericarp can occur in a few days at ambient temperature after chill storage. The discoloration has been linked to the degradation of anthocyanins by anthocyanase, and POD or PPO activities (Jiang et al., 2004; Zhang et al., 2005). PPO and POD activities were highest in the epicarp and as browning was not observed when both enzymes were selectively inhibited, it was postulated that they could both be associated with litchi browning. Evidence has now been presented suggesting that litchi PPO directly oxidized (-)-epicatechin and products of this oxidation catalyzed anthocyanin degradation, resulting in pericarp browning (Ruenroengklin et al., 2009).

2.3.1.4   Strawberry browning

Using slices of ripe strawberries, it has been found that added H_2O_2 stimulated anthocyanin and catechin degradation and it was concluded that POD may play an important role in brown polymer formation (López-Serrano and Ros Barceló, 1999). However, using purified strawberry POD and PPO enzymes has suggested that brown polymer formation in strawberries is mainly due to PPO, though POD also plays an important role (López-Serrano and Ros Barceló, 2002). Both PPO and POD activities in strawberries have been found to correlate well with measured color parameters, indicating that the browning of the fruit may be a function of both enzyme activities (Chisari et al., 2007).

2.3.2   Betalain degradation

The red betacyanins and yellow betaxanthins in red beetroot can both undergo enzymatic degradation. Decolorization of the violet red betacyanin, betanin, has been observed in model systems containing a red beet cell wall preparation that had the capacity to generate H_2O_2 (Wasserman and Guilfoy, 1983). It was concluded that the reaction proceeded via a peroxidatic mechanism. Both membrane-bound and cell wall-bound POD were identified in red beet (Wasserman and Guilfoy, 1984), and it was found that betanin was enzymatically decolorized at a greater rate than betaxanthin pigments in the presence of phenolic compounds (Wasserman et al., 1984). The betaxanthins were, however, more prone to chemical oxidation by H_2O_2.

PPO has been found in a latent state in beetroot (Escribano et al., 2002) though it can apparently be activated as a result of steam-peeling beetroots, leading to a "black-ring" discoloration (Im et al., 1990).

2.3.3   Carotenoid degradation

Carotenoid pigments in fruits and vegetables are generally relatively stable although potentially they can be degraded to yellow-orange and colorless products in a coupled oxidation with polyunsaturated fatty acids catalyzed by Type 2 lipoxygenases. However, the evidence for such reactions is based largely on studies with purified enzymes and pigments. Lipoxygenase activity, if present, can be inhibited by phenolic compounds with flavans having a stronger inhibitory effect on the enzyme than flavonols followed by ferulic and *p*-coumaric acids (Oszmianski and Lee, 1990). Carotenoid degradation has also been linked with PPO activity, the lycopene in tomato apparently being oxidized by quinones formed in the enzymatic reaction (Spagna et al., 2005).

2.3.4   Chlorophyll degradation

Chlorophylls can be degraded into green, olive-brown, and colorless products. The green compounds still contain the central magnesium (or the magnesium has been replaced by another metal such as zinc or copper), the olive-brown products have lost the metal ion, and the colorless products contain open tetrapyrrole rings. Although the fine details are not fully comprehended, the main enzymatic steps in the pathway of chlorophyll catabolism in green tissue undergoing true senescence are known to involve chlorophyllase, reductases, magnesium-dechelatase (although magnesium removal is possibly caused by a non-enzymatic factor), pheophorbide a oxygenase and red chlorophyll catabolite reductase (Figure 2.5). Loss of chlorophyll by reactions that lie outside this physiological pathway has been termed *pseudosenescent* behavior (Ougham et al, 2008). In practice, it is difficult to distinguish between true and pseudosenescence and both may co-occur in fruits and vegetables. The post-harvest rate of chlorophyll degradation varies considerably from species to

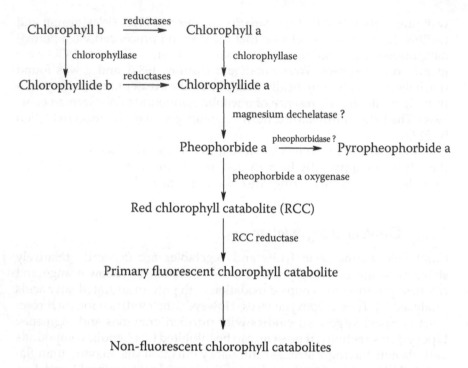

Figure 2.5 Chlorophyll degradation during senescence of green tissue (adapted from Ougham et al., 2008).

species, the rate being enhanced under conditions that lead to ethylene formation and lowered by oxidative defense mechanisms such as endogenous ascorbate.

2.3.4.1 Broccoli yellowing

Yellowing of raw broccoli florets can occur rapidly on storage in air at 15–20°C due to chlorophyll degradation. In green vegetables, it has been proposed that POD utilizes H_2O_2 to catalyze oxidation of phenolic compounds to phenoxy radicals that then oxidize the chlorophyll and its derivatives to colorless low molecular weight compounds through the formation of 13^2-hydroxychlorophyll a, a fluorescent chlorophyll catabolite and a bilirubin-like intermediate (Yamauchi et al., 2004). In addition to the phenoxy radical, superoxide anion, formed in the POD-catalyzed reaction, may be involved in chlorophyll oxidation. External application of ethylene to broccoli has been shown to accelerate chlorophyll degradation, putatively via chlorophyllase, Mg dechelatase, and POD activities, whereas a cytokinin-related compound reduced the rate of chlorophyll breakdown (Costa et al., 2005). Mild heat treatment at 50°C of broccoli has been found to inhibit chlorophyll degradation and to suppress the enhancement of POD isoenzyme activities occurring on storage (Funamoto et al., 2006). This suggested that suppression of POD activities could inhibit floret senescence. Alternatively, H_2O_2 generated by the heat treatment may have induced activation of enzymes related to the ascorbate–glutathione cycle, and the enhanced action of the cycle could have led to the suppression of senescence in heat-treated broccoli florets.

2.3.4.2   Olive fruit chlorophyll changes

During the development of the olive, two stages in chlorophyll evolution can be distinguished: one of synthesis when the fruit is in the growth phase, and a degradative stage which begins when the fruit is completely developed (Minguez-Mosquera and Gallardo-Guerrero, 1996). Although the enzyme chlorophyllase was present in olives throughout their development cycle, its activity reached a maximum at both the beginning and end of the vegetative growth phase, suggesting that chlorophyllase may be involved in chlorophyll synthesis. POD activity has been detected in thylakoid membranes of olives resulting in the formation of 13^2-hydroxychlorophylls *a* and *b* and degradation of carotenoids (Gandul-Rojas et al., 2004). In some cultivars of olive fruit, the early cleavage of the macro-ring of the chlorophyll molecule has indicated a rapid loss of chlorophylls before the synthesis of anthocyanins, probably via the pheophorbide *a* oxygenase pathway (Roca et al., 2007).

2.3.4.3   Frozen green vegetable chlorophyll loss

The available evidence suggests that chlorophyll degradation in frozen, unblanched green vegetables is linked to chlorophyllase and POD activity, and to enzymatically catalyzed lipid degradation (Kidmose et al., 2002). The latter mechanism involves hydrolytic breakdown of lipids by lipases to polyunsaturated fatty acids, and oxygenation of the fatty acids by lipoxygenase to hydroperoxides that then oxidize the chlorophyll to khaki/yellow or colorless products.

2.4   Conclusions

The enzymatic reactions associated with color changes in fruits and vegetables have been discussed with examples in two areas of discoloration: discoloring pigment formation, and degradation of naturally occurring pigments. It is concluded that the identity of enzymes and substrates involved is fully comprehended in relatively few cases and this has led to empirical techniques being employed to control color changes. Further knowledge of the enzymatic reactions involved in fruit and vegetable color is therefore desirable and, in particular, of the identity of the enzymes involved in browning and the factors controlling enzyme activity.

38   J. Brian Adams

Abbreviations

CA: controlled atmosphere
GGT: gamma-glutamyl transpeptidase
PAL: phenylalanine ammonia lyase
POD: peroxidase
PPO: polyphenol oxidase
SOD: superoxide dismutase

References

Adams, J.B. and H.M. Brown. 2007. Discoloration in raw and processed fruits and vegetables. *Critical Reviews in Food Science and Nutrition* 47: 319–333.

Ahn, T., G. Paliyath, and D.M. Murr. 2007. Antioxidant enzyme activities in apple varieties and resistance to superficial scald development. *Food Research International* 40: 1012–1019.

Akissoé, N., C. Mestres, J. Hounhouigan, and M. Nago. 2005. Biochemical origin of browning during the processing of fresh yam (*Dioscorea spp.*) into dried product. *Journal of Agricultural and Food Chemistry* 53: 2552–2557.

Amiot, M.J., M.Tacchini, S.Y. Aubert, and W. Oleszek. 1995. Influence of cultivar, maturity stage, and storage conditions on phenolic composition and enzymatic browning of pear fruits. *Journal of Agricultural and Food Chemistry* 43:1132–1137.

Aquino-Bolaños, E.N. and E. Mercado-Silva. 2004. Effects of polyphenol oxidase and peroxidase activity, phenolics and lignin content on the browning of cut jicama. *Postharvest Biology and Technology* 33: 275–283.

Bachem, C.W., G.J. Speckmann, P.C.G. Van der Linde, F.T.M. Verheggen, M.D. Hunt, J.C. Steffens, and M. Zabeau. 1994. Antisense expression of polyphenol oxidase genes inhibits enzymatic browning in potato tubers. *Bio/Technology* 12: 1101–1105.

Barbagallo, R.N., M. Chisari, F. Branca, and G. Spagna. 2008. Pectin methylesterase, polyphenol oxidase and physicochemical properties of typical long-storage cherry tomatoes cultivated under water stress regime. *Journal of the Science of Food and Agriculture* 88: 389–396.

Bauchot, A.D., S.J. Reid, G.S. Ross, and D.M. Burmeister. 1999. Induction of apple scald by anaerobiosis has similar characteristics to naturally occurring superficial scald in "Granny Smith" apple fruit. *Postharvest Biology and Technology* 16: 9–14.

Brandelli, A. and C.H.G.L. Lopez. 2005. Polyphenoloxidase activity, browning potential and phenolic content of peaches during postharvest ripening. *Journal of Food Biochemistry* 29: 624–637.

Casado-Vela, J., S. Sellés, and R. Bru. 2005. Purification and kinetic characterization of polyphenol oxidase from tomato fruits (*Lycopersicon esculentum* cv. Muchamiel). *Journal of Food Biochemistry* 29: 381–401.

Cheynier, V., H. Fulcrand, S. Guyot, J. Oszmianski, and M. Moutounet. 1995. Reactions of enzymically generated quinones in relation to browning in grape musts and wines. *ACS Symposium Series No.600*, 130–143.

Chisari, M., R.N. Barbagallo, and G. Spagna. 2007. Characterization of polyphenol oxidase and peroxidase and influence on browning of cold stored strawberry fruit. *Journal of Agricultural and Food Chemistry* 55: 3469–3476.

Choi, Y-J., F.A. Tomás-Barberán, and M.E. Saltveit. 2005. Wound-induced phenolic accumulation and browning in lettuce (*Lactuca sativa* L.) leaf tissue is reduced by exposure to n-alcohols. *Postharvest Biology and Technology* 37: 47–55.

Costa, M.L., P.M. Civello, A.R. Chaves, and G.A. Martínez. 2005. Effect of ethephon and 6-benzylaminopurine on chlorophyll degrading enzymes and a peroxidase-linked chlorophyll bleaching during post-harvest senescence of broccoli (*Brassica oleracea* L.) at 20°C. *Postharvest Biology and Technology* 35: 191–199.

de Castro, E., D.M. Barrett, J. Jobling, and E.J. Mitcham. 2008. Biochemical factors associated with a CO_2-induced flesh browning disorder of Pink Lady apples. *Postharvest Biology and Technology* 48: 182–191.

Decock, P.C., D. Vaughan, A. Hall, and B.G. Ord. 1980. Biochemical studies on blossom end rot of tomatoes. *Physiologia Plantarum* 48: 312–316.

Degl'Innocenti, E., L. Guidi, A. Pardossi, and F. Tognoni. 2005. Biochemical study of leaf browning in minimally processed leaves of lettuce (*Lactuca sativa* L. var. *Acephala*). *Journal of Agricultural and Food Chemistry.* 53: 9980–9984.

Escribano, J., F. Gandía-Herrero, N. Caballero, and M. Angeles Pedreño. 2002. Subcellular localization and isoenzyme pattern of peroxidase and polyphenol oxidase in beet root (*Beta vulgaris* L.). *Journal of Agricultural and Food Chemistry* 50: 6123–6129.

Fernandez-Trujillo, J.P., J.F. Nock, E.M. Kupferman, S.K. Brown, and C.B. Watkins. 2003. Peroxidase activity and superficial scald development in apple fruit. *Journal of Agricultural and Food Chemistry* 51: 7182–7186.

Franck, C., J. Lammertyn, Q.T. Ho, P. Verboven, B. Verlinden, and B.M. Nicolai. 2007. Browning disorders in pear fruit. *Postharvest Biology and Technology* 43: 1–13.

Funamoto, Y., N. Yamauchi, and M. Shigyo. 2006. Control of isoperoxidases involved in chlorophyll degradation of stored broccoli (*Brassica oleracea*) florets by heat treatment. *Journal of Plant Physiology* 163: 141–146.

Gandul-Rojas, B., M. Roca, and M.I. Mínguez-Mosquera. 2004. Chlorophyll and carotenoid degradation mediated by thylakoid-associated peroxidative activity in olives (*Olea europaea*) cv. Hojiblanca. *Journal of Plant Physiology* 161: 499–507.

Gong, Y. and J.P. Mattheis. 2003a. Physiological responses of "Braeburn" and "Gala" apples to high CO_2. *Hortscience* 38: 859.

Gong, Y. and J.P. Mattheis. 2003b. Effects of low oxygen on active oxygen metabolism and internal browning in "Braeburn" apple fruit. *ISHS Acta Horticulturae* 628: 533–539.

Hershkovitz, V., S.I. Saguy, and E. Pesis. 2005. Postharvest application of 1-MCP to improve the quality of various avocado cultivars. *Postharvest Biology and Technology* 37: 252–264.

Howard, L.R. and L.E. Griffin. 1993. Lignin formation and surface discoloration of minimally processed carrot sticks. *Journal of Food Science* 58: 1065–1067.

Im, J-S., K.L. Parkin, and J.H. von Elbe. 1990. Endogenous polyphenoloxidase activity associated with the "black ring" defect in canned beet (*Beta vulgaris* L.) root slices. *Journal of Food Science* 55: 1042–1045, 1059.

Imai, S., K. Akita, M. Tomotake, and H. Sawada. 2006. Identification of two novel pigment precursors and a reddish-purple pigment involved in the blue-green discoloration of onion and garlic. *Journal of Agricultural and Food Chemistry* 54: 843–847.

Jiang, Y., X. Duan, D. Joyce, Z. Zhang, and J. Li. 2004. Advances in understanding of enzymatic browning in harvested litchi fruit. *Food Chemistry* 88: 443–446.

Jimenez-Atienzar, M., M. A. Pedreno, N. Caballero, J. Cabanes, and F. Garcia-Carmona. 2007. Characterization of polyphenol oxidase and peroxidase from peach mesocarp (*Prunus persica* L. cv. Babygold). *Journal of the Science of Food and Agriculture* 87: 1682–1690.

Johnson, S.M., S.J. Doherty, and R R.D. Croy. 2003. Biphasic superoxide generation in potato tubers. A self-amplifying response to stress. *Plant Physiology* 131: 1440–1449.

Kader, F., M. Irmouli, J.P. Nicolas, and M. Metche. 2002. Involvement of blueberry peroxidase in the mechanisms of anthocyanin degradation in blueberry juice. *Journal of Food Science* 67: 910–915.

Kader, F., B. Rovel, M. Girardin, and M. Metche. 1997. Mechanism of browning in fresh highbush blueberry fruit (*Vaccinium corymbosum* L). Role of blueberry polyphenol oxidase, chlorogenic acid and anthocyanins. *Journal of Agricultural and Food Chemistry* 74: 31–34.

Kidmose, U., M. Edelenbos, R. Nørbæk, and L.P. Christensen. 2002. Color stability in vegetables. In: *Color in Food, Improving Quality*. pp. 179–232. MacDougall, D.B., Ed., Woodhead Publishing, UK and CRC Press, USA.

Kubec, R., M. Hrbáčová, R.A. Musah, and J. Velíšek. 2004. Allium discoloration: precursors involved in onion pinking and garlic greening. *Journal of Agricultural and Food Chemistry* 52: 5089–5094.

Lærke, P.E., E.R. Brierley, and A.H. Cobb. 2000. Impact-induced blackspots and membrane deterioration in potato (*Solanum tuberosum* L.) tubers. *Journal of Agricultural and Food Chemistry* 80: 1332–1338.

Lærke, P. E., J. Christiansen, and B. Veierskov. 2002. Color of blackspot bruises in potato tubers during growth and storage compared to their discoloration potential. *Postharvest Biology and Technology* 26: 99–111.

Larrigaudiere, C., I. Lentheric, E. Pintó, and M. Vendrell, 2001. Short-term effects of air and controlled atmosphere storage on antioxidant metabolism in Conference pears. *Journal of Plant Physiology* 158: 1015–1022.

Le Bourvellec, C., J-M. Le Quéré, P. Sanoner, J-F. Drilleau, and S. Guyot. 2004. Inhibition of apple polyphenol oxidase activity by procyanidins and polyphenol oxidation products. *Journal of Agricultural and Food Chemistry* 52: 122–130.

Lee, E-J., J-E. Cho, J-H. Kim, and S-K. Lee. 2007. Green pigment in crushed garlic (*Allium sativum* L.) cloves: Purification and partial characterization. *Food Chemistry* 101: 1709–1718.

Li, L., D. Hu, Y. Jiang, F. Chen, X. Hu, and G. Zhao. 2008. Relationship between γ-glutamyl transpeptidase activity and garlic greening, as controlled by temperature. *Journal of Agricultural and Food Chemistry* 56: 941–945.

López-Serrano, M. and A. Ros Barceló. 1999. H_2O_2-mediated pigment decay in strawberry as a model system for studying color alterations in processed plant foods. *Journal of Agricultural and Food Chemistry* 47: 824–827.

López-Serrano, M. and A. Ros Barceló. 2002. Comparative study of the products of the peroxidase-catalyzed and the polyphenoloxidase-catalyzed (+)-catechin oxidation. Their possible implications in strawberry (*Fragaria* × *ananassa*) browning reactions. *Journal of Agricultural and Food Chemistry* 50: 1218–1224.

Luza, J.G., R. van Gorsel, V.S. Polito, and A.A. Kader. 1992. Chilling injury in peaches: a cytochemical and ultra-structural cell wall study. *Journal of the American Society of Horticultural Science* 117: 881–886.

Macheix, J-J., J-C. Sapis, and A. Fleuriet. 1991. Phenolic compounds and polyphenoloxidase in relation to browning in grapes and wines. *Critical Reviews in Food Science and Nutrition* 30: 441–486.

Matsuoka, H., A. Takahashi, Y. Ozawa, Y. Yamada, Y. Uda, and S. Kawakishi. 2002. 2-[3-(2-Thioxopyrrolidin-3-ylidene)methyl]-tryptophan, a novel yellow pigment in salted radish roots. *Bioscience, Biotechnology and Biochemistry* 66: 1450–1454.

Mayer, A.M. 2006. Polyphenol oxidases in plants and fungi: Going places? A review. *Phytochemistry* 67: 2318–2331.

McGarry, A., C.C. Hole, R.L.K. Drew, and N. Parsons. 1996. Internal damage in potato tubers: A critical review. *Postharvest Biology and Technology* 8: 239–258.

Minguez-Mosquera, M.I. and L. Gallardo-Guerrero. 1996. Role of chlorophyllase in chlorophyll metabolism in olives cv. Gordal. *Phytochemistry* 41: 691–697.

Muneta, P. and G. Kalbfleisch. 1987. Heat-induced contact discoloration: A different discoloration in boiled potatoes. *American Journal of Potato Research* 64: 11–15.

Munshi, C. B. 1994. Effect of mineral nutrition, sprout inhibitors, and naturally-occurring toxicants on potato quality. *Dissertation Abstracts International-B* 54: 4979.

Murata, M., M. Sugiura, Y. Sonokawa, T. Shimamura, and S. Homma. 2002. Properties of chlorogenic acid quinone: relationship between browning and the formation of hydrogen peroxide from a quinone solution. *Bioscience, Biotechnology and Biochemistry* 66: 2525–2530.

Nicolas, J.J., F.C. Richard-Forget, P.M. Goupy, M.J. Amiot, and S.Y. Aubert. 1994. Enzymatic browning reactions in apple and apple products. *Critical Reviews in Food Science and Nutrition* 34:109–157.

Omidiji, O. and J. Okpuzor. 1996. Time course of PPO-related browning of yams. *Journal of the Science of Food and Agriculture* 70: 190–196.

Ougham, H., S. Hörtensteiner, I. Armstead, I. Donnison, I. King, H. Thomas, and L. Mur. 2008. The control of chlorophyll catabolism and the status of yellowing as a biomarker of leaf senescence. *Plant Biology* 10(Suppl. 1): 4–14.

Oszmianski, J. and C.Y. Lee. 1990. Inhibitory effect of phenolics on carotene bleaching in vegetables. *Journal of Agricultural and Food Chemistry* 38: 688–690.

Ozawa, Y., Y. Uda, and S. Kawakishi. 1993. Effects of pH, metal ions and ascorbic acid on formation of yellow pigments from tetrahydro-beta-carboline derivatives. *Journal of Japanese Society for Food Science and Technology* 40: 528–531.

Partington, J.C., C. Smith, and G.P. Bolwell. 1999. Changes in the location of polyphenol oxidase in potato (*Solanum tuberosum* L.) tuber during cell death in response to impact injury: Comparison with wound tissue. *Planta* 207: 449–460.

Pedreschi, R., C. Franck, J. Lammertyn, A. Erban, J. Kopka, M. Hertog, B. Verlinden, and B. Nicolaï. 2009. Metabolic profiling of "Conference" pears under low oxygen stress. *Postharvest Biology and Technology* 51: 123–130.

Piretti, M.V., G. Gallerani, and U. Brodnik. 1996. Polyphenol polymerisation involvement in apple superficial scald. *Postharvest Biology and Technology* 8: 11–18.

Richard-Forget, F.C. and F.A. Gauillard. 1997. Oxidation of chlorogenic acid, catechins, and 4-methylcatechol in model solutions by combinations of pear (*Pyrus communis* Cv. Williams) polyphenol oxidase and peroxidase: A possible involvement of peroxidase in enzymatic browning. *Journal of Agricultural and Food Chemistry* 45: 2472–2476.

Roca, M., B. Beatriz Gandul-Rojas, and M.I. Minguez-Mosquera. 2007. Varietal differences in catabolic intermediates of chlorophylls in *Olea europaea* (L.) fruit cvs. Arbequina and Blanqueta. *Postharvest Biology and Technology* 44: 150–156.

Ruenroengklin, N., J. Sun, J. Shi, S.J. Xue, and Y. Jiang. 2009. Role of endogenous and exogenous phenolics in litchi anthocyanin degradation caused by polyphenol oxidase. *Food Chemistry.* 115: 1253–1256.

Sabba, R.P. and B.B. Dean. 1994. Sources of tyrosine in genotypes of *Solanum tuberosum* L. differing in capacity to produce melanin pigments. *Journal of the American Society for Horticultural Science* 119: 770–774.

Saltveit, M.E. 2000. Wound induced changes in phenolic metabolism and tissue browning are altered by heat shock. *Postharvest Biology and Technology* 21: 61–69.

Schaller, K. and A. Amberger. 1974. The correlation between constituents responsible for blackspots of potatoes. *Qualitas Plantarum: Plant Foods for Human Nutrition* 24: 191–198.

Segovia-Bravo, K.A., M. Jarén-Galán, P. García-García, A. Garrido-Fernández. 2009. Browning reactions in olives: Mechanism and polyphenols involved. *Food Chemistry* 114: 1380–1385.

Sim, S. K., S.M. Ohmann, and C.B. Tong. 1997. Comparison of polyphenol oxidase in tubers of *Solanum tuberosum* and the non-browning tubers of *S. hjertingii.* *American Potato Journal* 74: 1–13.

Spagna, G., R.N. Barbagello, M. Chisari, and F. Branca. 2005. Characterization of a tomato polyphenol oxidase and its role in browning and lycopene content. *Journal of Agricultural and Food Chemistry* 53: 2032–2038.

Stevens, L. H., E. Davelaar, R.M. Kolb, E.J.M. Pennings, and N.P.M. Smits. 1998. Tyrosine and cysteine are substrates for blackspot synthesis in potato. *Phytochemistry* 49: 703–707.

Takahama, U. 2004. Oxidation of vacuolar and apoplastic phenolic substrates by peroxidase: Physiological significance of the oxidation reactions. *Phytochemistry Reviews* 3: 207–219.

Tomás-Barberán, F.A., J. Loaiza-Velarde, A. Bonfanti, and M.E. Saltveit. 1997. Early wound- and ethylene-induced changes in phenylpropanoid metabolism in harvested lettuce. *Journal of the American Society for Horticultural Science* 122: 399–404.

Wang, D., H. Nanding, N. Han, F. Chen, and G. Zhao. 2008. 2-(1H- pyrrolyl)carboxylic acids as pigment precursors in garlic greening. *Journal of Agricultural and Food Chemistry* 56: 1495–1500.

Wasserman, B.P. and M.P. Guilfoy. 1983. Peroxidative properties of betanin decolorization by cell walls of red beet. *Phytochemistry.* 22: 2653–2656.

Wasserman, B.P. and M.P. Guilfoy. 1984. Solubilisation of the red beet cell wall betanin decolorizing enzyme. *Journal of Food Science.* 49: 1075–1077.

Wasserman, B.P., L.L. Eiberger, and M.P. Guilfoy. 1984. Effect of hydrogen peroxide and phenolic compounds on horseradish peroxidase-catalyzed decolorization of betalain pigments. *Journal of Food Science.* 49: 536–538.

Weaver, M. L. and E. Hautala. 1970. Study of hydrogen peroxide, potato enzymes and blackspot. *American Potato Journal* 47: 457–468.

Werij, J., B. Kloosterman, C. Celis-Gamboa, C. de Vos, T. America, R. Visser, and C. Bachem. 2007. Unravelling enzymatic discoloration in potato through a combined approach of candidate genes, QTL, and expression analysis. *Theoretical and Applied Genetics* 115: 245–252 (8).

Yamauchi, N., Y. Funamoto, and M. Shigyo. 2004. Peroxidase-mediated chlorophyll degradation in horticultural crops. *Phytochemistry Reviews* 3: 221–228.

Zhang, C. and S. Tian. 2009. Crucial contribution of membrane lipids' unsaturation to acquisition of chilling-tolerance in peach fruit stored at 0 °C. *Food Chemistry* 115: 405–411.

Zhang, Z., X. Pang, D. Xuewu, Z. Ji, and Y. Jiang. 2005. Role of peroxidase in anthocyanin degradation in litchi fruit pericarp. *Food Chemistry.* 90: 47–52.

Zhou, Y., J.M. Dahler, S.J.R. Underhill, and R.B.H. Wills. 2003. Enzymes associated with blackheart development in pineapple fruit. *Food Chemistry.* 80: 565–572.

chapter three

Major enzymes of flavor volatiles production and regulation in fresh fruits and vegetables

Jun Song

Contents

3.1 Introduction

Fruits and vegetables play an important role in the human diet. In addition to their nutritional benefits, fruits are noted for their delicious flavor. Fresh fruit quality is dependent on many aspects, including appearance, color, texture, flavor, and nutritional value (Dull and Hulme, 1971). Among them, flavor is one of the most important quality traits for most fruits. Development of new technologies for the breeding, production, and postharvest handling of fresh fruits has extended their availability through improvements in yield, appearance, and decay resistance, but has often neglected flavor quality. Therefore, improvements in fruit flavor and nutrition are needed to satisfy consumer demands (Kader, 2004; Morris and Sands, 2006). Improving the flavor properties of fresh fruit would add value, increase consumption, and create new markets for these commodities.

In general, flavor quality can be defined as aroma/odor and taste including sweetness, acidity, bitterness, and freshness. Agriculture and food technology research has a long history of studying fruit and vegetable flavor. Development of analytical instruments and concepts of sensory science have provided insights on the formation and development of flavor compounds that contribute to fruit eating quality. Despite many exciting developments in flavor research, such as chemistry, biochemistry, analytical techniques, and sensory evaluation, challenges still remain. One of the biggest challenges is to understand the fundamental mechanisms controlling changes in flavor quality, and biochemical pathways that determine this quality trait. Another challenge is to reveal how horticultural practices, including breeding, production, postharvest, and processing, influence the consumer's perception of flavor.

It is well known that thousands of primary and secondary metabolites are produced during fruit ripening, which results from many metabolic pathways involving many enzymes. Several reviews have been published on flavor volatiles and fruit quality of apples (Yahia, 1994; Dixon and Hewett, 2000; Fellman et al., 2000), strawberry (Forney et al., 2000), tomato (Baldwin et al., 2007), apple, strawberry, and melons (Song and Forney, 2008). The complex chemistry of flavor volatile, technical challenges, and development of sensory science have been discussed. In this chapter, major enzymes influencing formation and regulation of volatile/aroma compounds in most popular fruit such as apple, strawberry, tomatoes, banana, and melons will be described. In addition, major flavor volatiles such as terpenes will be discussed using carrot as an example. The changes of flavor chemistry occurring during ripening, senescence, and postharvest handling will be the primary focus. Most recent publications in relation to these enzymes will be briefly introduced, leaving scientific detail in the reference papers. Comments on future research needed to understand enzymes and flavor biosynthesis in fruit will be discussed. The goal is to update our understanding of the mechanisms of flavor development, provide an update on fruit flavor research, and identify new opportunities to enhance the flavor of fresh fruits and vegetables.

3.2   *Aroma volatile compounds in fruits*

To better understand the enzymes influencing fruit volatile production, it is necessary to briefly outline volatile production and the metabolic pathways for volatile biosynthesis in fruits, including apple, banana, strawberry, and melon. Fruits produce distinct volatile compounds during ripening that impart the characteristic fresh fruit flavor consumers desire. The importance of volatile production and factors influencing it in fruit has been intensively discussed (Song and Forney, 2008). Apples, strawberries, banana, tomato, and melons represent a diversity of fruit ripening physiology. Apples, tomato, and banana are climacteric fruit that demonstrate a burst in respiration and ethylene production in association with the onset of ripening and volatile production during fruit ripening. Strawberries, on the other hand, are non-climacteric fruit and ripen without any apparent increase in respiration or ethylene production (Dull and Hulme, 1971). Melons are a diverse group that express both climacteric and non-climacteric characteristics (Kendall and Ng, 1988). More than 400 volatile compounds are produced by apple, banana, strawberry, tomato and melon fruit, which are comprised of diverse classes of chemicals, including esters, alcohols, aldehydes, ketones, and terpenes (Dirinck et al., 1989; Larsen and Poll, 1992; Beaulieu, 2006; Baldwin et al., 2007). Some volatile compounds are only produced in certain fruit (Song and Forney, 2008). Many of these volatile compounds are produced in trace amounts, which are below the thresholds of most analytical instruments, but can be detected by human olfaction. Human perception of volatile compounds is determined by two primary factors: fruit volatile concentration and the human aroma perception threshold. The aroma thresholds of volatile compounds help to relate their physical-chemical properties with human perception. Aroma thresholds were first applied to tomatoes to identify 16 important compounds such as cis-hexenal, ß-ionone, hexanal, ß-damascenone, 1-penten-3-one, 2+3 mehtylbutanal, and so on, contributing to tomato aroma (Buttery et al., 1989). Unlike tomatoes, most fruit such as apple, banana, strawberry, and melons produce esters which are the major aroma components. Based on the quantitative abundance and olfactory thresholds of esters, a fraction of these compounds have been identified as fruit flavor impact compounds (Cunningham et al., 1986; Shieberler et al., 1990; Wyllie et al., 1995). For example, using aroma thresholds, ethyl butanoate, ethyl 2-methylbutanoate, 2-methylbutyl acetate, ethyl hexanoate, and hexyl acetate were identified as important volatiles in Fuji apples (Lara et al., 2006).

In spite of the fact that overwhelming numbers of chemical compounds have been identified as volatile compounds in fresh fruit, many volatile compounds have similar chemical structure and can be classified into specific chemical groups. Their similar chemical structure may also be helpful to link the origin of these compounds, which may have similar biosynthesis mechanisms. Regardless of their contribution to the odor or aroma threshold, most of the volatile compounds in fruit have systematic structures with C_2, C_3, C_4, C_5, and C_6 carbon chains as building blocks, either with straight or branched chains. For example, butyl acetate and hexyl acetate are dominantly produced in apples, while 3-methyl butyl acetate is predominant in banana. Despite the diversity of volatile compounds among these fruit, the basic biosynthetic pathway may be similar. Even with intensive efforts using chemistry, biochemistry, and molecular biology, most of the pathways leading to fruit flavor volatile biosynthesis have not yet been determined. Presently it is believed that the metabolism of fatty acids and branched amino acids may serve as precursors for the biosynthesis of aroma volatiles in most fruit (Fellman et al., 2000; Perez et al., 2002; Rudell et al., 2002). Fatty acids play a major role in ester synthesis providing 2, 4, and 6 carbon straight chains. Straight-chain esters in whole fruit arise predominantly through ß-oxidation of fatty acids (Rowan et al., 1999). It is widely assumed that lipoxygenase (LOX) may contribute to the breakdown of long-chain fatty acids to C_6 aldehydes, which are converted to alcohols by aldehyde dehydrogenase (Rowan et al., 1999; Fellman et al., 2000). *De novo* synthesized free fatty acids also contribute to the formation of straight chain esters in apples (Song and Bangerth, 2003).

Branched amino acids leucine and isoleucine are important substrates for branched chain volatiles such as 2-methylbutyl acetate and ethyl-2-methylbutanoate in apple or 3-methylbutyl acetate in banana (Tressl and Drawert, 1973; Fellman et al., 2000). Feeding studies show this pathway may be present in many fruit such as apple, banana, and strawberry (Tressl and Drawert, 1973; Rowan et al., 1996; Fellman et al., 2000; Perez et al., 2002). In melons, both fatty acid and branched amino acid biosynthesis are important contributors to volatile formation. The amino acids alanine, valine, leucine, iso-leucine, and methionine increase during fruit ripening, in close association with volatile production and are believed to supply the carbon chains for four groups of esters, ethyl acetate, 2-methylpropyl, 2-methylbutyl and thioether ester, respectively (Wyllie et al., 1995; Wang et al., 1996). When a comparison of amino acid content was made between the highly aromatic melon Makdimon and the low-aroma melon Alice, no significant difference in volatile concentration was found. Therefore, the difference in volatile concentrations in melon is not due to the availability of amino acid substrates, but rather is dependent on other biosynthetic pathways (Wyllie et al., 1995).

3.3   Major enzymes of flavor volatiles production and regulation in fresh fruits

3.3.1   Alcohol acetyl transferase (AAT)

More than 80% of the volatiles produced by ripe apple, strawberry, banana, and melon are esters (Dirinck et al., 1989; Song and Forney, 2008). Most esters have sweet and fruity odors and relative low aroma thresholds. These chemical characteristics make ester compounds important contributors to fruit flavor and they usually comprise the major flavor impact compounds, being largely responsible for consumer perception of fruit flavor. Therefore, production of these compounds has been the major focus of research in the past a few years. The enzyme that is responsible for the final step of ester formation is acyl CoA alcohol transferase AAT (EC 2.3.1.84), which combines alcohols and acyl CoAs to form esters. AAT is a common enzyme existing in many fruit and belongs to the gene family BAHD (**B**EAT, **A**HCTs, **H**CBT, and **D**AT). Plants contain a large number of acyl transferases with approximately 88 found in *Arabidopsis* and more than 40 in rice. Only a few in *Arabidopsis* have been characterized for biochemical function (St-Pierre and De Luca, 2000). To date, a few genes encoding AATs have been cloned in apples, banana, strawberries, and melons. The gene encoding for enzyme AAT (MpAAT1) gene was successfully cloned in apples (*Malus pumila* cv. Royal Gala) (Souleyre et al., 2005). The MpAAT1 gene was expressed in leaves, flowers, and fruit. This gene produced a protein that contains features similar to other plant acyl transferases. It has the ability to utilize a broad range of substrates including C_3-C_{10} straight chain alcohols, aromatic alcohols and branched chain alcohols. However, the binding of the alcohol substrates is the rate-limiting step compared with the binding of the CoA substrates. Another gene encoding for enzyme AAT, MdAAT2 was cloned in "Golden Delicious" apples (Li et al., 2006). It had high sequence identity (93.4%) to MpAAT1 but low sequence homology with other AATs such as strawberry and melon. In contrast to other apple cultivars, MdAAT2 of "Golden Delicious" was exclusively expressed in the fruit tissue. The MdAAT2 protein is 47.9 kD and is localized primarily in the fruit peel. The expression of the MdAAT2 protein was regulated at the transcription level in the fruit peel and showed the highest activity at late developmental stages and was inhibited by treatment with 1-MCP, an ethylene action inhibitor. When apple AAT2 gene was induced in tobacco leaves, the transgenic tobacco leaves did not produce any apple-like volatile esters; rather, they produced compounds such as methyl caprylate, methyl caprate, and methyl dodecanoate, and the concentration of methyl benzoate and methyl tetradecanoate were also significantly increased (Li, et al., 2008). These results indicated that AAT2 reacts with a very broad range of substrates including both alcohols and acyl CoAs, which may be the case for all AATs.

In strawberry, AAT is one of the most studied genes in volatile biosynthesis. Both wild and cultivated strawberry fruit produce linear esters such as ethyl butanoate, ethyl hexanoate, octyl acetate, and hexyl butyrate, in contrast to banana fruit, which produce predominantly isoamyl acetate and butyl acetate. The expression of these genes is related to the onset of the green and breaker stages of ripening (Nam et al. 1999). Micro-array analysis of gene expression in strawberry demonstrated that AAT is exclusively expressed in fruit tissue and demonstrates a 16-fold increase in activity

from the pink to the full red stage of ripeness (Aharoni, 2004). Both AAT
in cultivated and wild strawberry are related to a group of genes that
encodes enzymes with unrelated substrates. The amino acid sequence
comparison among the banana, strawberry and wild strawberry reveals
that there are 86% identical with two regions primarily around position 60
and 430 (Beekwilder et al., 2004). All these acyltransferase genes expressed
in fruit indicate the ability to produce a number of esters using a variety
of combinations of alcohols and acetyl-CoAs. Attempts to engineer ester
production in plants such as petunia by using the strawberry AAT gene
resulted in the AAT enzyme activity but failed to increase ester production or increase benzylbenznoate. This indicted that AAT alone is not the
limiting factor for ester formation, but rather alcohol availability may be
limiting. If the goal is to increase ester production, increasing alcohol production should be considered (Beekwilder, 2004).

In melons, four genes encoding CM-AATs (CM termed for *Cucumis
melo*) have been isolated and characterized (Shalit et al., 2001; El-Sharkawy
et al., 2005). Those genes encoding enzymes comprise significantly different functions that contribute to volatile ester formation. All AAT gene-
encoded proteins, except for CM-AAT-2, were enzymatically active upon
expression in yeast and demonstrated substrate preferences to produce different ester compounds (El-Sharkawy, 2005). Further screening of amino
acids that are unique to melon AATs confirmed that at least four amino
acid residues were unique to the melon Cm-AATs in general or to specific
melons, including an aromatic amino acid, phenylalanine at the position
of 49 (F49) in Cm-AAT1, alanine at the position 61 (A61) for Cm-AAT3,
glutamine at 135 (Q135), and leucine at 339 (L339) in Cm-AAT4. Replacing
F49 by leucine (L) in Cm-AAT1 induced a change in stereoisomer recognition of the preferred substrates and a reduction of the range of the esters
produced, while replacing A61 by V in Cm-AAT3 greatly extended the
range of substrates accepted by the enzyme (Lucchetta et al., 2007). By
comparing the amino acid sequence of all AATs, it appears that alanine
268 is unique to Cm-AAT2 while other AATs have a threonine at this
position. The enzyme activity of Cm-AAT1 can be significantly reduced
if the threonine is replaced by alanine, indicating that one single amino
acid can play a decisive role in determining AAT activity. After mutation of Cm-AAT2, by replacing alanine with threonine and Cm-AAT1 by
replacing threonine with alanine, CmAAT-2 was capable of producing a
wide range of esters, while Cm-AAT1 enzyme activity was reduced significantly. This result clearly indicated that AAT is important in forming
esters. All melon AAT genes are expressed in the fruit at the ripe stage
and are inhibited by antisense ACC oxidase and treatment with 1-MCP
(El-Sharkawy, 2005). Characterization of AATs in fruit and other plant tissues demonstrated a highly variable range of Km for acetyl CoA from 0.02
mM to 2 mM, while the Km for alcohol can be up to 46.5 mM (Lucchetta,

2007; Shalit, 2001). It is suggested that the AATs in melons is in tetrameric form with a molecular weight around 200 kDa. This enzyme is strongly regulated by the product of the reaction and dependent on the concentration of CoA-SH, which can be either stimulatory or inhibitory to the ester-forming activity (Lucchetta, 2007). It is well known that the ester formation in melons is also dependent on substrate availability.

Feeding alcohols to the fruit has been used to study ester formation (AATs) in many fruit such as apples, bananas, and strawberry (Song, 1994; Jayanty et al., 2002; Ferenczi et al., 2006; Khanom and Ueda, 2008; Song and Forney, 2008). The conversion of alcohols into their corresponding esters demonstrated that the bottleneck of ester production in fruit is the availability of alcohols and substrates, which limits fruit volatile biosynthesis. Background levels of AAT in apples are present even at early developmental stages prior to ripening (Song, 1994; Li et al., 2006). In contrast, AAT in banana was up-regulated in banana pulp after ripening was initiated by ethylene (Medina -Suarez et al., 1997). When feeding apple fruit disks with alcohols at different developmental stages, differences in ester formation was found, indicating that the variety and quantity of acetyl-CoA play an important role in apple ester production (Li et al., 2006). Data collected from substrates studies shed light on these AATs that they have broad substrate specificity for alcohols. For example, the study with different alcohol substrates in combination with acetyl CoA, butyryl-CoA, and hexanoyl-CoA demonstrated a high affinity of strawberry AAT for geraniol in combination with acetyl-CoA, 1-octanol with butyryl CoA and 1-nonanol for hexyanoyl-CoA. 1-Butanol weakly reacted with acetyl CoA, but it readily combined with butyryl CoA and hexanoyl CoA. It is well known that butyl acetate is only produced in small amounts in strawberry (Khanom and Ueda, 2008). Similar results were found in apples, where the preference of MpAAT1 for alcohol substrates is dependent on substrate concentration, which therefore determines the aroma profiles of apple fruit (Souleyre et al., 2005).

In a study using melon pulp slices of two cultivars that were incubated with both aliphatic and aromatic alcohols for 48 h at 30°C, production of corresponding esters were found. Major amounts of esters such as hexyl acetate, isobutyl acetate, isoamyl acetate, as well as benzyl acetate and 3-phenyl–propyl acetate were found (Khanom and Ueda, 2008). This result indicates that AAT actively converted both aliphatic and aromatic alcohols with acetyl CoA in melon pulp tissue. This is further supported by the functional characterizations of AATs in banana and strawberry, which indicate that the aroma profiles in a given fruit species are determined by the supply of precursors (Beekwilder, 2004).

Different AATs exist among apple cultivars and tissues and their levels are differentially controlled during fruit development (Holland et al., 2005). As with the AAT in melons and strawberry, substrate availability rather than AAT activity is the limiting factor for ester formation in apples (Souleyre et al., 2005). It is also interesting to look at the evolution of AAT in fruit as well as in other plant tissues. Apparently, AATs are present in many plant tissues and serve as detoxifiers for plants to remove the unwanted metabolites/compounds or substrates from the cell (Gang, 2005). Ester formation can be a very effective method to vaporize unwanted compounds or transfer them to other forms to be used as intermediates. Pichersky and Gang (2000) suggested a special form of convergent evolution in which new enzymes with the same function evolve independently in separate plant lineages from a shared pool of related enzymes with similar but not identical functions. It was concluded that the metabolic diversity is also caused by low enzyme specificity and directly related to availability of suitable substrates. Therefore, future work should focus on the entire metabolic pathway rather than on a single enzyme at a time.

Physiology and biochemistry studies revealed that AAT is under control of fruit development and is regulated during fruit ripening. When apple fruit was treated with 1-methylcyclopropene (1-MCP), an ethylene action inhibitor, MdAAT2 was inhibited, but partially recovered with ethylene treatment 20 days after 1-MCP treatment, indicating that MdAAT2 is influenced by ethylene (Li, 2006). Using transgenic apple lines that block ethylene biosynthesis by reducing 1-aminocyclopropane 1-carboxylic acid (ACC) oxidase or ACC synthase (ACS), AAT activity was found to be regulated by ethylene, while other enzymes such as alcohol dehydrogenase (ADH) and lipoxygenase (LOX) were unaffected by ethylene modulation (Defilippi et al., 2005). Total ester production was inhibited by 65–70% in the transgenic apple fruit silenced for ethylene, while alcohol precursors were inhibited by 12–38% (Dandekar et al., 2004). No major differences were found in aldehyde production; however, a significant change in the ratio of hexanal/E-2-hexenal was observed. Similar results were found with apples treated with 1-MCP, where volatiles, organic acids, and sugar metabolism were found to be ethylene dependent (Defilippi et al., 2004). Other enzymes such as LOX, ADH, and pyruvate decarboxylase (PDC) are believed to be involved in the pathways to provide aldehydes and alcohols for ester synthesis (Fellman et al., 2000), although the source of alcohols and aldehydes for ester synthesis in fruit is not fully understood. A poor relationship between LOX activity and fruit volatile production was found in "Golden Delicious" apples at early and mid-maturity harvests. Late-harvested fruit demonstrated an increase of LOX activity associated with fruit senescence. It was suggested that LOX may not be directly involved in ester formation, but rather newly synthesized free fatty acids may serve as precursors for volatile biosynthesis (Song and Bangerth, 2003). However, using multivariate analysis of the biosynthesis of volatile compounds, no difference in AAT activity was found between controlled atmosphere (CA) and refrigerated air (RA) storages while, but rather different levels of LOX and PDC activity are responsible for the production of different volatiles occurring in CA and RA stored fruits (Lara et al., 2006).

3.3.2 Lipoxygenase

Lipoxygenase (LOX, Linoleate:oxygen reductase, E.C. 1.13.11.12) plays important roles in both plant defense and flavor formation. It is one of the most studied enzymes in fruits and vegetables (Matsui et al., 2001). LOX catalyzes the addition of molecular oxygen at either the C_9 or C_{13} residue of unsaturated fatty acids with a 1,4,-pentadiene structure. Linoleic acid and linolenic acid are the most abundant fatty acids in the lipid fraction of plant membranes and are the major substrates for LOX to form fatty acid hydroperoxides (HPOs). Cleavage of the HPOs forms short-chain aldehydes and oxo-acids through the action of fatty acid hydroperoxide lyase (HPL). The fate of fatty acids in this enzyme system is determined by the substrates, LOX specificity and HPL (Rosahl, 1996). It is well known that LOX produces C_6 and C_9 carbon aldehydes, which are the significant flavor compounds in many fruits and vegetables when tissues are mechanically damaged, homogenized, or chewed. In many fruit tissues, however, those C_6 and C_9 volatiles cannot be detected in intact tissues and they become the substrates for further flavor metabolism. Therefore, LOX can be seen as an enzyme responsible for secondary volatile generation that directly influences human flavor perception. Different fruits and vegetables have different volatile profiles resulting from the LOX pathway. For example, to produce C_6 aldehydes, 13-HPL acts on 13-HPO to form C_6 aldehydes as well as 12-oxo-(Z)-9-dodecenoic acid, which is found in tomatoes (Baldwin et al., 1998; Baldwin et al., 2007; Baldwin et al., 2008). In cucumbers and melons, however, both 9/13-HPL and 9-HPL were found to react with 9- and 13-HPOs to form C_9 aldehydes (Matsui et al., 2006). Interestingly, there have been a few attempts to modify the flavor composition in tomato fruit by expression of targeted genes. For example, tomato is generally believed to have high LOX activity, which results in the formation of 9-HPOs. However, when cucumber fatty acid hydroperoxide lyase, which acts on 9-HPO from fatty acids to form C_9 aldehydes, was introduced into tomato plants, it resulted in high HPL activity in both leaves and fruit of tomatoes. However, the production of short-chain volatile aldehydes and alcohols was not enhanced. When linoleic acid was added to a crude homogenate prepared from the transgenic tomato fruit, a high amount of C_9-aldehyde was formed, while no difference in C_6 aldehyde was evident. This result revealed that the formation of 13-HPO of fatty acids is preferably formed from endogenous substrates. In contrast, 9-HPO is formed from the exogenous fatty acid substrates (Matsui et al., 2001). Adding exogenous LOX and ADH to the tomato homogenate decreased the concentration of hexanal, cis-3-hexenal, and trans-2-hexenal (Yilmaz et al., 2001). Significant differences in enzyme activity of LOX, HPL, and ADH were also found at green, pink, and red stages of ripeness from 12 tomato selections. Unfortunately, there were no predictable patterns in volatile compounds formed as a function of the activity of these enzymes when volatile compounds were analyzed from the whole fruit tissues. These results suggest that either there is little or no direct relationship between these enzymes and volatile compounds formed during fruit ripening and/or direct headspace analysis of these volatile compounds did not reflect the activity of these enzymes and in fresh tomato fruit (Yilmaz et al., 2001; Yilmaz et al., 2002). In the same study, a significant difference in enzyme activity of 12 selections of tomatoes was found. It has been proposed that the synthesis of the C_6 volatiles is limited by the availability of non-esterified fatty acids to the chloroplast localized 13-lipoxygenase. The chloroplast-to-chromoplast transition and disruption of the thylakoid membrane resulted in bringing 13-lipoxygenase in contact with its fatty acid substrates (Mathieu et al., 2009). Using antisense reduction of one iso-form of the 13-lipoxygenase, TomloxC, resulted in almost complete loss of multiple C_6 compounds in tomato fruit. These data suggested that TomloxC is responsible for the formation of C_6 volatiles in tomato fruit (Chen et al., 2004). These studies point out that the modification of volatile formation has to consider the effect of enzyme localization and maceration of tissues. To improve total volatile production, efforts to improve availability of substrates are essential if substrates are the limiting factor of volatile synthesis, which confirmed the hypothesis that LOX is not a limiting factor for volatile production in tomato fruit (Griffiths et al., 1999).

Changes in LOX activity during fruit ripening has been reported by various studies conducted with apple and strawberry (Defilippi et al., 2005; Leone et al., 2006). In apple, a large number of LOXs have been found in a non-redundant EST sequence dataset, along with ADH and branched amino acid biosynthesis enzymes (Newcomb et al., 2006). LOX activity increased during fruit ripening; however, no close relationship between LOX activity and fruit respiration, ethylene production, and volatile production was found in apple fruit during ripening (Song and Bangerth, 2003). Combining enzyme activity with enzyme localization and immunolocalization analysis, researchers found that LOX forms may have specific locations in different cell compartments of strawberry fruit and their activity is temporally differentiated. Applying 2-dimensional plots, at least two mobility groups of LOXs were found with molecular weights of 100 and 98 kDa and pI ranging between 4.4 and 6.5. However, LOX activity in ripe strawberry fruit was not higher than that in unripe and turning stages fruit. The presence of different LOX isoforms in strawberry fruits and that the lipoxygenase-hydroperoxide lyase pathway plays role in converting lipid to C_6

volatiles during ripening (Leone et al., 2006). Using a transgenic apple line suppressed for ACC oxidase or ACC synthase, it was reported that LOX enzyme activity increased only slightly during ripening and responded to an exogenous ethylene treatment (Defilippi et al., 2005). Other enzymes other than LOX in the pathway, such as HPL, which has shown substrate specificity, may play a role in regulating the formation of aldehyde. Apparently, more research is needed to clarify the role of LOX in formation and regulation of volatiles. In order to identify common mechanisms, a standard analytical procedure for studying, both LOX and secondary volatile production should be employed to allow comparison between fruit tissues and laboratories.

3.3.3 Alcohol dehydrogenase (ADH) and pyruvate decarboxylase (PDC)

In fruit volatile production systems, LOX may not act alone. Another enzyme responsible for formation of aldehydes and alcohols is ADH (alcohol dehydrogenase, EC:1.1.1.1). It is well known that ADH converts aldehydes and alcohols back and forth depending on the condition of the tissue. The predominant role that ADH plays in the production of aldehydes and ethanol has been studied in fruits and vegetables under stress conditions such as high CO_2 or low oxygen (Ke et al., 1994; Prestage et al., 1999; Imahori et al., 2003; Saquet and Streif, 2008). Despite the common understanding of the role of ADH in fruit, a limited role has been found for ADH in ester formation of apple fruit during ripening and in response to ethylene treatment. It was reported that ADH activity increased at the beginning of fruit ripening and then decreased gradually in peel tissue and remained constant in flesh tissue. ADH activity was higher in the fruit of an ACO antisense line as compared to controls but did not change after exposure to ethylene. In addition, there is no association between ADH and alcohol concentration (Defilippi et al., 2005). Observation of volatile production and enzyme activity of LOX, PDC and ADH from apple fruit maturing on the tree indicated that both LOX and HPL may act together to control emission of volatile compounds. As fruit approached the climacteric stage, PDC activity increased in skin tissues concurrently with acetaldehyde production. It was found that acetaldehyde production preceded commercial harvest by about one week, immediately following an upsurge of LOX, HPL, and ADH. ADH activity, measured in both peel and flesh tissues, increased as fruit approached maturation, while production of ethanol decreased. These results demonstrated that high concentrations of acetaldehyde found in fruit of advanced maturation was related to PHL rather than PDC (Villatoro et al., 2008). It has been reported that acetaldehyde in fruit can be formed from pyruvic acid through the action of PDC, from fatty acids via the LC/HPL pathway, or from ethanol through enzyme oxidation by ADH. The decrease of ethanol production throughout maturation was believed to be due to the diversion to acetaldehyde rather than ethyl acetate (Villatoro et al., 2008).

Two highly divergent ADH genes (CmADH1 and CmADH2) are specially expressed in ripening melon fruit and are up-regulated by ethylene in Charentais cantaloupe melons (*Cucumis melo* var. *cantalupensis*). They encode proteins that operate preferentially as aldehyde reductase using acetaldehyde in the presence of NADPH. Besides aliphatic aldehydes, CmADH1 utilizes branched aldehydes such as 2- and 3-methylbutyraldehyde, 2-methylproponaldehyde, and aromatic aldehyde, but the activity for these types of substrates is much lower than that for acetaldehyde. Cm-ADH1 has a K_m for acetaldehyde that was 10 times lower than the K_m for ethanol. The respective V_{max} were of 2500 μmol mg protein^{-1}min^{-1} and 5000 μmol mg protein^{-1} min^{-1}. Induction of these enzymes is closely associated with fruit ethylene production and action, which is inhibited by 1-MCP, indicating that they are under the control of ethylene. Sequence analysis indicated that CmADH1 has 83% homology with apple Md-ADH (Manríquez et al., 2006).

Transformed tomato plants with fruit-specific expression of the transgene(s) of ADH displayed a range of enhanced ADH activities in the ripening fruit, but no suppression of ADH was observed (Speirs et al., 2002). Preliminary sensory evaluations indicated that fruit with elevated ADH activity and higher levels of alcohols were found to have a more intense "ripe fruit" flavor. Modified ADH levels in the ripening fruit influenced the balance between some of the aldehydes and the corresponding alcohols associated with flavor production. Hexanol and Z-3-hexenol levels were increased in fruit with increased ADH activity and reduced in fruit with low ADH activity (Speirs et al., 1998).

Pyruvate decarboxylase (E.C. 4.1.1.1) is one of the enzymes specially required for ethanol formation through fermentation and catalyses the decarboxlyation of pyruvate to acetaldehyde. In strawberry fruit, PDC gene expression correlates with that of AAT during fruit development. It is believed that PDC plays a role in volatile formation, especially with ethyl esters in strawberry fruit (Aharoni et al., 2000). Two PDC genes (Fapdc1 and Fapdc3) were reported in strawberry fruit, and their expression patterns were investigated. The Fapdc1 gene seemed to play a role in fruit ripening, aroma biosynthesis, and stress response, while Fapdc3 showed a consistent expression pattern throughout fruit ripening (Moyano et al., 2004). Further research on LOX, PDC, and ADH in fruit will enhance our understanding of volatile biosynthetic pathways and reveal mechanisms controlling substrate availability for ester formation (Moyano et al., 2004).

3.3.4 Other enzymes

In addition to AAT, LOX, ADH, and PDC, more enzymes, especially those that contribute to the metabolism of substrates for volatile seem to be involved in fruit volatile production. Global gene expression in an antisense ACC oxidase transgenic line of "Royal Gala" apple that produced no detectable levels of ethylene revealed that 17 genes from four biosynthetic pathways (fatty acid, branched amino acid [isoleucine], sesquiterpene, and phenylpropanoid biosynthesis) were found to be the control points for ethylene-induced aroma volatile production (Schaffer et al., 2007). Only certain points within the aroma biosynthesis pathways are controlled by ethylene. Often the first and final steps are important transcriptional regulation points within all four pathways. For example, genes related to fatty acid biosynthesis seemed not to respond to ethylene treatment, while only LOX1 and LOX7 were induced by ethylene. In the branched amino acid biosynthesis pathway, however, threonine deaminase (E.C. 4.3.1.19), the first step in isoleucine biosynthesis, is induced by ethylene treatment. Isoleucine, which is an important precursor for the 2-methyl branched esters in apples, increased in apple peel tissue in fruit ripening and in response to ethylene treatment (Defilippi et al., 2005). Exogenous ethylene treatment of immature apple fruit induced fruit ripening and volatile production, with preference to the branched esters (Song, 1994). Branched volatile compounds, especially 2-methylbutyl acetate, were not as strongly suppressed as other volatiles during CA storage in "Red Delicious" apple fruit. A related pathway from isoleucine to 2-methylbutylbutanoic acid was markedly reduced, resulting in lower volatile production (Rudell et al., 2002). The conversion of 2-methylbutanol to 2-methylbutyl acetate and 2-methylbutanoic acid to ethyl-2-methylbutanote and hexyl 2-methylbutanote was limited by the availability of 2-methylbutyl substrates which are derived from isoleucine, but not by acetyl CoA, ethanol, or hexanol (Matich and Rowan, 2007). In higher plants, a unique feature of branched amino acid biosynthesis is that Val and Ile are synthesized in chloroplast and with two parallel pathways each with a set of four identical enzymes catalyzing the reactions with different substrates. As an exception, threonine deaminase (TD) catalyzes the deamination of threonine (Thr.) to form α-ketobutylrate (KB) and ammonia. So far, TD has been expressed in young leaves and flowers and little is known about its expression in fruit. Biosynthetic TD activity is involved in isoleucine synthesis and is subject to feedback inhibition by this amino acid. Developmentally regulated, the biosynthetic enzyme TD is one of the most abundant proteins in mature tomato fruit. TD mRNA level has been reported to be 2–5 times higher in young fruit than in leaves (Samach et al., 1991). Its gene expression is induced by treatments of abscisic acid (ABA) or methyl jasmonate or wounding. The precise details of the pathway for all these branched volatiles synthesis and the enzymes responsible have not been identified in plants, but a pathway has been described in yeast (Dickinson et al., 2000; Dickinson et al., 2003). On the basis of structural considerations, it is assumed that 3-methylbutanal and 3-methylbutanol are derived from leucine and 2-methylbutanal and 2-methylbutanol from isoleucine. The biosynthesis of higher alcohols involves the decarboxylation of keto acids to form aldehydes, followed by a reduction of the aldehydes to produce the alcohols. Therefore the branched amino acids are synthesized from the keto acids and during fermentation via the branched amino acids metabolic pathway by decarboxylation and reduction (Dickinson, et al., 2000). The proposed pathway begins with the action of branched chain aminotransferase (BCAT) that removes the amino groups from the respective amino acids to form the corresponding α-keto acids. PDC then converts the resulting α-keto acids to the corresponding branched chain aldehyde. The final step of amino acid catabolism (for conversion of an aldehyde to an alcohol) can be accomplished by ADH or by the SFA1-encoded formaldehyde dehydrogenase (Dickinson et al., 2003).

In apples, a large number of LOXs have been found in a non-redundant EST sequence dataset, along with ADH and all branched amino acid biosynthesis enzymes (Newcomb et al., 2006). Applying EST frequency analysis with this EST database revealed that EST clusters from fruit-derived tissue show strong sequence homology with biochemically characterized enzymes that are involved in ester biosynthesis including acyl-CoA dehydrogenase, acyl-CoA oxidase, enoyl-CoA hydratase, malonyl-CoA:ACP transacylase, LOXs, 3-ketoacyl-CoA thiolase, acyl-CoA synthetase, and acyl carrier proteins (Park et al., 2006). These findings add new clues for the identification of enzymes responsible for aroma volatile biosynthesis beyond AAT, LOX, and fatty acid ß-oxidation, and may open new opportunities for aroma volatile research to identify unknown pathways related to fruit volatile biosynthesis. Therefore, the enzymes regulating the branched amino acid pathway during fruit ripening deserve more in-depth research.

Another important aroma compound is α-farnesene, which is synthesized through the sesquiterpene biosynthesis pathway. In this pathway, 3-hydroxy-3-methylglutaryl (HMG)-CoA synthase and α-farnesene synthase1 biosynthesis were strongly induced by ethylene treatment (Schaffer et al., 2007). The gene of α-farnesene synthase 1 was cloned in apple tissue, which is closely related to fruit ripening and under the control of ethylene (Pechous and Whitaker, 2004; Lucchetta et al., 2007; Zhu et al., 2008). In the phenylpropanoid pathway, Phe ammonia lyase and O-methyltransferase were induced by ethylene. Both enzymes may be involved in the biosynthesis of estragole in apples (Schaffer et al., 2007).

The enzyme O-methyltransferase (OMT, E.C.2.1.1.x), which is responsible for methylation of 2,5-dimethyl-4-hydroxy-3(2H) furanone (DMHF) to 2,5-dimethyl-4-methyoxy-3(2H) furanone (DMMF) in strawberries, was reported by Wein et al. (2002). Northern-hybridization indicated that the strawberry–OMT specific transcripts accumulated in fruit during ripening, but were absent in leaf, petiole, root, and flower tissues. The protein exhibited substrate specificity for catechol, caffeic acid, protocatechuic aldehyde, caffeoyl CoA, and DMHF, with a common structural feature of the accepted substrates with an o-diphenolic structure. Based on the expression pattern of strawberry OMT and enzyme activity at different fruit maturities, it was concluded that the strawberry OMT plays an important role in the biosynthesis of strawberry volatiles such as vanillin and DMMF (Wein et al., 2002). Furaneol, methoxyfuraneol, and O-methyltransferase activity increased significantly at the ripe stage in strawberry fruit, and more than one O-methyltransferase may be present in the fruit (Lavid et al., 2002). DMHF and its methoxy derivative are also important volatile compounds contributing to strawberry fruit flavor (Bood and Zabetakis, 2002). A *Fragaria × ananassa* quinine oxidoreductase (FaQR) was identified and shown to form HDMF, an important volatile compound in strawberry (Raab et al., 2006). FaQR is ripening related and showed maximum activity at the full red stage. Both HDMF and DMMF biosynthetic enzymes offer potential targets to engineer the flavor of strawberry from breeding populations.

A cDNA termed CmCCD1 (carotenoids cleavage dioxygenase) has been isolated from Tam Dew melon fruit, and its coding protein is similar to other plant carotenoid cleavage dioxygenase enzymes with a broad substrates specificity for the carotenoids precursors, but specific for the site of cleavage. CmCCD1 gene expression in melons is up-regulated during fruit ripening. Since the Tam Dew melon is a white-fleshed cultivar that does not produce any carotenoids or apocarotene volatiles, it is believed that the accumulation of ß-ionone, an important flavor compound in melons, is limited by the availability of carotenoids. This discovery suggests that color formation and biosynthesis of certain volatiles may share the same metabolic pathways and precursors during fruit ripening (Ibdah, 2006).

Using transgenic techniques to express *Ocimum basilicum* geraniol synthase under the control of a tomato ripening specific polygalacturonase promoter resulted in monoterpene accumulation and increase of geraniol and derivative compounds such as geraniol, geranial, geranyl acetate, citronellal, and rose oxide. Although the fruits were evaluated by untrained panelists as preferred, the transgenic fruit failed to develop deep red color due to the decrease of lycopene by 50%. In addition, phytoene content was also reduced by 70–90% (Davidovich-Rikanati et al., 2007). This result indicated that a transgenic approach may be feasible to improve fruit volatile production; however, multiple pathways may need to be considered as a systematic approach to balance the substrate pools (Davidovich-Rikanati et al., 2007). Surprisingly, transgenic fruit produced not only geraniol and geranial but also geranyl acetate and citronellyl acetate, implying the involvement of endogenous acetyl-CoA: alcohol acetyl transferase, contrary to the common belief that there is no AAT in tomato fruit. This result further suggests that flavor biosynthesis enzymes, especially AAT, could be induced by substrates. This hypothesis can be supported by the characterization of strawberry OMT, which indicates that the evolution of secondary metabolites does not proceed step by step. Due to the presence of multifunctional enzymes, new plant metabolites can be converted by several existing biocatalysts. Formation of DMHF probably led to the immediate production of DMMF and its corresponding glucoside. Thus, at a specific moment in evolution three new products evolved simultaneously (Wein et al., 2002).

3.4   Enzymes of flavor volatiles production and regulation in vegetables—carrot

Most vegetables are vegetative tissues and do not have the synthetic mechanisms to produce esters, while many contain ADH, PDC, and LOXs. In contrast to most fruits, vegetables contain a large group of terpenes; these are important volatile compounds that play an important role in influencing the sensory qualities that are measured as taste and flavor. Fresh carrot presents a diverse volatile profile from the most fruit that is worthwhile to be discussed.

The majority of volatile constitutes emitted from fresh carrots are mono- and sesquiterpenes. These compounds can comprise up to 97% of the total volatile compounds. Based on the GC-olfactory (GC-O) description, in general, volatile of carrots can be divided into three groups that represent the overall volatile characteristics of carrots: *carrot top, fruity,* and *spicy-wood* (Kjeldsen et al., 2003). It was reported that characteristic carrot-top odor was related to α-pinene, ß-pinene, sabinene, α-phellandrene, ß-myrcene, and p-cymene. The monoterpenes, such as limonene, γ-terpinene, and terpinolene, contribute to the citrus-like and fruity notes, while the major sesquiterpenes, ß-caryophyllene, α-humulene, ß-bisabolene, and (E)-and (Z)-y-biabolene, were found to be related to spicy and woody notes. Therefore, increasing the metabolisms related to monoterpenes and reducing the sesquiterpenes may improve the overall flavor of carrots.

Bitter and harsh taste in carrots often causes consumer rejection and is one of the main reasons for low preference scores when carrots are evaluated by sensory method.

The harsh bitterness of carrots was first described by Simon et al. (1980), who described it as a strong burning turbintin-like flavor most clearly perceived at the back of the throat during chewing. Harsh flavor seemed to be correlated with high content of volatile and low sugars. A strong relationship was found between the sensory attributes terpene flavor, burning aftertaste, and bitterness and the level of p-cymene, γ-terpinene, limonene, (E)-ß-farnesene, α-phellandrene, and terpinolene. Terpene aroma, green aroma, and carrot aroma were positively related with ß-myrcene, α-terpinene, and ß-phellandrene (Kreutzmann et al., 2008). However, it was impossible to predict the flavor sensation from individual terpene.

Based on the identification of volatile compounds, the volatile biosynthesis in carrots is derived from terpene metabolism. A feeding experiment with labeled precursors such as [5,5-^{2}H]- mevalonic acid lactone (d$_2$-MVL) and [5,5-^{2}H] -1-deoxy-D-xylulose (d$_2$-DOX) demonstrated independent *de novo* biosynthesis of terpenoids in carrot roots and leaves. In both tissues, monoterpenes are synthesized exclusively via the 1-deoxy-D-xylulose/2-C-methyl-D-erythritol-4-phosphate (DOXP/MEP) pathway, whereas sesquiterpenes are generated by the classical mevalonic acid pathway (MVA) as well as by the DOX/MEP route (Hampel et al., 2005). This study proved that the biosynthesis of terpenes of carrot roots is mainly localized in the phloem, while a de novo biosynthesis of ß-caryophellene in the xylem is also present. Both mono-and sesquiterpenes are biosynthesized via the novel DOX/MEP route, while the MVA pathway is exclusively reserved for the biosynthesis of sesquiterpenes.

3.5 Conclusions

Production of flavor volatile compounds in fruits and vegetables is the result of complex metabolic networks involving many pathways and control mechanisms. The metabolomic diversity is also caused by low enzyme specificity and is directly related to the availability of substrates. Future work must therefore take into account the entire metabolic pathway rather than a single enzyme. While precursor studies of early harvested or 1-MCP treated fruit indicate that AAT is not the limiting factor in volatile production, it may also imply that both fatty acids and branched amino acid substrates were physically unavailable to the enzyme. Due to the overwhelming evidence that substrates may be the bottleneck of volatile production in most fruit, it becomes very important to clarify the compartmentalization of volatile biosynthesis within the fruit cell to understand the biosynthesis of those substrates or precursors and their compartmentalization and transport in the cell.

Substrates are important for volatile production, but they may not be the only limiting factor. Metabolic energy in forms such as ATP may also limit volatile production, especially in early harvested, 1-MCP treated, or long-term CA stored fruit. All of these conditions reduce fruit metabolism and respiration (Rudell et al., 2002), which may result in limited energy supply. In CA-stored apple fruit, low rates of respiration decreased ATP as well as the ATP:ADP ratio, which was associated with decreased fatty acid biosynthesis, while other co-factors such as pyridine nucleotides (NADH and NADPH) were not impaired (Saquet et al., 2003). Volatile biosynthesis in both climacteric and non-climacteric fruit is highly integrated with fruit ripening and senescence. Therefore, a better understanding of fruit ripening and its triggers will help in our understanding of fruit volatile production (Alexander and Grierson, 2002).

It is also necessary to remember that volatile biosynthesis pathways are only a part of the complex network of fruit metabolism and are influenced by many factors, such as genetics, production practices, and postharvest handling. To fully understand volatile biosynthesis, more detailed biochemical and molecular studies are needed to investigate aroma volatile biosynthesis and its regulation. Immunolocalization, immunoblot, and fluorescence imaging techniques could help to determine protein expression, localization, activity state, and cell compartmentalization (Giepmans et al., 2006). An understanding of where the substrates are produced and where the enzymes are localized will improve our understanding of mechanisms of volatile biosynthesis regulation. Investigations at the genomic, proteomic, and metabolomic levels will bring major breakthroughs to understand the flavor biosynthesis pathways. Investigations on fatty acid and branched amino acid biosynthesis and their interactions during fruit ripening may clarify limiting factors of volatile synthesis.

Acknowledgment

The author would like to thank Dr. C. Forney at AAFC for his critical review of this manuscript and his constructive suggestions.

Abbreviations

1-MCP	1-methylcyclopropene
AAT	alcohol acyltransferase
ACC	1-aminocyclopropane 1-carboxylic acid
ADH	alcohol dehydrogenase
AVG	aminovinylglycine
CA	controlled atmosphere
DMHF	2,5-dimethyl-4-hydroxy-3(2H)–furanone (furaneol)
DMMF	2,5-dimethylene-4-methyoxy-3(2H)-furanone (mesifurane)
DOXP/MEP	1-deoxy-D-xylulose/2-C-methyl-D-erythritol-4-phosphate

EST	expressed sequence tags
GC	gas chromatography
HDMF	4-hydroxy-2,5-dimethylene-3(2H)-furanone
LOX	lipoxygenase
MJ	methyl jasmonate
MS	mass spectrometry
MVA	mevalonic acid
PDC	pyruvate decarboxylase

References

Aharoni, A. 2004. Functional characterization of enzymes forming volatile esters from strawberry and banana. *Plant Physiology* 135: 1865–1878.

Aharoni, A., L.C.P. Keizer, H.J. Bouwmeester, Z. Sun, M. Alvarez-Huerta, H.A. Verhoeven, J. Blaas, A.M.M.L. van Houwelingen, R.C.H. De Vos, H. van der Voet, H.R.C. Hansen, M. Guis, J. Mol, R. Davis, M. Schena, A.J. van Turnen, and A. O'Connell. 2000. Identification of the SAAT gene involved in strawberry flavor biogenesis by use of DNA microarrays. *Plant Cell* 12: 647–662.

Alexander, L., and D. Grierson. 2002. Ethylene biosynthesis and action in tomato: A model for climacteric fruit ripening. *Journal of Experimental Botany* 53: 2039–2055.

Baldwin, E.A., K. Goodner, and A. Plotto. 2008. Interaction of volatiles, sugars, and acids on perception of tomato aroma and flavor descriptors. *Journal of Food Science* 73: S294–S307.

Baldwin, E.A., A. Plotto, and K. Goodner. 2007. Shelf-life versus flavour-life for fruits and vegetables: how to evaluate this complex trait. *Stewart Postharvest Review* 1: 1–10.

Baldwin, E.A., J.W. Scott, M.A. Einstein, T.M.M. Malundo, B.T. Carr, R.L. Shewfelt, and K.S. Tandon. 1998. Relationship between sensory and instrumental analysis of tomato flavor. *Journal of the American Society for Horticultural Science* 123: 906–915.

Beaulieu, J.C. 2006. Volatile changes in cantaloupe during growth, maturation, and in stored fresh-cuts prepared from fruit harvested at various maturities. *Journal of the American Society of Horticultural Science* 131: 127–139.

Beekwilder, J., M. Alvarez-Huerta, E. Neef, F.A. Verstappen, H.J. Bouwmeester, and A. Aharoni. 2004. Functional characterization of enzymes forming volatile esters from strawberry and banana. *Plant Physiolology* 135: 1865–1878.

Bood, K.G., and I. Zabetakis. 2002. The biosynthesis of strawberry flavour. (II) Biosynthesis and molecular biology studies. *Journal of Food Science* 67: 2–8.

Buttery, R.G., R. Teranishi, R.A. Flath, and L.C. Ling. 1989. Fresh tomato volatiles: Composition and sensory studies. In R. Teranishi, R. Buttery, and F. Shahidi, eds., *Flavor Chemistry: Trends and Development*. Amer. Chem. Soc, Washington, DC.

Chen, G., R. Hackett, D. Walker, A. Taylor, Z. Lin, and D. Grierson. 2004. Identification of a specific isoform of tomato lipoxygenase (TomloxC) involved in the generation of fatty acid-derived flavor compounds. *Plant Physiology* 136: 2641–2651.

Cunningham, D.G., T.E. Acree, J. Barnard, R.M. Butts, and P.A. Braell. 1986. Charm analysis of apple volatiles. *Food Chemistry* 19: 147–147.

Dandekar, A.M., G. Teo, B.G. Defilippi, S.L. Uratsu, A.J. Passey, A.A. Kader, J.R. Stow, R.J. Colgan, and D. James. 2004. Effect of down-regulation of ethylene biosynthesis on fruit flavour complex in apple fruit. *Transgenic Research* 13: 373–384.

Davidovich-Rikanati, R., Y. Sitrit, Y. Tadmor, Y. Iijima, N. Bilenko, E. Bar, B. Carmona, E. Fallik, N. Dudai, J.E. Simon, E. Pichersky, and E. Lewinsohn. 2007. Enrichment of tomato flavor by diversion of the early plastidial terpenoid pathway. *Nature Biotechnology* 25: 899–901.

Defilippi, B.G., A.M. Dandekar, and A.A. Kader. 2004. Impact of suppression of ethylene action or biosynthesis on flavour metabolites in apple (Malus domestica Borkh) fruits. *Journal of Agricultural and Food Chemistry* 52: 5694–5701.

Defilippi, B.G., A.M. Dandekar AM, and A.A. Kader. 2005. Relationship of ethylene biosynthesis to volatile production, related enzymes, and precursor availability in apple peel and flesh tissues. *Journal of Agricultural and Food Chemistry* 53: 3133–3141.

Defilippi, B.G., A.A. Kader, and A.M. Dandekar. 2005. Apple aroma: Alcohol acyltransferase, a rate limiting step for ester biosynthesis, is regulated by ethylene. *Plant Science* 168: 1199–1210.

Dickinson, J.R., S.J. Harrison, J.A. Dickinson, and M.J.E. Hewlins. 2000. An investigation of the metabolism of isoleucine to active amyl alcohol in *Saccharomyces cerevisiae*. *Journal of Biological Chemistry* 275: 10937–10942.

Dickinson, J.R., L.E.J. Salgado, and M.J.E. Hewlins. 2003. The catabolism of amino acids to long chain and complex alcohols in *Saccharomyces cerevisiae*. *Journal of Biological Chemistry* 278: 8028–8034.

Dirinck, P., H. De Pooter, and N. Schamp. 1989. Aroma development in ripening fruits. In R. Teranishi, R. Buttery, and F. Shahidi, eds., *Flavor Chemistry: Trends and Development*. Amer. Chem. Soc., Washington, DC, pp 24–34.

Dixon, J., and E.W. Hewett. 2000. Factors affecting apple aroma/flavour volatile concentration: A review. *New Zealand Journal of Crop and Horticultural Science* 28: 155–173.

Dull, G.G., and A.C. Hulme. 1971. Quality. In AC Hulme, ed., *The Biochemistry of Fruits and Their Products*. Academic Press, London, pp 721–725.

El-Sharkawy, I., D. Manríquez, F.B. Flores, F. Regad, M. Bouzayen, A. Latché, and J.C. Pech. 2005. Functional characterization of a melon alcohol acyl-transferase gene family involved in the biosynthesis of ester volatiles. Identification of the crucial role of a threonine residue for enzyme activity. *Plant Molecular Biology* 59: 345–362.

Fellman, J.K., T.W. Miller, D.S. Mattinson, and J.P. Mattheis. 2000. Factors that influence biosynthesis of volatile flavor compound in apple fruits. *HortScience* 35: 1026–1033.

Ferenczi, A., J. Song, M. Tian, K. Vlachonasios, D. Dilley, and R. Beaudry. 2006. Volatile ester suppression and recovery following 1-methylcyclopropene application to apple fruit. *Journal of the American Society for Horticultural Science* 131: 691–701.

Forney, C.F., W. Kalt, and M.A. Jordan. 2000. The composition of strawberry aroma is influenced by cultivar, maturity and storage. *HortScience* 35.

Gang, D.R. 2005. Evolution of flavors and scents. *Annual Review of Plant Biology* 56: 301–325.

Giepmans, B.N., S.R. Adams, M.H. Ellisman, and R.Y. Tsien. 2006. The fluorescent toolbox for assessing protein location and function. *Science* 312: 217–223.

Griffiths, A., S. Prestage, R. Linforth, J. Zhang, A. Taylor, and D. Grierson. 1999. Fruit-specific lipoxygenase suppression in antisense-transgenic tomatoes. *Postharvest Biology and Technology* 17: 163–173.

Hampel, D., A. Mosandl, and M. Wüst. 2005. Biosynthesis of mono- and sesquiterpenes in carrot roots and leaves (*Daucus carota* L.): Metabolic cross talk of cytosolic mevalonate and plastidial methylerythritol phosphate pathways. *Phytochemistry* 66, (3), 305–311.

Holland, D., O. Larkov, E. Bar-Ya'akov Bar, A. Zax, E. Brandeis, U. Ravid, E. Lewinstsohn. 2005. Development and varietal differences in volatile ester formation and acetyl-CoA: Alcohol acetryl tranferase activities in apple (Malus domestica Borkh.) fruit. *Journal of Agricultural and Food Chemistry* 53: 7198–7203.

Ibdah, M., Y. Azulay, V. Portnoy, B. Wassserman, E. Bar, A. Mier, Y. Burger, J. Hirschberg, A.A. Schafer, N. Katzir, Y. Tadmor, and E. Lewinsogn. 2006. Functional characterization of CmCCD1, a carotenoid cleavage dioxygenase from melon. *Phytochemistry* 67: 1579–1589.

Imahori, Y., K. Matushita, M. Kota, Y. Ueda, M. Ishimaru, and K. Chachin. 2003. Regulation of fermentative metabolism in tomato fruit under low oxygen stress. *Journal of Horticultural Science and Biotechnology* 78: 386–393.

Jayanty, S., J. Song, N.M. Rubinstein, A. Chong, and R.M. Beaudry. 2002. Temporal relationship between ester biosynthesis and ripening events in bananas. *Journal of the American Society for Horticultural Science* 127: 998–1005.

Kader, A.A. 2004. Perspective on postharvest horticulture (1978–2003). *HortScience* 38: 759–761.

Ke, D., L. Zhou, and A.A. Kader. 1994. Mode of oxygen and carbon dioxide action on strawberry ester biosynthesis *Journal of the American Society for Horticultural Science.* 119: 971–975.

Kendall, S.A., and T.J. Ng. 1988. Genetic variation of ethylene production in harvested]muskmelon fruits. *HortScience* 23: 759–761.

Khanom, M.M. and Y. Ueda. 2008. Bioconversion of aliphatic and aromatic alcohols to their corresponding esters in melons (*Cucumis melo* L. cv. Prince melon and cv. Earl's favorite melon). *Postharvest Biology and Technology* 50: 18–24.

Kjeldsen, F., L.P. Christensen, and M. Edelenbos. 2003. Changes in volatile compounds of carrots (*Daucus carota* L.) during refrigerated and frozen storage. *Journal of Agricultural and Food Chemistry* 51, (18), 5400–5407.

Kreutzmann, S., A.K. Thybo, M. Edelenbos, M., L.P. Christensen. 2008. The role of volatile compounds on aroma and flavour perception in coloured raw carrot genotypes. *International Journal of Food Science and Technology* 43, (9), 1619–1627.

Lara, I., J. Graell, M.L. López, and G. Echeverría. 2006. Multivariate analysis of modifications in biosynthesis of volatile compounds after CA storage of "Fuji" apples. *Postharvest Biology and Technology* 39: 19–28.

Larsen, M., and L. Poll. 1992. Odour thresholds of some important aroma compounds in strawberries. *Zeitschrift für Lebensittel-. Untersuchung und Forschung* 195: 120–123.

Lavid, N., W. Schwab, E. Kafkas, M. Koch-Dean, E. Bar, O. Larkov, U. Ravid, and E. Lewinsohn. 2002. Aroma biosynthesis in strawberry: S-adenosylmethionine: furaneol O-methyltranferase activity in ripening fruits. *Journal of Agricultural and Food Chemistry* 50: 4025–4030.

Leone, A., T. Bleve-Zacheo, C. Gerardi, M.T. Melillo, L. Leo, and G. Zacheo. 2006. Lipoxygenase involvement in ripening strawberry. *Journal of Agricultural and Food Chemistry* 54: 6835–6844.

Li, D., Y. Xu, G. Xu, L. Gu, D. Li, and H. Shu. 2006. Molecular cloning and expression of a gene encoding alcohol acyltransferase (MdAAT2) from apple (cv. Golden Delicious). *Phytochemistry* 67: 658–667.

Lucchetta, L., D. Manriquez, I. El-Sharkawy, F.B. Flores, P. Sanchez-Bel, M. Zouine, C. Ginies, M. Bouzayen, C. Rombaldi, J.C. Pech, and A. Latche. 2007. Biochemical and catalytic properties of three recombinant alcohol acyltransferases of melon. Sulfur-containing ester formation, regulatory role of CoA-SH in activity, and sequence elements conferring substrate preference. *Journal of Agricultural and Food Chemistry* 55: 5213–5220.

Manríquez, D., I. El-Sharkawy, F.B. Flores, F. El-Yahyaoui, F. Regad, M. Bouzayen, A. Latché, and J.C. Pech. 2006. Two highly divergent alcohol dehydrogenases of melon exhibit fruit ripening-specific expression and distinct biochemical characteristics. *Plant Molecular Biology* 61: 675–685.

Mathieu, S., V.D. Cin, Z. Fei, H. Li, P. Bliss, M.G. Taylor, H.J. Klee, and D.M. Tieman. 2009. Flavour compounds in tomato fruits: Identification of loci and potential pathways affecting volatile composition. *Journal of Experimental Botany* 60: 325–337.

Matich, A. and D. Rowan. 2007. Pathway analysis of branched-chain ester biosynthesis in apple using deuterium labeling and enantioselective gas chromatography-mass spectrometry. *Journal of Agricultural and Food Chemistry* 55: 2727–2735.

Matsui, K., S. Fukutomi, J. Wilkinson, B. Hiatt, V. Knauf, and T. Kajwara. 2001. Effect of overexpression of fatty acid 9-hydroperoxide lyase in tomatoes (*Lycopersicon esculentum* Mill.). *Journal of Agricultural and Food Chemistry* 49: 5418–5424.

Matsui, K., A. Minami, E. Hornung, H. Shibata, K. Kishimoto, V. Ahnert, H. Kindl, T. Kajiwara, and I. Feussner. 2006. Biosynthesis of fatty acid derived aldehydes is induced upon mechanical wounding and its products show fungicidal activities in cucumber. *Phytochemistry* 67: 649–657.

Medina-Suarez, R., K. Manning, J. Fletcher, J. Aked, C.R. Bird, and G.B. Seymour. 1997. Gene expression in the pulp of ripening bananas. Two-dimensional sodium dodecyl sulfate-polyacrylamide gel electrophoresis of in vitro translation products and cDNA cloning of 25 different ripening-related mRNAs. *Plant Physiology*. 115(2):453–461.

Morris, C.E. and D.C. Sands. 2006. The breeder's dilemma-yield ot nutrition. *Nature Biotechnology* 24: 1078–1080.

Moyano, E., S. Encinas-Villarejo, J.A. López-Ráez, J. Redondo-Nevado, R. Blanco-Portales, M.L. Bellido, C. Sanz, J.L. Caballero, J. Muñoz-Blanco. 2004. Comparative study between two strawberry pyruvate decarboxylase genes along fruit development and ripening, post-harvest and stress conditions. *Plant Science* 166: 835–845.

Newcomb, R.D., R.N. Crowhurst, A.P. Gleave, E.H.A. Rikkerink, A.C. Allan, L.L. Beuning, J.H. Bowen, E. Gera, K.R. Jamieson, B.J. Janssen, W.A. Laing, S. McArtney, B. Nain, G.S. Ross, K.C. Snowden, E.J.F. Souleyre, E.F. Walton, and Y.K. Yauk. 2006. Analyses of expressed sequence tags from apple. *Plant Physiology* 141: 147–166.

Park, S., N. Sugimoto, M.D. Larson, R. Beaudry, and S. van Nocker. 2006. Identification of genes with potential roles in apple fruit development and biochemistry through large-scale statistical analysis of expressed sequence tags. *Plant Physiology* 141: 811–824.

Pechous, S.W. and B.D. Whitaker. 2004. Cloning and functional expression of an (E,E)-α-farnesene synthase cDNA from peel tissue of apple fruit. *Planta* 219: 84–94.

Perez, A.G., R. Olias, P. Lucaces, and C. Sanz. 2002. Biosynthesis of strawberry aroma compounds through amino acid metabolism. *Journal of Agricultural and Food Chemistry* 50: 4037–4042.

Pichersky, E. and D.R. Gang. 2000. Genetics and biochemistry of secondary metabolisms in plants: an evolutionary perspective. *Trends in Plant Science.* 5: 439–445.

Prestage, S., R.S.T. Linforth, A.J. Taylor, E. Lee, J. Speirs, and W. Schuch. 1999. Volatile production in tomato fruit with modified alcohol dehydrogenase activity. *Journal of the Science of Food and Agriculture* 79: 131–136.

Raab, T., J.A. Lopez-Raez, D. Klein, J.L. Caballero, E. Moyano, W. Schwab, and J. Munnoz-Blanco. 2006. FaQR, required for the biosynthesis of the strawberry flavour compound 4-hydroxy-2,5-dimethyl-3-(2H)-furanone, encodes an enone oxidoreductase. *Plant Cell* 18: 1023–1037.

Rosahl, S. 1996. Lipoxygenases in plants: Their role in development and stress response. *Zeitschrift fur Naturforschung. C, Journal of Biosciences* 51: 123–138.

Rowan D.D., J.M. Allen, S.Fielder, and M.B. Hunt. 1999. Biosynthesis of straight-chain ester volatiles in red delicious and granny smith apples using deuterium-labelled precursors. *Journal of Agricultural and Food Chemistry* 47: 2553–2562.

Rowan, D.D., H.P. Lane, J.M. Allen, S. Fielder, and M.B. Hunt. 1996. Biosynthesis of 2-methylbutyl, 2-methyl-2-butenyl, and 2-methylbutanoate esters in "Red Delicious" and "Granny Smith" apples using deuterium-labelled substrates. *Journal of Agricultural and Food Chemistry* 44: 3276–3285.

Rudell, D.R., D.S. Mattinson, J.P. Mattheis, S.G. Wyllie, and J.K. Fellman. 2002. Investigation of aroma volatile biosynthesis under anoxic conditions and in different tissues of "Redchief Delicious" apple fruit (Malus domestica Borkh.). *Journal of Agricultural and Food Chemistry* 50: 2627–2632.

Samach, A., D. Hareven, T. Gutfinger, S. Ken-Dror, and E. Lifschitz. 1991. Biosynthetic threonine deaminase gene of tomato: Isolation, structure, and upregulation in floral organs. *Proceedings of the National Academy of Sciences of the United States of America* 88: 2678–2682.

Saquet, A.A. and J. Streif. 2008. Fermentative metabolism in "Jonagold" apples under controlled atmosphere storage. *European Journal of Horticultural Science* 73: 43–46.

Saquet, A.A., J. Streif, and F. Bangerth. 2003. Impaired aroma production of CA-stored "Jonagold" apples as affected by adenine and pyridine nucleotide levels and fatty acid concentrations. *Journal of Horicultural Science and Biotechnology* 78: 695–705.

Schaffer, R.J., E.N. Friel, E.J.F. Souleyre, K. Bolitho, K. Thodey, S. Ledger, J.H. Bowen, J.H. Ma, B. Nain, D. Cohen, A.P. Gleave, R.N. Crowhurst, B.J. Janssen, J.L. Yao, and R.D. Newcomb. 2007. A genomics approach reveals that aroma production in apple is controlled by ethylene predominantly at the final step in each biosynthetic pathway. *Plant Physiology* 144: 1899–1912.

Shalit, M., N. Katzir, Y. Tadmor, O. Larkov, Y. Burger, F. Shalekhet, E. Lastochkin, U. Ravid, O. Amar, M. Eldelstein, Z. Karchi, and E. Lewinsohn. 2001. Acetyl-CoA: alcohol acetryltransferase activity and aroma formation in ripening melon fruits. *Journal of Agricultural and Food Chemistry* 49: 794–799.

Shieberler, P., S. Ofner, and W. Grosch. 1990. Evaluation of potent odorants in cucumbers and muskmelons by aroma extract dilution analysis. *Journal of Food Science* 55: 193–195.

Simon, P., C.E. Peterson, and R.C. Lindsay. 1980 Genetic and environmental influence on carrot flavor. *Journal of the American Society for Horticultural Science* 105, 416–420.

Song, J. 1994. Influence of different harvest maturities on fruit ripening with the special consideration of aroma volatile production of apples, tomatoes and strawberries. Publisher Ulrich Grauer, Stuttgart, Germany.

Song, J. and F. Bangerth. 2003. Fatty acids as precursors for aroma volatile biosynthesis in pre-climacteric and climacteric apple fruit. *Postharvest Biology and Technology*. 20: 113–121.

Song, J. and C. Forney. 2008. Flavour volatile biosynthesis and regulation in fruit. *Canadian Journal of Plant Science* 88: 537–550.

Souleyre, E.J.F., D.R. Greenwood, E.N. Friel, S. Karunairetnam, and R. Newcomb. 2005. An alcohol acyl transferase from apple (cv. Royal Gala), MpAAT1, produces esters involved in apple fruit flavor. *FEBS Journal* 272: 3132–3144.

Speirs, J., R. Correll, and P. Cain. 2002. Relationship between ADH activity, ripeness and softness in six tomato cultivars. *Scientia Horticulturae* 93: 137–142.

Speirs, J., E. Lee, K. Holt, K. Yong-Duk, N.S. Scott, B. Loveys, and W. Schuch. 1998. Genetic manipulation of alcohol dehydrogenase levels in ripening tomato fruit affects the balance of some flavor aldehydes and alcohols. *Plant Physiology* 117: 1047–1058.

Tressl, R. and F. Drawert. 1973. Biogenesis of banana volatiles. *Journal of Agricultural and Food Chemistry* 21: 560–565.

Villatoro, C., R. Altisent, G. Echeverría, J. Graell, M.L. López, and I. Lara. 2008. Changes in biosynthesis of aroma volatile compounds during on-tree maturation of "Pink Lady" apples. *Postharvest Biology and Technology* 47: 286–295.

Villatoro, C., G. Echeverría, J. Graell, M.L. López, and I. Lara. 2008. Long-term storage of "Pink Lady" apples modifies volatile-involved enzyme activities: Consequences on production of volatile esters. *Journal of Agricultural and Food Chemistry* 56: 9166–9174.

Wang, Y., S.G. Wyllie, and D.N. Leach. 1996. Chemical changes during the development and ripening of the fruit Cucumis melo (cv. Makdimon). *Journal of Agricultural and Food Chemistry* 44: 21–216.

Wein, M., N. Lavid, S. Lunkenbein, E. Lewinsohn, W. Schwab, and R. Kaldenhof. 2002. Isolation, cloning and expression of a multifunctional O-methyltransferase capable of forming 2,5-dimethyl-4-methoxy-3(2H)-furanone, one of the key aroma compounds in strawberry fruits. *Plant Journal* 31: 755–765.

Wyllie, S.G., D.N. Leach, Y. Wang, and R. Shewfelt. 1995. Key aroma compounds in melons: their development and cultivar dependence. In R.L. Rouseff, and Leahy, M.M., eds., *Fruit Flavour Biogenesis, Characterization and Authentication.* Amer. Chem. Soc., Wash. DC, pp 248–257.

Yahia, E.M. 1994. Apple flavour. *Horticulture Reviews* 16: 197–234.

Yilmaz, E., E.A. Baldwin, and R.L. Shewfelt. 2002. Enzymatic modification of tomato homogenate and its effect on volatile flavor compounds. *Journal of Food Science* 67: 2122–2125.

Yilmaz, E., K.S. Tandon, J.W. Scott, E.A. Baldwin, and R.L. Shewfelt. 2001. Absence of a clear relationship between lipid pathway enzymes and volatile compounds in fresh tomatoes. *Journal of Plant Physiology* 158: 1111–1116.

Zhu, Y., Rudell, D.R., and J.P. Mattheis. 2008. Characterization of cultivar differences in alcohol acyltransferase and 1-aminocyclopropane-1-carboxylate synthase gene expression and volatile ester emission during apple fruit maturation and ripening. *Postharvest Biology and Technology* 49: 330–339.

chapter four

Effect of enzymatic reactions on texture of fruits and vegetables

Luis F. Goulao, Domingos P. F. Almeida, and Cristina M. Oliveira

Contents

4.1   Introduction

4.1.1   Describing texture

Textural characteristics of fruits and vegetables are important in determining consumer acceptance, and even minor deviations from the expected texture can result in produce rejection. The textural properties of a food are the group of physical characteristics that arise from its structural elements, are sensed by touch, and are related to the deformations, fracture, disintegration, and flow under force; these physical properties are measured objectively and expressed as functions of mass, time, and distance (Bourne, 2002). Although the term is used widely and loosely, texture is not a single, well-defined attribute. It is a multi-trait attribute encompassing individual characteristics described by terms like firm, stiff, breakdown, crisp, granular, hard, juicy, spongy, melty, floury, or gritty (Harker et al., 1997ab; King et al., 2000). Each of the mentioned attributes is likely to reflect particular facets of cell wall structure, especially cell wall strength and cell-to-cell adhesion. Hence, texture should be defined as a collective term that encompasses the structural and mechanical properties of a plant organ and their sensory perception by the consumer (Figure 4.1). Depending on the specific fruit or vegetable, one or a few textural attributes are appropriate to define its texture for the purpose of quality control throughout the supply chain.

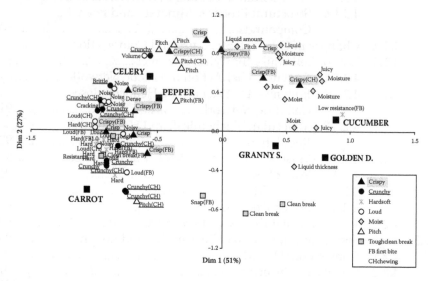

Figure 4.1 Plotted map of consumer perception of representative types of fruit and vegetable textures. (Fillion and Kilcast, 2002. Consumer perception of crispness and crunchiness in fruits and vegetables. *Food Quality Preference* 1:23–29.)

4.1.2   Textural Properties of fruits and vegetables

The texture of plant organs is determined by the organ archestructure at different levels of organization. Structural polymers, their organization into macromolecular complexes in the cell wall, cell size and geometry, and their organization into tissues, further organized into organs with defined geometry and structure, all contribute to the textural properties of fruits and vegetables. Moreover, the texture of living fruits and vegetables changes during development including the developmental stages occurring during postharvest storage (e.g., ripening and senescence).

The metabolic events responsible for the textural changes in fruits are believed to involve loss in turgor pressure, physiological changes in membrane composition, modifications in the symplast/apoplast relations, degradation of starch, and modifications in the cell wall dynamics. Although the relative contribution of each event for fruit softening is still under debate, changes in cell wall structure have been considered as the most important factor (Fischer and Bennett, 1991; Hadfield and Bennett, 1998). Softening has been mainly associated with changes in the primary cell walls of the parenchyma cells, including the middle lamella. During the decline in firmness, the first change observed in a ripening fruit is the dissolution of the middle lamellae (Ben-Arie et al., 1979) and a decrease in intercellular adhesion, generally accompanied by a reduced area of intercellular adhesion. It is followed by solubilization and/or depolymerization of pectic and hemicellulosic polysaccharides and, in some instances, wall swelling. Texture depends upon the geometric characteristics of these cells, including shape, size, thickness, and strength of the wall, cell turgor pressure, the manner in which they bind to form a tissue, and the presence of fibers or air pockets. Cell size and packing patterns determine the volume of intercellular space affecting cell adhesion by determining the extent of cell-to-cell contact.

Textural properties of fleshy fruits vary among species and genotypes and, in many instances, undergo dramatic changes during ripening. Based on their softening behavior, fruits can be divided into two categories (Bourne, 1979): those that soften greatly to a melting texture as they ripen [e.g., tomato (*Solanum lycopersicum* L.), peach (*Prunus persica* L.), strawberry (*Fragaria ananassa* Dutch) or kiwifruit (*Actinidia deliciosa* A.Chev.)] and those that soften moderately to a crisp, fracturable texture [e.g., apple (*Malus* x *domestica* Borkh.), nashi pear (*Pyrus pyrifolia* Nakai) or cranberry (*Vaccinium macrocarpon* Aiton.)]. In fruits belonging to the first category, softening is accompanied by cell wall swelling, which results from penetration of water into larger microfibrilar spaces created by cell separation of cells in synchrony with pectic solubilization (Crookes and Grierson, 1983; Hallett et al., 1992; Redgwell et al., 1997b).

In contrast with fruits, vegetables are made of a wider range of morphological structures with different biological roles. Our understanding of textural changes related to developmental processes in vegetative plant organs is limited compared with that of fleshy fruits. The limited research attention given to the physiology of texture in these commodities is understandable considering that their textural changes are generally less dramatic than those of fleshy fruit. In contrast, considerable research is available on the effects of processing on the textural properties of vegetables (Adams, 2004). Textural properties are but one of the quality determinants in most raw non-fruit vegetables, but rank high in the quality attributes of certain commodities, like celery (*Apium graveolens* L.), asparagus (*Asparagus officinalis* L.), and snap peas (*Pisum sativum* L.).

In general, tissues of non-fruit vegetable are harder than ripe, fleshy fruit due to the higher proportion of thickened and lignified cell walls (Toivonen and Brummell, 2008). Textural properties of leafy vegetables are largely related to turgor pressure and usually regarded as a direct effect of water loss. From a sensorial perspective these changes are readily perceived by vision as appearance. Structural changes in cell walls of leaves have not been comprehensively analyzed and the role of aquaporins in regulating cell water content has not been explored in relation to the texture of leafy vegetables (Maurel, 1997, 2007).

Roots and storage organs may undergo textural changes during development as the proportion of different tissues changes and geometric features are altered (Reeve et al., 1973a,b), although textural changes during postharvest storage are not striking in this class of vegetables.

Asparagus spears are stem vegetables that undergo rapid hardening after harvest associated with a decrease in uronic acid concentration and lignin deposition (Rodríguez et al., 1999). Stem hardening also occurs in the inflorescence vegetables, broccoli (*Brassica oleracea* L. Italica Group) (Serrano et al., 2006), and cauliflower (*Brassica oleracea* L. Botrytis Group) (Simón et al., 2008).

The rapid textural changes of seeds of legume vegetables and immature kernels of sweet corn are highly detrimental to their quality. Texture in these vegetables is primarily determined by the levels of water-soluble sugars, prior to starch formation, and moisture content.

Fresh-cut fruits and vegetables are a convenient and fast-growing segment of horticultural produce. Textural properties are key quality parameters in fresh-cut produce, and changes in textural attributes are often the limiting factor of shelf-life. Textural properties of fresh-cut fruit are, to some extent, related to those of whole fruit used as raw materials for processing. Nonetheless, fresh-cut processing involves operations that substantially alter the ripening-related textural changes occurring in whole fruit (see discussion below). In non-fruit vegetables, textural changes are generally associated with water loss or lignification

(Viña and Chaves, 2003), whereas water loss and cell wall disassembly are major determinants of textural changes in fresh-cut fruits. Excessive softening, a major problem in several fresh-cut fruit, has implications beyond the perception of texture, affecting juice retention and flavor perception (Beaulieu and Gorny, 2001).

4.2   Biochemical bases for textural changes

4.2.1   Cell wall structure and metabolism

4.2.1.1   Representations of the primary cell wall

Plant cell walls consist of a complex and highly variable combination of polysaccharides and other polymers that are secreted by the cell and assembled in organized networks linked together by covalent and non-covalent bonds. The plant cell wall of dicotyledonous species is composed of approximately 90% polysaccharides (McNeil et al., 1984) that can be classified into three main groups: cellulose, hemicellulose, and pectin, representing respectively, 35%, 15%, and 40% of the cell walls mass of fruits and vegetables (Brett and Waldron, 1996). Structural glycoproteins, phenolic esters, minerals, and enzymes are also present and interact to allow genetically determined modifications in the wall's physical and chemical properties. Knowledge about the structural complexity of these individual cell wall components and the different ways by which they are linked together is fundamental to understand the significance of the enzyme-driven action in the polysaccharide backbones or side groups during softening. Several models have been proposed to explain the architecture of the primary cell wall. The cell wall is viewed as a three-dimensional network containing fluid-filled pores that interconnect to form pathways for solutes through the walls (Harker et al., 2000). The mechanical properties of fruit primary cell walls are mostly determined by a unique mixture of matrix (pectic and hemicellulosic) and fibrous (cellulose) polysaccharides. The network of pectic polymers appears to have the finest mesh size and determines apoplastic porosity (Read and Bacic, 1996). The interactions of polysaccharide polymers depend upon their cross-linkages, molecular size, and hydrogen-bonding characteristics, and determine the rigidity, cohesiveness, and shear properties of the cell wall that define texture.

In the "covalently cross-linked" model of Keegstra et al. (1973), the wall matrix polymers xyloglucans, pectic polysaccharides, and glycoproteins are covalently linked to one another and xyloglucan binds to cellulose microfibrils by hydrogen-bonding, resulting in a non-covalently cross-linked network that provides tensile strength to the wall. Even though the interaction among the matrix polymers proposed by this simplistic model has been generally considered out of date, recent evidence that a small amount of xyloglucan is attached to pectic polysaccharides (Thompson and Fry, 2000; Popper and Fry, 2005), sustains the concept put forward by the model. The "tether-network" model (Fry, 1989) has been widely used in the last years (Carpita and Gibeaut, 1993). In this model, xyloglucan is proposed to form hydrogen bonds with cellulose microfibril, acting as a tether between the microfibrils, which reinforces the cell wall. Xyloglucans not only bind to the surface of cellulose microfibrils, but are also woven into the amorphous regions. This enables enhanced binding, since its location both in the inner and outer surfaces of microfibrils allows binding of adjacent microfibrils. The cross-linking between perpendicular fibrils may function as a bracket. Another hypothesis explaining such patterns of distribution among microfibrils of cellulose is to prevent hydrogen bonding between cellulose microfibrils, allowing each microfibril to slide during cell enlargement. The xyloglucan-cellulose framework is embedded, but not covalently-bond, in an amorphous pectin matrix together with a domain of other less abundant components, including structural glycoproteins. In the "diffuse layer" or "multicoat" model (Talbott and Ray, 1992), xyloglucan molecules are proposed to be hydrogen-bonded to the cellulose microfibrils without directly cross-linking them. This tightly bound xyloglucan is coated with layers of progressively less-tightly bound polysaccharides, and linkages between microfibrils occur indirectly by lateral, non-covalent associations between each different polysaccharide layer. Also in this model, cellulose and xyloglucan are embedded in a pectic matrix. The "stratified layer" model (Ha et al., 1997) suggests xyloglucan molecules hydrogen-bonded to and cross-linking cellulose microfibrils. This cellulose-xyloglucan lamella would be separated by strata of pectic polysaccharides, responsible for the control of wall thickness and slippage between the cellulose-hemicellulose layers. Additional evidence suggests the existence of xyloglucan-RG-I (rhamnogalacturonan-I) conjugates (Popper and Fry, 2005), with RG-I very firmly integrated into the wall.

More recently, based on ^{13}C Nuclear Magnetic Resonance, a new representation was proposed (Bootten et al., 2004) in which only a relatively short length of each xyloglucan molecule is actually adsorbed to cellulose, and only a small proportion of the total surface area of the cellulose microfibrils has xyloglucan adsorbed onto it. In this model, the partly rigid xyloglucan cross-links adjacent cellulose microfibrils and/or cellulose microfibrils and other non-cellulosic polysaccharides, such as pectins. Moreover, a new model in which the different classes of pectin are covalently cross-linked, with HG (homogalacturonan) and rhamnogalacturonan-II (RG-II) representing different regions of the same molecule and network has been suggested (Vincken et al., 2003).

The different models described emphasize aspects of the cell wall structure and are helpful in providing a mind-map of this complex structure. There seems to be no definitive evidence favoring a given model over the others (Cosgrove, 2001), therefore all models should be considered for the interpretation of cell wall modifications. However, all of the current models of the primary cell wall of higher plants describe the wall as a network of structurally independent but interacting networks: the cellulose-hemicellulose network, the pectin network and, in some tissues, the extensin network, and in all models, cellulose microfibrils are coated with xyloglucan (Carpita and Gibeaut, 1993; Cosgrove, 2001).

4.2.1.2   Pectic matrix

Pectins are a class of heterogeneous macromolecules that constitute the most abundant polysaccharides within the cell wall matrix, forming hydrophilic gels that impose important mechanical features to the wall. Pectins are a family of acidic polysaccharides containing 1,4-linked α-D-galacturonic acid residues, assembled with a range of modifications and substitutions that include methyl- and acetyl-esterified structural domains with variable degrees of ramification by single sugars or complex side-chains (Voragen et al., 1995). The backbone of pectins can be estimated to be more than 500 residues long and the degree of methyl-esterification of the galacturonate residues can vary over a wide range. The term *pectic acid* still prevails in some literature to refer to the low-methoxyl pectic fraction being the term *pectin* reserved to the highly methylated fraction. This distinction based on the solubility characteristics of the polymers has little physiological significance, since pectins are synthesized and deposited in the wall with the uronate residues methyl-esterified (Roberts, 1990) and deesterification occurs in the cell wall through the action of PMEs (pectin methylesterases). Homogalacturonan (or polygalacturonic acid; HG), rhamnogalacturonan-I (RG-I), and the substituted galacturonan referred to as rhamnogalacturonan-II (RG-II) are the predominant pectic polysaccharides present in the primary cell walls (Seymour et al., 1990; Carpita and Gibeaut, 1993; Brett and Waldron, 1996; Ridley et al., 2001), while the middle lamella is composed almost exclusively of HG and some structural proteins. These three structures are covalently linked to one another to form the pectic matrix, envisioned as a unique and complex macromolecule (Ridley et al., 2001; Vincken et al., 2003; Coenen et al., 2007), although the nature of their covalent arrangements is still unclear (O'Neill et al., 2004). Pectin polymers can be covalently linked by diferulic acid bonds (Fry, 1986) that are proposed to link together neutral pectins via their terminal galactose residues. The linkages that integrate the pectin superstructure in the wall include calcium bridges (egg-box) and borate di-esters of RG-II monomers (see below). Pectins are also described in terms of "smooth" (corresponding to HGs) and "hairy" (which include RG-I and -II) blocks that may reside as components of a single pectin polymer. "Smooth" blocks are linear polymers of α-(1 $\rightarrow$ 4)-linked galacturonic acid and its methyl ester, and long "smooth" regions are interspersed with stretches of "hairy" backbone carrying few side-chains, since inserted within the "smooth" HG polymer are α-(1 → 2)-linked rhamnosyl residues at regular and determined spaces. These rhamnosyl residues delineate HG domains for methyl esterification or de-esterification, thus enabling calcium cross-linking at regular intervals. Rhamnosyl residues also serve as attachment sites for arabinose-rich and galactose-rich side-chains (reviewed in Fischer and Bennett, 1991). Therefore, the backbone, rich in galacturonic acid and rhamnose, bears numerous side-chains rich in (1 → 5)-α-arabinan and (1 → 4)-β-galactan components (O'Neill et al., 1990). Pectin side-chains may also contain fucose, methylfucose, methylxylose, apiose, glucuronic acid, aceric acid, keto-deoxy-octulosonic acid, and/or glucose (Fischer and Bennett, 1991). The role of RG side-chains on cell wall assembly and mechanical properties remains poorly understood but modifications in their structure can have a strong impact on the morphology of the tissue (Oomen et al., 2002).

RG-II is composed of at least eight 1,4-linked α-D-galacturonic acid backbone highly branched which contains at least eleven different sugars forming an extremely complex pattern of linkages. It exists in the primary cell wall as a dimmer cross-linked by a borate diester cross-link (Kobayashi et al., 1996; O'Neill et al., 1996; Ishii et al., 1999) and is stabilized by the presence of calcium (Kobayashi et al., 1999). RG-II and HG have been proposed to be linked covalently to one another (Ishii and Matsunaga, 1996). Dimmer formation results in cross-linking of two HG chains upon which the RG-II molecules are constructed and the calcium-dependent ionic cross-linking of HG is required for the formation of the three-dimensional pectin network *in muro* (O'Neill et al., 2004).

HG is the principal constituent of the middle lamella and is thought to be responsible for cell-to-cell adhesion that holds together the primary wall and adjacent cells through the formation of calcium cross-links between adjacent chains of HG (Thompson et al., 1999), and for the porosity of the cell wall to macromolecules (Carpita and Gibeaut, 1993). The pore size in the cell wall is established by a combination of the frequency and length of the junction zones, the degree of methyl esterification and the length of arabinans, galactans, and arabinogalactans attached to RG-I that extend into the pores (Carpita and McCann, 2000).

Pectin molecules are proposed to be linked together non-covalently via a structure denominated the "egg-box" (Grant et al., 1973) that is thought to stabilize the middle lamella (Fry, 1986). In this structure, calcium ions are chelated by de-esterified galacturonic acid residues on adjacent polymers, resulting in supramolecular assemblies and gels that add rigidity to the wall (Jarvis, 1984; Willats et al., 2001; Jarvis et al., 2003). However, homogalacturonic acid (HGA) is synthesized and secreted in a completely or highly methyesterified form (Roberts, 1990; Doong et al., 1995), requiring subsequent de-esterification of HGAs *in muro*, by PME action (Willats et al., 2001; Jarvis et al., 2003), which increases the negative charge density of the cell wall environment, making HG prone to be cross-linked by divalent cations, such as calcium.

In contrast with the middle lamella, pectins of the primary wall are more highly branched and possess longer side-chains. The carboxylic groups of the molecules of the backbone are extensively methylesterified, reducing the potential for calcium cross-bridging. Therefore, ester linkages should be involved in cross-linking this pectic gel (Steele et al., 1997).

Substituted galacturonans include several different polysaccharides. Xylogalacturonans are pectic polysaccharides with unknown function, composed by a galacturonan backbone with β-D-Xyl$p$ residues attached its C-3 position. Xylogalacturonans are known to be present only in reproductive tissues, and this polysaccharide was identified in apple fruits (Schols et al., 1995).

4.2.1.3   Hemicellulose-cellulose network

Hemicelluloses (also denominated cellulose-linking glycans) are a group of polysaccharides organized in highly branched structures, composed of 1®4-linked-β-D-hexosyl residues in which O-4 is in the equatorial orientation. Hemicelluloses include xyloglucan, xylan, glucuronoxylan, arabinan, arabinoxylan, mannan, glucomannan, galactomannan, and galactan (Carpita and Gibeaut, 1993; Brett and Waldron, 1996). Xyloglucan comprises 15-25% of the primary walls of dicotyledonous (Carpita and Gibeaut, 1993), except for some *Solanaceae* species. The $1 \to 4$-β-glucan backbone of xyloglucans is probably composed of repeating heptasaccharide units to which variable amounts of sugar residues are added during synthesis (Hayashi, 1989). The xyloglucan family displays a large heterogeneity as the result of the attachment of short side-chains containing xylose, galactose, and, in certain cases, a terminal fucose, attached to about 75% of the β-(1 $\to$ 4)-D-Glc$p$ residues of the backbone (Hayashi, 1989). The α-xylosyl residues are attached to the 6-position of the β-glucosyl residues, terminal galactose attached to the 2-position of the xylosyl residues by β-linkage, while L-fucose is attached by α-linkage to the 2-position of the galactosyl residues. The molar ratio of D-glucose, D-xylose, and D-galactose of all xyloglucans is 4:3:1 (Hayashi, 1989). Due to their structural similarity, hemicelluloses are characteristically hydrogen-bound to the surface of the cellulose microfibrils, forming tethers that form the major load-bearing structure in the primary wall, or may act as a slippery coating to prevent direct microfibril-microfibril contact (Taiz and Zieger, 1998). Since xyloglucans are longer (about 50 to 500 nm) than the spaces between cellulose microfibrils (20 to 40 nm), they have the potential to link several microfibrils together (Taiz and Zieger, 1998).

After xyloglucan, the other major matrix glycans are glucuronoarabinoxylan and glucomannan. Arabinogalactans or arabinans are also present as free macromolecules, possibly forming a layer surrounding the xyloglucan/cellulose network (Talbott and Ray, 1992).

Cellulose is made up of linear chains of cellobiose, a $1 \rightarrow 4$-linked β-D-glucose disaccharide (Taiz and Zieger, 1998), in which every other glucose residue is related 180° with respect to its neighbor (Brown et al., 1996) leading to the formation of long, unbranched polymers. Each cellulose molecule contains 3,000 to more than 25,000 glucose units (Brown et al., 1996) and cellulose molecules are grouped together in microfibrils by hydrogen bonding of adjacent glucose units. In higher plants each microfibril is about 10 to 15 nm wide and contains approximately 50-60 cellulose molecules. The extremely high number of hydrogen bonds within a microfibril of cellulose gives it a great tensile strength, and the individual polysaccharide chains become closely aligned and bonded to each other to make crystalline or paracrystalline arrays of glucan chains that are relatively inaccessible to enzymatic attack. Furthermore, the microfibrils wind together to form fine threads that coil around one another, resulting in about 0.5 μm width and 4 μm length structures named macrofibrils.

4.2.1.4   *Structural proteins, minerals, and phenolic compounds*

In addition to the polysaccharides described, primary cell walls also contain structural proteins. Five classes of structural apoplastic proteins have been described: extensins, glycine-rich proteins (GRPs), proline-rich proteins (PRPs), arabinogalactan proteins (AGPs), and solanaceous lectins (Showalter, 1993). Although associations between these structural proteins and fruit ripening have been less reported, mRNA for structural proteins was shown to be constitutive or up-regulated during ripening of peaches [*Prunus persica* (L.) Batsch] (Trainotti et al., 2003). AGPs have been implicated in cell adhesion (Majewska-Sawka and Nothnagel, 2000). The carbohydrate moieties of AGPs have a common structure of β-$(1 \rightarrow 3)$-galactosyl backbones to which side-chains of β-$(1 \rightarrow 6)$-linked galactosyl residues are attached through O-6. The β-$(1 \rightarrow 6)$-linked galactosyl chains are further substituted with L-arabinofuranose and lesser amounts with other sugars. Minerals, including calcium and boron, and phenolic esters such as ferulic and coumaric acids can be also present and important to maintain the structure of the polysaccharides in the cell wall (Brett and Waldron, 1996).

4.2.2   *Changes in cell wall structure and composition*

Changes in the structure of the cell wall are associated with dissolution of the middle lamella and modifications of the primary cell wall (Crookes and Grierson, 1983), where modifications in pectin, hemicellulose, and cellulose together are assumed to be responsible for the alteration of cell wall structure during ripening-related loss of firmness (Huber, 1983, Seymour et al., 1990). Structural changes common to all fleshy fruit involve loosening of the primary cell walls and loss of cell cohesion, which can be or not accompanied by actual cell wall degradation.

Events initiated early in fruit softening include loss of neutral sugars (in particular galactan and arabinan) from side chains of RG-I (Gross and Sams, 1984), de-methyesterification of HGs and solubilization of polyuronides (Brummel, 2006; Vicente et al., 2007b) (Figure 4.2). These events are considered to be a universal feature of softening. Pectin solubilization may result from un-cross-linking of pectin molecules with each other as the result of loss of neutral sugars in the form of neutral galactose-rich side-chains of RG-I (Seymour et al., 1990, Redgwell et al., 1992). These changes result in an apparent dissolution of the pectin-rich middle lamella region, and as ripening progresses the cell wall becomes increasingly hydrated. The changes in cohesion of the pectin gel govern the ease with which a cell can be separated from another, which in turn affects the final texture of the ripe fruit. In some species, solubilized pectins are subsequently (Redgwell et al, 1992; Brummell et al., 2004; Vicente et al., 2007b)

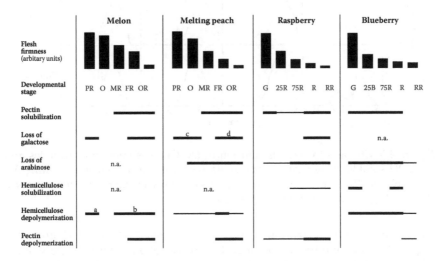

Figure 4.2 Proposed models for temporal sequence of cell wall changes occurring during maturation and ripening in fruits (redrawn from Rose et al., 1998; Brummell, 2006; Vicente et al., 2007bc). PR, pre-ripe; O, onset; MR, mature-ripe; FR, full-ripe; OR, overripe; G, green; 25R(B), 25% red(blue); 50%R(B), 50% red(blue); R, red; B, blue; R(B)R, red(blue) ripe; n.a., data not available; a, from xylose-rich polymer; b, from KOH-soluble xyloglucan; c, from KOH extract; d, from CDTA extract.

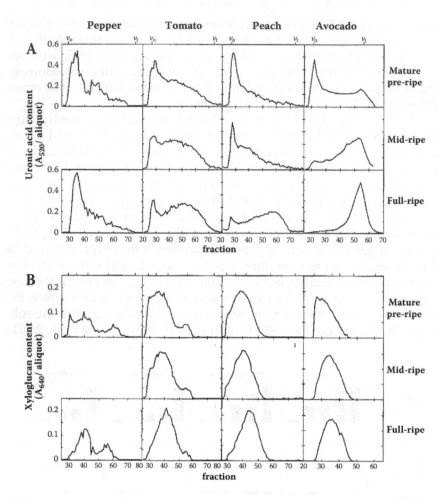

Figure 4.3 Changes in size distribution in (A) CDTA-soluble polyuronides and (B) KOH-soluble xyloglucan in representative fruit species. Vo, void volume; Vt, total volume. (Brummell, 2006. Cell wall disassembly in ripening fruit. *Functional Plant Biology* 33:103–119. With permission.)

subject to depolymerization in the later stages of ripening (Figure 4.2; Figure 4.3) through the action of endo- and/or exo-acting PGs (polygalacturonases) (Dawson et al., 1992). However, PG-mediated depolymerization can occur in calcium-bound pectins prior to solubilization (Almeida and Huber, 2007). The highly branched status of much of the wall-bound pectins in unripe fruits presumably limits its attack by endo-PGs, unless the removal of side chains makes the molecule more labile to this enzyme. However, this hypothesis has been challenged. A correlation between cell wall swelling and pectin solubilization, but not with galactose loss, has been established in kiwifruit (Redgwell and Percy, 1992; Redgwell et al., 1997b), and inhibiting the action of endogenous β-gal (endogenous β-galactosidase) prevents galactose loss but does not affect the rate of softening or the degree of pectin solubilization (Redgwell and Harker, 1995). Moreover, some pectin solubilization precedes galactose loss early in softening (Redgwell et al., 1992) and neither depolymerization nor removal of pectic side chains is necessary for the initial solubilization of some pectic polymers in kiwifruit (Redgwell et al., 1992). Also, in tomato, softening occurs before β-gal activity or galactose loss is detected (Carrington and Pressey, 1996). Besides affecting hydration status and solubilization, pectin de-methylesterification modifies conditions in the apoplast, namely pH and ion balance, and creates charged surfaces that may restrict the movement and action of some enzymes in the wall matrix. Another early event in fruit softening in some species is the regulated disassembly of hemicelluloses. Xyloglucan does not decline in amount during ripening, suggesting that small molecules are not generated and xyloglucan is mainly cleaved by the ends in regions attached to cellulose or pectin. Although the amounts of hemicellulose (mainly xyloglucan) remain constant during ripening (Maclachlan and Brady, 1994), a significant decrease in the relative molecular weight of the tightly bound xyloglucan fraction was noticed, coincident with the onset of softening (Maclachlan and Brady, 1994) (Figure 4.3).

Mid and later softening events that accompany overripe deterioration are associated with pectin disassembly (Rose et al., 1998; Brummell et al., 2004; Vicente et al., 2007b) in the species where it occurs. This depolymerization of polyuronides is initiated at mid-softening in some fruits like avocado or tomato, or at late-softening in other fruits, like melon (*Cucumis melo* L.) and peach (reviewed in Brummell, 2006).

It should be noted that, as stated, depending on the fruit species, different modifications of the polysaccharides may occur and may occur at distinct extents (Figure 4.3). Some species soften without detectable depolymerization of polyuronides (apple and strawberry) and/or matrix glycans [apple, papaya (*Carica papaya* L.), raspberry (*Rubus idaeus* L.)], whereas ripening-related depolymerization of these compounds appears to occur in others (reviewed in Goulao and Oliveira, 2008).

Studies in potato (*Solanum tuberosum* L.) tubers with altered RG-I structure and content show that arabinan or galactan side chains are important to assure tuber resistance to fracture (Ulvskov et al., 2005).

The modifications occurring in cell wall of fruits and vegetables are largely imposed by the action of enzymes that target the pectic and the hemicellulose-cellulose networks. The cited modifications, particularly those that occur early in the process, may result in creating the conditions for a series of such enzymes to enter into the ripening cell wall and act on their specific linkages.

4.2.3   Role of cell wall-modifying enzymes

Members from several enzyme families act in a coordinated way to alter the physical properties of the wall, namely the hydration status and the viscosity of the matrix, or to cleave the backbone and modify the pattern of ligation of the polysaccharides that compose the wall. A large body of literature has reported many enzymes whose proposed mode of action is consistent with some of the biochemical modifications that occur during softening (see previous subsection), with increased mRNA, protein and/or *in vitro* activity during fruit ripening and concomitant softening (recently reviewed by Prasanna et al., 2007, and Goulao and Oliveira, 2008). The strongest candidates are endoglycanases, glycosidases, trans-glycosylases, esterases, and acetylases, and include endo- (EC 3.2.1.15) and exo-polygalacturonases (EC 3.2.1.67), pectate lyases (EC 4.2.2.2), pectin methylesterases (EC 3.1.1.11), pectin acetylesterases (EC 3.1.1.-), β-galactosidases (3.2.1.23), α-L-arabinofuranosidases (3.2.1.55), endo-1,4-β-glucanases (3.2.1.4), endo-1,4-β-xylanases (EC 3.2.1.8), β-xylanases (EC 3.2.1.37), xyloglucan endotransglycosylase/hydrolases (EC 2.4.1.207), and expansins. Other less studied candidate enzymes have been suggested, such as rhamnogalacturonan hydrolases, yieldins, lipid transfer proteins, or deoxyhypusine synthases (reviewed in Vicente et al., 2006).

The production of overexpressing and antisense lines strongly down-regulated in the expression of specific members of some of these candidate families have provided additional *in muro* substantiation of the direct effect of specific members on their target polysaccharides. However, a phenotype that fails to soften was never obtained from suppression of an individual enzyme and only in few cases a reduction in fruit softening progression was obtained (Table 4.1).

The wide variability among fruit species concerning protein and poly-saccharide composition, wall architecture, metabolism, pattern of growing and ripening, and softening behavior makes it difficult to propose an unambiguous model for enzymatic action from this largely incomplete and sparse information reported using distinct species (Goulao and Oliveira, 2008). Therefore, in this chapter we will focus on the enzymatic role attributed to softening based on works conducted by comparing fruit from the same species that exhibits differential softening patterns, like naturally occurring mutations, different cultivars, or genetically modified lines.

4.2.3.1   Enzymes that act on the pectic network

Tomato fruit softening is accompanied by a massive increase in PG mRNA accumulation, detectable protein, and enzyme activity (Brady et al., 1982; Grierson and Tucker, 1983; DellaPenna et al., 1986; Campbell et al., 1990). Antisense tomatoes expressing 1% of wild-type PG activity (Sheehy et al., 1988; Smith et al., 1988) displayed inhibited pectin depolymerization in

Table 4.1 Summary of Experimental Results Cited in the Literature Which Report Effects on Polysaccharides and Softening Behavior after Suppression and Overexpression of Individual Cell Wall–Modifying Enzymes

Enzyme	Modification	Species	Effect on polysaccharides	Effect on firmness and softening	References
PG	suppression	tomato	Inhibition of pectin depolymerization	Normal or slightly reduced softening	Smith et al., 1988, 1990; Brummell and Labavitch, 1997; Cooley and Yoder, 1998
*Fa*PG1	suppression	strawberry	Reduction in pectin solubilization increase in pectins covalently bounded to the wall	Increased firmness	Quesada et al., 2009
PME1	suppression	tomato	n.d.	Faster softening rate	Phan et al., 2007
PME2	suppression	tomato	Decreased pectin depolymerization Increased pectin methylesterification	Little or no effect on fruit softening	Tieman et al., 1992; Gaffe et al., 1984; Hall et al., 1993

(continued)

Table 4.1 Summary of Experimental Results Cited in the Literature Which Report Effects on Polysaccharides and Softening Behavior after Suppression and Overexpression of Individual Cell Wall–Modifying Enzymes (Continued)

Enzyme	Modification	Species	Effect on polysaccharides	Effect on firmness and softening	References
PL	suppression	strawberry	Reduction in pectin solubility and decreased depolymerization of more tightly bound polyuronides; Lower degree of *in vitro* swelling; Less ionically bound pectins; Reduced cell-to-cell adhesion	Increased fruit firmness	Jiménez-Bermúdez et al., 2002; Santiago-Doménech et al., 2008
TBG1	suppression	tomato	No effect on the levels of cell wall galactosyl residues	No effect on fruit firmness	Carey et al., 2001
TBG4	suppression	tomato	Reduced free galactose levels at mature green stage; normal levels at ripening; No changes in total cell wall galactosyl content	Increased firmness	Smith et al., 2002
TBG6	suppression	tomato	Increased fruit cracking, reduced locular space Increased thickness of cuticle	No effect on fruit firmness	Moctezuma et al., 2003
*Le*Cel1	suppression	tomato	n.d.	No effect on fruit firmness	Lashbrook et al., 1998
*Le*Cel2	suppression	tomato	n.d.	No effect on fruit firmness	Brummell et al., 1999
EGase	suppression	pepper	Normal xyloglucan depolymerization	No effect on fruit firmness	Harpster et al., 2002a

Enzyme	Treatment	Fruit	Cell wall effect	Firmness effect	Reference
EGase	overexpression	pepper	Normal xyloglucan depolymerization	No effect on fruit firmness	Harpster et al., 2002b
*Fa*Cel1	suppression	strawberry		No effect on fruit firmness	Woolley et al., 2001; Palomer et al., 2006
XTH	suppression	tomato	n.d.	No effect on fruit firmness	Referred in Brummell and Harpster, 2001 and da Silva et al., 1994
*Le*Exp1	suppression	tomato	Reduced polyuronide depolymerization Normal hemicellulose depolymerization	Lower softening rate	Brummell et al. 1999a, 2002
*Le*Exp1	overexpression	tomato	Normal polyuronide depolymerization Increased hemicellulose depolymerization	Higher softening rate	Brummell et al., 1999a; Brummell et al., 2002

Note: n.d., not determined.

tissue homogenates, providing evidence that PG is the primary determinant of cell wall polyuronide degradation in tomato fruit. However, these fruits sustained almost normal solubilization and fruit softening was not prevented or the fruits became just slightly firmer than controls (Smith et al., 1990; Brummell and Labavitch, 1997). Similarly, in the mutant *rin* (ripening-inhibitor) overexpressing PG up to 60% of its normal activity, solubilization and depolymerization of polygalacturonan occurred *in vivo*, but softening was not restored (Giovannoni et al., 1989; DellaPenna et al., 1990). In another experiment, transposon-tagged tomato lines with an insertion in the PG gene also resulted in fruits with normal softening behavior (Cooley and Yoder, 1998). Thus, although necessary for polyuronide degradation in tomato, PG itself is neither sufficient nor necessary to promote fruit softening. However, in strawberry, a fruit in that softens without major pectin depolymerization (Huber, 1984), antisense down-regulation of a ripening related PG (*FaPG1*) resulted in fruits with reduced pectin solubilization, less covalently bounded to the wall, and with increased firmness (Quesada et al., 2009), suggesting a central role of PGs in ripening, even in species in which pectin degradation has not been considered as a determinant for fruit softening. In contrast with ripening-related exo-PG, which is found in both melting and non-melting peaches, endo-PG activity increases only in ripening melting varieties, coincident with the melting phase (Pressey and Avants, 1978; Orr and Brady, 1993; Lester et al., 1994; Callahan et al., 2004; Manganaris et al., 2006; Murayama et al., 2009). In apple, the expression pattern of a PG gene is cultivar-dependent, occurring first and with significantly higher abundance in Royal Gala and later in cultivars with lower softening rates like Braeburn and Granny Smith (Atkinson et al., 1998; Goulao et al., 2008). In strawberry cultivars with contrasting softening rates, total PG activity and expression of two PG genes were accessed and revealed that the softest cultivar showed the higher total PG activity in all ripening stages analyzed. The expression of a gene that encodes an inactive PG protein (*T-PG*) is favored in firmer cultivars in relation to the expression of the gene that encodes active PG (*FaPG1*), while the opposite occurred in the softest cultivar (Villarreal et al., 2008). These results support a role for PG in the process of pectin solubilization or depolymerization attributed to fruit softening.

Removal by PME of methanol esterified to the carboxyl group of uronic acids affects the properties of the cell wall in several ways. This deesterification is necessary to generate susceptible sites for PG action (Pressey and Avants, 1982; Koch and Nevins, 1989), relaxing the wall, it generates increased binding trough calcium bridges, strengthening the wall, it promotes a localized reduction in the pH, which controls the activity of other enzymes (Fischer and Bennett, 1991) and it generates free COO- which facilitates the formation of new cross-linkages in cell walls (Fry, 1986).

Antisense suppression of PME mRNA abundance and activity by 90% in tomato (Tieman et al., 1992; Hall et al., 1993; Gaffe et al., 1994; Watson et al., 1994) and wild strawberry (*Fragaria vesca*) (Osorio et al., 2008) resulted in fruit with decreased pectin degradation, and an increase in methylesterification, although with little effect on fruit firmness. Low PME activity in the antisense tomato fruit modified both accumulation and partitioning of cations between soluble and bound forms (Tieman and Handa, 1994). The absence of phenotype during tomato ripening may be the result of the fact that loss of bound calcium and the ability to form cross-bridges that add strength to the polygalacturonate gels might have annulled any effects of reduced depolymerization of pectins on ripening-associated softening of transgenic fruits. It should be considered that these antisense lines only suppressed the activity of type 1 and had no affect on type 2, which are also present in fruit tissues (Gaffe et al., 1994), so they might represent an incomplete picture. On the other hand, antisense inhibition of PMEU1, an isoform with more significance before ripening, results in fruits that soften faster than controls (Phan et al., 2007), suggesting that this isoform may act to strengthen the cell wall, making fruits less susceptible to the ripening-induced softening process.

The middle lamella of the ripe tomato mutant *Cnr* (colorless non-ripening; a non-ripening, non-softening tomato that exhibits reduced cell-to-cell adhesion) lacks long un-methylesterified HGA blocks (Orfila et al., 2002), and unesterified HGA-containing pectic regions in *Cnr* cell walls appeared to have a different de-esterified block structure that may account for its ineffectiveness to bind calcium with consequences in cell adhesion (Orfila et al., 2002). The *Cnr* mutation appears to have effects on the solubility of HGA, and *Cnr* cell walls contain less chelator-soluble HGA, which is more susceptible to endo-PG degradation (Orfila et al., 2002). Supporting the role of PME in this behavior is the fact that the ripening-associated PME isoform is not active in *Cnr* fruits (Orfila et al., 2002). However, tomato mutants *nor* (non-ripening) and *Nr* (never-ripe) display PME activity similar to that of normally ripening genotypes (Harriman et al., 1991), but fail to soften. Considerable differences in PME activity occur between two raspberry cultivars with different patterns of fruit softening. While in Glen Prosen (firm cultivar), PME activity is maintained at lower levels throughout ripening, in Glen Clova (soft cultivar) an increase in PME activity occurs as the fruit softens (Iannetta et al., 1998). Similarly, the softest strawberry varieties display the highest PME and PG activities than those that remain firmer during ripening (Lefever et al., 2004), and the mRNAs of two PME genes in stony hard type peach are reduced during ripening as compared with these genes in the melting type peach (Murayama et al., 2009). Differences in the expression of PME genes may account, at least in part, for the differences in tuber texture among potato cultivars (Ducreux et al., 2008).

A pectate lyase (PL) gene has been isolated from ripe strawberry (Medina-Escobar et al., 1997) and banana (*Musa acuminata* Colla) (Dominguez-Puigjaner et al., 1997), with expression restricted to ripening fruits. Strawberry fruits suppressed in PL mRNA expression were significantly firmer than controls and their cell wall material showed a lower degree of *in vitro* swelling, less pectin solubilization, and a lower amount of ionically bound pectins, indicating a higher integrity of cell wall structure than control fruits (Jiménez-Bermúdez et al., 2002; Santiago-Doménech et al., 2008) and suggesting a direct role for this enzyme in fruit softening.

Three isoforms of β-gal were identified from ripening tomato fruits (Pressey, 1983) and six tomato β-gal genes are expressed during fruit ripening (Smith and Gross, 2000). In contrast with tomato β-galactosidases (TBGs) 1,2,3 and 5 which are present in *rin, nor* and *Nr* mutant fruits in a chronological pattern similar to that of wild-type accumulation (Smith and Gross, 2000), TBG4 transcript accumulation is significantly impaired in *rin* and *nor* mutants, and TBG6 persisted in mutants whereas it is not detected in wild fruit (Smith and Gross, 2000), suggesting a putative involvement in softening. TBG4 codes for β-Gal II, an enzyme known to possess both β-gal and exo-galactanase activities (Carey et al., 1995). Antisense suppression of TBG4 transcription was correlated with a reduction in extractable exo-galactanase activity (Smith et al., 2002), even though total β-gal activity and total cell wall galactosyl contents remain at the same levels as controls. Some lines were significantly firmer than untransformed fruits, suggesting that reduced levels of exo-galactanase during early ripening have an impact on fruit firmness during later stages of ripening, which may suggest that galactosyl-containing side-chains in the wall result in decreasing wall porosity, thereby obstructing access to wall components by other wall hydrolases (Brummell and Harpster, 2001). Suppression of TBG6 showed that this isoform may be important in early fruit growth and development but its role in fruit loss of texture is deemphasized (Moctezuma et al., 2003). A β-gal cDNA clone expresses earlier in apple cultivars that soften more rapidly when compared to firmer cultivars (Ross et al., 1994; Goulao et al., 2008), and softer raspberry cultivars have lower PG but higher β-gal activities than firmer cultivars (Iannetta et al., 1999).

Differences in the expression levels of three arabinofuranosidases (AFases) were detected between strawberry cultivars with different softening patterns (Rosli et al., 2009). In the softest cultivar, specific AFase activity and the expression of the three *Fa*Aras were higher at ripening Rosli et al., 2009). It should be noted that this enzyme catalyzes the hydrolysis of terminal non-reducing α-L-arabinofuranosil residues not only from various pectic polysaccharides, but also from hemicellulosic homo-(arabinans) and heteropolysaccharides (arabinogalactans, arabinoxylans, arabinoxyloglucans, glucuronoarabinoxylans, and so on).

4.2.3.2 Enzymes that act on the hemicellulose-cellulose network

For many years it was believed that endo-1,4-β-glucanases (EGases) were responsible for xyloglucan depolymerization. However, it has been shown, in studies using transgenic plants, that plant EGases have a minimal effect on xyloglucan's degree of polymerization (Harpster et al., 2002a,b). The mechanism is thought to be through hydrolysis of cross-linking xyloglucans, leading to loosening of each cellulose microfibril. However, the *in vivo* substrate of EGases remains unidentified.

The two known activities of xyloglucan endotransglucosylase/hydrolase (XTH) proteins are enzymologically described as xyloglucan endotransglucosylase activity (XET) and xyloglucan endohydrolase (XEH) activity (Nishitani and Tominaga, 1992; Fry et al., 1992; Farkas et al., 1992; Fanutti et al., 1993; Rose et al, 2002). Both transglucosylation and hydrolysis are suggested to be involved in the structural changes of xyloglucans (Nishitani and Tominaga, 1991). In the presence of suitable oligosaccharides, such as could arise during partial lyses of matrix polysaccharides during fruit ripening, XTHs can catalyze polysaccharide-to-oligosaccharide endotransglucosylation, causing a reduction in the relative molecular mass (*Mr*) of the xyloglucan (Lorences and Fry, 1993) or can depolymerize xyloglucan by acting as a strict hydrolyse (Farkas et al., 1992; Nishitani and Tominaga, 1992; da Silva et al., 1994). Endotransglucosylase activity measured from extracts of ripening tomato fruits was reported to be lower in the non-softening *rin* mutant than in wild-type fruits (Maclachlan and Brady, 1992, 1994).

A xylanase isoform expressed during fruit ripening in papaya correlates with the variation in softening patterns of different varieties with contrasting softening behavior (Chen and Paull, 2003). In strawberry, β-xylosidase mRNA accumulation, protein accumulation and activity, correlates with softening in two cultivars with contrasting firmness (Bustamante et al., 2006).

Expansins are thought to act transiently on the cellulose-hemicellulose linkages, allowing wall loosening. At least seven expansin isoforms are expressing with individual characteristic expression patterns during tomato growing and ripening, of which *Le*Exp1 is fruit-specific and ripening-related (Rose et al., 1997; Brummell et al., 1999b). In tomato, lines suppressed and lines overexpressed in *Le*EXP1 were analyzed and the results from both independent experiments suggest a function during softening by promoting the conditions to other hydrolases, namely EGases or XTHs, to act on their specific substrates. In fact, overexpression of *Le*Exp1 is correlated with cell wall hemicellulose depolymerization, and occurs even in mature green fruits before ripening (Brummell et al., 1999a). The direct effect of expansins on depolymerization of hemicelluloses is

challenged by the fact that suppression of *Le*Exp1 did not detectably affect their molecular mass profiles during ripening (Brummell et al., 1999a). Contrasting results were obtained when analyzing the effect on the pectic network. In antisense fruits suppressed in *Le*Exp1, polyuronide depolymerization was substantially arrested later in ripening (Brummell et al., 1999a), while overexpression did not produce any significant effect on its depolymerization (Brummer et al., 1999a). This reduction of polyuronide depolymerization in *Le*Exp1 suppressed lines was even larger than the one caused by inhibition of PG activity (Brummell and Labavitch, 1997; Brummel et al., 1999a). Since overexpression of *Le*Exp1 did not alter pectin depolymerization (Brummell et al., 1999a), the effect of expansins on the pectic network probably is derived from indirect effects and an inter-relationship between expansins and PG has been suggested. Genetic manipulation of *Le*Exp1 has significant effects on tomato firmness, with softening occurring at lower rates than in controls, particularly in the late ripening stages. In *Le*Exp1 overexpressing lines, fruits softened more rapidly, particularly in the early ripening stages (Brummell et al., 1999a; Brummell et al., 2002). The mRNA accumulation of a ripening-related expansin mRNA was different in a set of apple cultivars with different softening. While this gene is transcribed during ripening at relatively high levels in some cultivars, including Golden Delicious or Kitarou, in others, like Fuji and Ralls Janet it was below the experimental detection level, and in other cultivars it increases until commercial maturity and decreases thereafter, during ripening (e.g., Kotaro) (Wakasa et al., 2006). The same pattern of expression of expansins related with differences in firmness was reported in strawberry. Protein abundance (Dotto et al., 2006) and the expression of three out of five expansin mRNAs were correlated with fruit firmness in three cultivars that differ in fruit softening during ripening (Salentijn et al., 2003; Dotto et al., 2006). The mRNA for a peach expansin is detectable in Akatsuki peaches (a rapidly softening cultivar) but hardly detectable in the Manami cultivar (a cultivar that remains firm during postharvest storage) (Hayama et al., 2003).

In summary, some enzymatic activities have an effect on the wall through direct degradation of polysaccharides and loss of cohesion of the tissue. Others, in particular the ones that modify polysaccharide side-chains, may act by facilitating solubilization, promoting increases in hydration, modifying the pore size, which adjust the conditions for mobility, access to substrate and activity of other cell wall–modifying enzymes. The overall results from gene manipulation in transgenic fruits suggest that no single enzyme appears to be sufficient or necessary to account for the textural changes that occur during fruit ripening. This suggests a sequential and cooperative action between various enzymes, in which the action of one enzyme is necessary for subsequent and generalized degradation by other enzymes.

4.2.3.3 Molecular, biochemical, and hormonal regulation

Even with a softening-related pattern, the *in vivo* action of cell wall associated enzymes may not mimic their pattern of gene expression or enzymatic activity. Factors contributing to these differences include the apoplastic solubility or mobility of the enzymes (Bordenave and Goldberg, 1994), control and modulation by apoplastic pH and ionic conditions (Huber and O'Donoghue, 1993; Almeida and Huber, 1999; Fry, 2004), and steric influences (Ricard et al., 1981). Decreases in pH of the apoplast occur during ripening of peach, apricot (*Prunus armeniaca* L.; Ugalde et al., 1988), and tomato (Almeida and Huber, 1999). In tomato, the pH of an apoplastic fluid decreases from 6.7 at the mature-green stage to 4.4 in ripe fruit. This apoplastic acidification is accompanied by a 3-fold increase in potassium concentration (Almeida and Huber, 1999). Apoplastic acidification may be responsible for the solubilization of pectins via calcium displacement in fruits during ripening (Knee, 1982) or through influence in the catalytic behavior or activity of cell wall hydrolases, since ion composition and strength can influence protein binding and kinetics (Huber and O'Donoghue, 1993). The changes in apoplastic pH during fruit ripening provide a mechanism to regulate the activity of cell wall enzymes, which is superimposed on the transcriptional regulation. Figure 4.4 summarizes the pattern of transcription of major cell wall enzymes during ripening of tomato fruit and presents data on the effect of pH on enzyme activity. For instance, tomato endo-PG starts accumulating at the onset of ripening, when the apoplastic pH is unsuitable for its activity, and continues accumulating during ripening as pH becomes favorable. In contrast, PME expression is highest at the mature-green stage, when apoplastic pH is suitable for PME activity, and decreases during ripening as apoplastic acidification progresses and PME activity would be restricted by pH (Figure 4.4). Despite the potential for biochemical regulation of enzyme activity, the temporal and special changes in apoplastic pH remain largely uncharacterized.

The five classical hormones, namely, auxins, cytokinins, gibberellins, abscisic acid, and ethylene, are all known to modulate growth and development at various stages of the developing fruit (Ozga and Rienecke 2003). In addition, the roles of growth regulators such as polyamines, salicylic acid, jasmonic acid, and brassinosteroids on fruit development and ripening are being identified (Cohen 1998; Mehta et al., 2002; Sheng et al., 2003). Until recently, control of climacteric fruit ripening has been attributed to ethylene signaling. It is generally recognized that ethylene regulates the transcription and/or translation of ripening-related enzymes, including some of those that act on the cell wall. However, evidence that the ethylene peak may not always be the major determinant of fruit softening has been reported (e.g., Hallett et al., 1992; Crooks and Grierson, 1983), and the occurrence of both ethylene-dependent and ethylene-independent

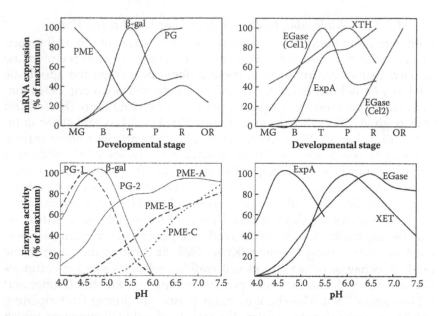

Figure 4.4 Expression of selected cell wall enzymes and correspondent catalytic activities at various pH levels. Expression of enzymes that act on the pectic matrix (A) and on the cellulose-hemicellulose network (B); effect of pH on the activity of enzymes that act on the pectic matrix (C) and on the cellulose-hemicellulose network (D). Redrawn from Almeida (1999), based on data derived from (A): Harriman et al. (1991), Smith et al. (1998), and DellaPenna et al. (1986); (B): Lashbrook et al. (1994), Gonzalez-Bosch et al. (1996), Arrowsmith and de Silva (1995), and Rose et al. (1997); (C): Warrilow and Jones (1995), Ross et al. (1993), Pressey and Avants (1973), and Chun and Huber (1998); (D): Purugganan et al. (1997), Hinton and Pressey (1974), McQueen-Mason et al. (1992), and Li et al. (1993).

pathways in the development of the climacteric ripening has been demonstrated in the last several years (Lelièvre et al., 1997; Hadfield et al., 2000; Itai et al., 2000; Trainotti et al., 2003; Johnston et al., 2009).

In addition to modifying the wall architecture by cutting and/or rearranging its polysaccharides, cell wall–modifying enzymes like PGs, PLs, or EGases may control ripening by generating oligosaccharide fragments (oligosaccharins) that will be involved in hormonal control (Aldington and Fry, 1993). Oligogalacturonide fragments derived from pectins consist of between two and twenty α-1,4-D-galactopyranoslyuronic acid residues, and it has been suggested that may induce and regulate the ethylene climacteric rise and ripening of tomato (Campbell and Labavitch, 1991; Dumville and Fry, 2000). An enhancement of ethylene production has been observed following treatment of fruit tissues with pectic fragments

produced by fungal enzymes (Tong et al., 1986), by acid hydrolysis
(Campbell and Labavitch, 1991) or with other fruit cell wall derived pectic
oligosaccharides released autolytically (Brecht and Huber, 1988; Melotto
et al., 1994). Short oligogalacturonide fragments are thought to induce
the expression of both allene oxide synthase, and enzyme involved in the
biosynthesis of jasmonates (Norman et al., 1999), and aminocyclopropane
1-carboxylic acid oxidase (Simpson et al., 1998). Even in non-climacteric
strawberry, a fruit-specific PG gene is suggested to be involved through
production of oligosaccharins rather than through pectin degradation
(Redondo-Nevado et al., 2001). Xyloglucans are another source of oligosaccharides (Pilling and Hofte, 2003). Persimmon (*Diospyrus kaki* L.) fruits
injected with a mixture of xyloglucan-derived oligosaccharides showed a
burst in ethylene production when compared to controls or fruits injected
with different monosaccharide solutions (Cutillas-Iturralde et al., 1998).
It is under debate the hypothesis that hydrolytic enzymes may function
as oligosaccharide generators that will induce the ethylene rise which in
turn regulates the expression of late-ripening genes (Baldwin and Pressey,
1988), or if their role is related with other relevant processes, including
proteinase inhibition (Bishop et al., 1984), or induction of responses at the
plasma membrane such as ion-flux or protein phosphorylation (Farmer
et al., 1991; Mathieu et al., 1991).

   Expression of genes encoding proteins other than non-classical cell
wall–modifying enzymes or other enzymes known to be involved in the
metabolic pathways of other events thought to contribute to fruit softening
has been proposed. A cDNA clone with similarity with GTPase proteins
is up-regulated during ripening of mangoes (*Mangifera indica* L.; Zainal
et al., 1996) and tomatoes (Lu et al., 2001), and this enzyme is thought to be
involved in trafficking of cell wall–modifying enzymes between different
cellular compartments, regulating its action (Lu et al., 2001). Tomato plants
expressing an antisense *Rab1 GTPase* gene show reduced levels of PG and
PME and reduced fruit softening (Lu et al., 2001). The PG β-subunit does
not possess any intrinsic activity, although antisense tomato plants suppressed for this protein exhibit increased fruit softening during ripening, higher extractable PG activity, and more polyuronide solubilization
(Brummell and Harpster, 2001). This suggests that this protein may be distributed throughout the cell wall to control PG diffusion or action during
ripening (Chun and Huber, 2000). A small heat-shock protein, *vis1*, was
shown to play a role in pectin depolymerization and juice viscosity during ripening of tomato (Ramakrishna et al., 2003). This protein acts by pro-
tecting enzymatic activities with disintegration of fruit components from
thermal denaturation caused by daily temperatures under field conditions.
Some proteins down-regulated during ripening may assume a regulatory
role on the activity of others. A glycoprotein inhibitor of PMEs has been
isolated from kiwifruit (Balestrieri et al., 1990) and its down-regulation

during ripening may modulate the de-esterification of pectins by allowing the activity of PME isoforms. In addition, and as stated above, silencing a growing-abundant PME isoform in tomato results in faster softening of the fruits (Phan et al., 2007). A basic protein that inhibits the hydrolytic activity of endoglucanases in tomato and its corresponding cDNA were purified, cloned, and characterized (Qin et al., 2003). This transcript was shown to be widely expressed in tomato vegetative tissues and is during expanding fruit, but it is down-regulated during ripening (Qin et al., 2003). A class III chitinase homologue with no enzymatic activity is present in young developing banana fruits and is down-regulated during ripening (Clendennen et al., 1998; Peumans et al., 2002). This protein behaves as a transient vegetative storage protein and its degradation is thought to be a source of amino acids for the synthesis of ripening-related proteins (Peumans et al., 2002).

Transcription factors control the expression or the posttranslational modifications needed to give the enzyme its catalytic properties. Comparisons between *Arabidopsis thaliana* fruit ripening (non-fleshy fruit), tomato (melty, climacteric fleshy fruit), strawberry (melty, non-climacteric fleshy fruit), and apple (fracturable fleshy fruit) have been made to disclose general mechanisms responsible for the control of fruit ripening. The results suggest that ripening is regulated, at least in part, by MADS-box transcription factors (Vrebalov et al., 2002; Sung and An, 1997; Yao et al., 1999; Sung et al., 2000). Both the *rin* and *Cnr* tomato mutants are related to mutations in MADS-box transcription factors or their targets (Seymour et al., 2008). The evidence that MADS-box regulation of ripening occurs in both tomato and strawberry also suggests that may be part of a common regulatory mechanism that operates in both climacteric and non-climacteric species (Vrebalov et al., 2002).

4.2.4   Turgor pressure and related biochemical changes

Although biochemical changes in the cell wall polysaccharides are a major mechanism underlying textural changes in fleshy fruit, turgor pressure also accounts for texture of plant organs, especially when texture is accessed as resistance to compression.

4.2.4.1   Water loss

Turgor pressure is an important determinant of fruit softening (Shackel et al., 1991; Tong et al., 1999). Turgor pressure decreases during ripening (Thomas et al., 2008) as ripening-related modifications of structural cell wall polymers supporting the cell tensile strength will influence the magnitude of turgor pressure. However, even in the absence of wall modifications, turgor pressure is substantially reduced by the presence of apoplastic solutes (Wada et al., 2008). Since turgor pressure in plant cells depends on the osmotic potential and the influx of water into the cell (Cosgrove, 1986) the turgor decline in ripening fruit is likely associated with efflux of solutes to the apoplast (Shackel et al., 1991; Almeida and Huber, 1999) and accompanying efflux of water from the protoplasm. Turgor pressure results from the dilution of the cell content and/or the direct loss of water from the fruit (Harker et al., 1997a), leading to a redistribution of solutes and an increase in cell volume that may occur when the cell wall elastic modulus changes as the result of turnover and breakdown of its structure, and by transpiration. Its influence in tissue strength derives from swelling, which reduces cell-to-cell contact area (Glenn and Poovaiah, 1990; Harker and Hallett, 1992). Ripening *DFD* (Delayed Fruit Deterioration) tomato fruits show minimal transpirational water loss and substantially elevated cell turgor. In this mutant, wall disassembly and genes expression appear similar to controls, but it possesses unusual features in its cuticle that suffer different modifications during ripening (Saladié et al., 2007). Therefore, at least in tomatoes, fruit softening results from an early decline in turgor pressure as a result of wall relaxation, followed by substantial water transpiration that occurs in parallel with a reduction in intercellular adhesion (Saladié et al., 2007). The water binding capacity also has a significant impact on texture. In kiwifruit, cell wall hydration represents a major change in tissue structure as the fruit softens (Harker and Hallett, 1994) and in stone fruit, the loss of juiciness is thought to occur when pectins bind water into a gel-like structure within the cell wall (Ben-Arie and Lavee, 1971). Therefore, studies regarding the distribution of water and osmolites within the tissues and solute transporters may contribute to our understanding of textural changes.

4.2.4.2 Loss of membrane integrity

Evidence points to the idea that although fruit cells do not experience a massive breakdown in cellular compartmentation during ripening, an increased leakiness of membranes may occur. Changes in the membrane composition that occur during fruit ripening lead to the increase of their "apparent free space" and result in leaking of ions, namely potassium, amino nitrogen, sugars, acids, and other solutes. The phenomenon has been reported in fruits like bananas (Brady et al., 1970), avocados (*Persea americana* Mill.; Sacher, 1962), tomatoes (Vickery and Bruinsma, 1973), and apples (Bartley, 1984). Changes occur in the structure of mitochondria, endoplasmic reticulum, and plasmalemma during ripening. At least in avocado, these changes are associated with the export of the cell wall–modifying enzymes (Platt-Aloia and Thomson, 1981). Since the transport of cell wall materials and enzymes occurs across the plasma membrane, this organelle can influence texture through the regulation of the ionic composition of the extracellular solution. For instance, acidification could be responsible for calcium removal from binding sites within the cell wall, promoting its weakening, or for the regulation of the activity of cell wall–modifying enzymes, contributing to the regulation of cell wall disassembly and solubilization, as stated before. In addition, changes in membrane permeability increase leakage of intracellular solutes into the extracellular fluid, reducing turgor pressure, and hydrating the apoplastic spaces. These events may cause cell-to-cell separation and consequent softening. However, it should be noted that the rate at which the cells leak solute into water depends not only on membrane permeability but also on the solute concentration gradient across their membranes and the rates of active uptake and efflux of the ion.

4.2.4.3   Starch degradation

Starch is a glucose homopolymer with $\alpha(1 \rightarrow 4)$ glycosidic linkages in the polymer chain and $\alpha(1 \rightarrow 6)$ linkages at the branchpoints. Unlike cell wall polysaccharides, starch is a storage polymer with no structural function. However, due to the high levels accumulated in certain plant organs, starch metabolism can affect texture. Soluble sugars accumulating as a result of starch hydrolysis increase the osmotic pressure within the cells, therefore contributing to the turgor pressure.

Most fruits that accumulate significant levels of starch during development (e.g., banana, kiwifruit, and apple) undergo extensive starch degradation during ripening. In contrast, in starchy vegetables such as potato, starch levels remain high during postharvest storage although starch reserves undergo mobilization (Hajirezaei et al., 2003).

Differences in texture within the potato germplasm have been associated with differences in starch structure (Taylor et al., 2007) and expression of α-amylase (Ducreux et al., 2008). Also, in banana starch degradation, concomitant with the disassembly of cell walls, has been implicated in fruit softening (Kojima et al., 1994).

4.2.5   Wounding and the case of fresh-cut produce

Fresh-cut processing invariably involves cutting operations that deeply affect tissue metabolism. Wounding enhances respiration rate, ethylene synthesis and increases area-to-volume ratio, favors water loss, and destroys structural barriers to the entry of microorganisms (King and Bolin, 1989; Watada et al., 1990; Watada and Qi, 1999).

Cell wall metabolism in wounded tissues differs in some aspects from that of whole fruit, whose general pattern was described above. The activity of pectic enzymes in fresh-cut fruit is altered in relation to their normal levels in whole fruit. The levels of PG and β-gal in papaya are increased after cutting and remain higher than in intact fruit for the remaining storage period (Karakurt and Huber, 2003). In ripening tomato, however, wounding has the opposite effect (Chung et al., 2006). Fresh-cut prepared from ripening tomato fruit had lower PG and β-gal activities than intact fruit, but once the fruit reaches a fully ripe stage wounding no longer alters the activities of pectic enzymes (Chung et al., 2006).

Cutting, by itself, may induce undesirable changes in texture by a mechanism related to the biochemical control of enzymatic activity. Tissue wounding, during the processing of fresh-cut fruit, exposes the apoplast in the vicinity of the cut area to acidic vacuolar contents modifying the apoplastic environment, therefore altering the activity of cell wall enzymes.

Studies showing that infiltrating tomato pericarp disks with buffers at pH 7.0 helps firmness retention when compared with the extensive softening induced at pH 4.5 prove the concept that apoplastic pH affects the texture of fresh-cut fruit (Pinheiro and Almeida, 2008). Significant linear relationships were found between pectin solubilization, which is strongly affected by pH, and the softening of tomato pericarp disks (Pinheiro and Almeida, 2008). It is known that PG hydrolyses only HG, the uronic acid residues of which have been previously demethylated via action of PMEs (Brummell and Harpster, 2001). Since pectins are deposited fully methylated on the cell wall (Staehelin and Moore, 2005), PME-generated negative charges in HGs are necessary to allow calcium binding and thus bring about its firming effects. Thus, the modulation of apoplastic pH can launch the basis for improved technologies aimed at firmness retention of fresh-cut fruit. The ability to maintain an apoplastic pH favorable to catalytic activity of PME, yet unfavorable to catalytic activity of PG (e.g., pH 7.0) associated with calcium supply, will likely improve the texture of fresh-cut fruit.

Moreover, treatments with acidic solutions are often applied to fresh-cut fruit to prevent browning and control microbial growth (Beaulieu and Gorny, 2001), providing an additional mechanism for apoplastic acidification in fresh-cut fruit.

4.3   Technological manipulation of texture-related enzymes

4.3.1   Conventional breeding

Plant breeding benefits from understanding the specific gene functions and the way they relate to a desired characteristic to be introduced in a new cultivar. Genomic tools have been useful in identifying the genes and the biochemical and molecular basis of texture and cell wall metabolism. In particular, genetic mapping can be explored based on the recognition of a strong genetic contribution to fruit texture (King et al., 2000). Flesh firmness is considered as a quantitative trait, being affected by several biochemical and physiological factors, and genetic variation from texture results from the segregation of numerous interacting Quantitative Trait Loci (QTL) (Seymour et al., 2002). Since fruit cultivars of several species vary considerably in their intrinsic textural properties, displaying a cultivar-specific softening behavior, breeders can take advantage of this natural source of variation to generate introgression line populations from crosses between cultivars with superior and poor softening rates and storage ability. This allows obtaining a powerful source of phenotypic variation in a characterized genetic background and facilitates the identification of QTLs related to several fruit quality attributes. Tools such as EST (Expressed Sequence Tag) and microarray analyses with subsequent PCR or *in vivo* expression analysis may facilitate the identification of a considerably high number of QTLs in several species, including tree fruits (Arús and Gardiner, 2008).

The availability of appropriate segregation populations and physical and linkage mapping programs opens the possibility of hypothesizing corresponding gene function based on linkage with characterized ripening mutants as in the case of tomato (Giovannoni et al., 1999), through linkage or co-linearity studies. Functionally linked genes are co-regulated and occur in proximity to each other in the chromosome, so the occurrence of an uncharacterized gene under the same operon of genes of known function provides information about its putative function. Co-linearity with other genomes is another powerful tool for identification of other uncharacterized candidate genes, such as ones involved in up- or downstream events and signaling, or coordination mechanisms involved in the general initiation and regulation of ripening. For instance, genes responsible for such physiological disorders as woolliness of peaches and nectarines, which involve changes in the expression of genes associated with cell wall metabolism and endo-membrane trafficking, have been revealed recently (González-Agüero et al., 2008).

To date, several programs aimed to identify QTLs related to texture have been reported in different fruits, such as melon (Moreno et al., 2008), tomato (Chaib et al., 2007), and apple (King et al., 2000; Kenis et al., 2008). Moreover, sensory and rheological assessments were successfully used in associating QTLs representing different attributes of fruit texture. In tomato, several genetic linkage maps were developed to study diverse fruit developmental, ripening, and quality loci (Bucheli et al., 1999; Grandillo et al., 1999; Doganlar et al., 2000; Ku et al., 2000; Chaib et al., 2007). QTL related to important quality traits including fruit mass, pH, soluble solid concentration (Paterson et al., 1991), shape, size, ripening time (Grandillo et al., 1999; Doganlar et al., 2000; Ku et al., 2000), organoleptic qualities (Causse et al., 2002), and fruit vitamin content (Rousseaux et al., 2005) are already available. Recently, a functional marker encoding an apoplastic invertase was identified as a QTL for increased soluble solids content (Fridman et al., 2004).

Identified useful QTLs can be transformed into genetic markers for convenient marker-assisted selection (MAS) routines. The identification of chromosomal regions contributing to major attributes of fruit texture is the first stage in developing selectable fully informative markers for the early selection of desirable genotypes, and it was shown in apple, based on their position in genetic linkage maps, that specific regions of the genome can account for significant proportions of the genetic variation related to aspects of texture (King et al., 2000; Chaib et al., 2007). Similarly, in peach, QTLs for several fruit quality traits including acidity and soluble solid content have been located in the same regions of just two linkage groups (Dirlewanger et al., 1999). A genetic marker for a softening trait would be especially desirable because the trait shows only when fruits are mature.

4.3.2   Genetic engineering

Knowledge of the genes involved in fruit softening may be used to produce genetically modified plants, overexpressing or suppressed in a specific isozyme. Advances in our understanding of the molecular and enzymatic basis of texture would allow the release of genetically modified fruits with improved storage and texture characteristics. Genetically modified fruits have become a commercial reality with the release of the transgenic tomato FlavrSavr™, in which a ripening-specific PG was strongly down-regulated. Due to its higher viscosity under cold-break processing, processed FlavrSavr™ tomatoes could be introduced in the market at lower costs. However, the poor genetic background of the genotype used in the transformation makes these tomatoes a failure in the marketplace. The public debate around genetically modified foods has, so far, hindered the commercial exploration of cultivars obtained by genetic engineering, particularly in Europe. Non-commercial genetically modified lines of tomato, strawberry, and pepper (*Capsicum annuum* L.) fruits suppressed in individual members of cell wall–modifying enzymes were produced for functional analysis studies (see above and Table 4.1) and are available. Despite the fact that a non-softening fruit has never been obtained, improvements in important physical properties that have implications in postharvest characteristics, including extended shelf life, reduction of the susceptibility to postharvest pathogens, and increased viscosity of juice and paste, were achieved in tomatoes down-regulated in PG (Kramer et al., 1992; Langley et al., 1994), PME (Tieman and Handa, 1994; Tieman et al., 1992, 1995; Thakur et al., 1996), and expansin (Brummell et al., 2002) without affecting other quality attributes.

Little research has been conducted on the characterization of promoters of ripening-induced genes. It should be informative to isolate fruit-specific promoters, which will be necessary to regulate the expression of genes in a tissue- and temporal-specific manner. Some promoters from fruit-specific, ripening-induced genes have been identified to date. Among those are the tomato PG (Montgomery et al., 1993; Nicholass et al., 1995), E8 (Deikman et al., 1992), 2A11 (Van Haaren and Houck, 1993), and 1-aminocyclopropane-1-carboxylate oxidase (ACO1) (Blume and Grierson, 1997). Genes corresponding to factors that result in the observed promoter binding activities remain unknown, thus limiting knowledge relative to specific genetic regulatory mechanisms that control expression of fruit-specific and ripening-related genes (Giovannoni, 2001). As stated before, no single enzyme has been identified as being solely responsible for fruit softening. Therefore, the identification of a regulatory mechanism that controls the transcription, translation, or activity of the sequential action of the genes and enzymes involved in the process opens the possibility of genetic manipulation with the aim of understanding and controlling the ripening process. Down-regulation of a gene encoding a tomato Auxin Response Factor (ARF), whose expression is regulated by ethylene in the fruit, was shown to result in fruits with enhanced firmness at the red-ripe stage (Jones et al., 2002), which is related to differences in pectin fine structure associated with changes in tissue architecture (Guillon et al., 2008). The suppression of deoxyhypusine synthase (DHS) was also shown to delay tomato postharvest softening (Wang et al., 2005). By controlling the expression of softening-related transcription factors, it should be possible to manipulate more factors rather than inhibit the expression of a single gene individually, and it should be possible to identify the proteins that are being regulated by these factors. A cDNA clone with similarity with *GTPase* proteins was found to be up-regulated during ripening of mangoes (Zainal et al., 1996) and tomatoes (Lu et al., 2001). This class of enzymes is thought to be involved in trafficking of cell wall–modifying enzymes. Antisense *GTPase* tomato fruits do not exhibit changes in color but failed to soften normally (Lu et al., 2001), accompanied by reduced levels of PME and PG activities (Lu et al., 2001). The strategy of suppressing transcription factors for biotechnological breeding in agricultural crops is a promising field for technological manipulation of texture-related enzymes.

4.3.3   Additives to interfere with texture-related enzymes

Calcium salts are used to improve firmness retention in fresh fruits and vegetables. Calcium is involved in cell wall metabolism in a variety of ways, including structural stabilization of polysaccharides, regulation of the activity of cell wall enzymes, and ion exchange properties of the wall (Demarty et al., 1984). Examples of the correlation between calcium content in fruit tissues and delayed softening are well documented (Poovaiah et al., 1988). The firming effect of calcium has been attributed to its role in cell wall metabolism. However, the role of calcium in maintaining fruit turgor potential may also contribute to firmness retention (Saftner et al., 1998).

The structural role of calcium in the cell wall derives from its ability to cross-link adjacent polygalacturonate chains. Since pectins are methylesterified when secreted into the cell wall (Staehelin and Moore, 2005), the action of PMEs *in muro* is required to generate the anionic groups in the galactosyluronate residues where Ca^{2+} can bind in an arrangement described as the "egg-box" (Grant et al., 1973). In addition, calcium contributes to the formation of the pectic matrix through the formation and stabilization of boron-RG-II complexes (Kobayashi et al., 1999).

Calcium is present in the cell wall in two different "compartments": a fraction is bound to fixed negative charges, whereas the remainder is soluble in the water-free space. The levels of calcium in an apoplastic fluid, of ripening tomato fruit presumably originating in water-free space, are 4 to 5 mM (Almeida and Huber, 1999). These calcium concentrations strongly inhibit *in vitro* cell wall autolysis (Rushing and Huber, 1984), and are likely to restrict ripening-related pectin solubilization *in vivo* (Almeida and Huber, 2007, 2008).

Increasing fruit calcium content enhances the stability of pectic polymers in the cell wall, decreasing the fraction of pectins that are water-soluble (Mignani et al., 1995). By pressure-infiltrating apple fruits with $CaCl_2$ solutions of concentrations ranging from 70 to 340 mM, Saftner et al. (1998) showed a close linear relation between the Ca^{2+} concentration in the solution and the concentration of Ca^{2+} in the cortical parenchyma (1 to 4 mm below the peel). Higher tissue Ca^{2+} contents correlated with increased firmness retention, increased turgor-induced stress, and decreased internal air space. The increased tissue Ca^{2+} content is distributed between cell wall bound and Ca^{2+} soluble in the water-free space. Ca^{2+}-binding sites in the apple cortex cell walls are saturated at 200 mM, whereas soluble calcium in the cortex apoplast increases linearly with the concentration of the bathing solution between 0 and 270 mM. Soluble apoplastic calcium affects tissue osmotic potential. A calcium-induced increase in turgor and decrease in interstitial air space may add to the mechanical strength of the tissue, helping to maintain firmness (Saftner et al., 1998). The overwhelming evidence that calcium positively impacts on fruit texture as well as other quality attributes (e.g., resistance to decay) provided the motivation for developing methods to improve fruit calcium content. Briefly, these methods include pre- and postharvest approaches as presented in Table 4.2 (Almeida, 2005).

Table 4.2 A Classification of Methods for Applying Calcuim
to Fruits and Vegetables

Time of application	Method of application	Notes
Preharvest	To the soil	Poor correlation between calcium fertilization and calcium content in fruits. Often dismissed as an unreliable or ineffective way of providing calcium to fruits, especially in woody crops.
	To the canopy	Effective in increasing fruit calcium content if applied directly to the fruit surface; may require multiple application during fruit development; risk of skin injury.
Preharvest	Dipping or drenching	Easy and inexpensive means to increase calcium content in the outer cell layers of fruit; may cause injury and enhance decay.
	Pressure (positive or negative) infiltration	Susceptibility to calcium toxicity and pressure-induced injury and technological complexity hinders commercial application.
	Coating	Release of calcium from a coating to the fruit outer cells.

Source: Adapted from Almeida, D.P.F. 2005. Managing calcium in the soil-plant-fruit system. In *Crops: Growth, Quality and Biotechnology,* Ed. R. Dris, 448–459. Helsinki: WFL.

4.4   Conclusion

Texture is a complex attribute of fruits and vegetables determined at various levels of plant architecture. In this review we have focused on the biochemical level, emphasizing the roles of polymer structure and composition, polymer integration on cell walls, and enzyme-mediated changes in the individual polymers and their interactions.

Two decades after the first attempts at using molecular tools to manipulate the expression of individual enzymes implicated in cell wall metabolism, it is becoming clear that no single enzyme is responsible for the textural changes in ripening fruit and that ripening-related changes in texture are not alike in the various species. Softening of fleshy fruits is likely to be such a crucial feature for seed dispersal and species survival that evolution assured that fruit cell wall architecture could be dismantled by various mechanisms. Textural changes in most non-fruit vegetables are limited compared with the dramatic changes occurring in many fleshy fruits and have, therefore, received comparatively less attention.

Despite the advances in our understanding of textural changes, manipulating the texture of fresh fruits and vegetables remains an important goal for reducing produce losses and improving produce quality throughout the supply chain. Meeting the challenge will require further research efforts at the fundamental level to obtain a better model of cell wall structure, and a comprehensive model for the regulation of cell wall metabolism and at the applied level to devise techniques of molecular and biochemical engineering of textural properties.

Abbreviations

β-gal	*β*-galactosidase
ACO1	1-aminocyclopropane-1-carboxylate oxidase
AFase	Arabinofuranosidase
AGP	arabinogalactan protein
ARF	Auxin Response Factor
Cnr	colorless non-ripening
DFD	delayed fruit deterioration
DHS	deoxyhypusine synthase
EGase	endo-1,4-*β*-glucanase
EST	Expressed Sequence Tag
GRP	glycine-rich protein
HG	Homogalacturonan
HGA	homogalacturonic acid
MAS	marker-assisted selection
Mr	relative molecular mass
nor	non-ripening
Nr	never-ripe
PCR	Polymerase Chain Reaction
PG	Polygalacturonase
PL	pectate lyase
PME	pectin methylesterase
PRP	proline-rich protein
QTL	Quantitative Trait Loci
RG-I	rhamnogalacturonan-I
RG-II	rhamnogalacturonan-II
rin	ripening-inhibitor
TBG	tomato *β*-galactosidase
XET	xyloglucan endotransglucosylase activity
XHE	xyloglucan endohydrolase activity
XTH	xyloglucan endotransglycosylase/hydrolase

References

Adams, J.B. 2004. Raw materials quality and the texture of processed vegetables. In *Texture in Food: Solid Foods*, Ed. D. Kilcast, 342–363. Boca Raton: CRC Press.

Aldington, S. and S.C. Fry. 1993. Oligosaccharins. *Advances in Botanical Research* 19:1–101.

Almeida, D.P.F. 1999. Cell wall metabolism in ripening and chilled tomato fruit, as related to the pH and mineral composition of the apoplast. PhD dissertation, Univ. Florida.

Almeida, D.P.F. 2005. Managing calcium in the soil-plant-fruit system. In *Crops: Growth, Quality and Biotechnology*, Ed. R. Dris, 448–459. Helsinki: WFL.

Almeida, D.P.F. and D.J. Huber. 1999. Apoplastic pH and inorganic ion levels in tomato fruit: A potential means for regulation of cell wall metabolism during ripening. *Physiologia Plantarum* 105:506–512.

Almeida, D.P.F. and D.J. Huber. 2007. Polygalacturonase-mediated dissolution and depolymerization of pectins in solutions mimicking the pH and mineral composition of tomato fruit apoplast. *Plant Science* 172:1087–1094.

Almeida, D.P.F. and D.J. Huber. 2008. *In vivo* pectin solubility in ripening and chill-injured tomato fruit. *Plant Science* 174:174–182.

Arrowsmith, D.A., and J. de Silva. 1995. Characterization of two tomato fruit-expressed cDNAs encoding xyloglucan endo-transglycosylase. *Plant Molecular Biology* 28:391–403.

Arús, P. and S. Gardiner. 2008. Genomics for improvement of *Rosaceae* temperate tree fruits. In *Genomics-Assisted Crop Improvement*, vol. 2: *Genomics Applications in Crops*, eds. R. Varshney and R. Tuberosa, 357–398. Dordrecht: Springer.

Atkinson, R.G., K.M. Bolitho, M.A. Wright, T. Iturriagagoitia-Bueno, S.J. Reid, and G.S. Ross. 1998. Apple ACC-oxidase and polygalacturonase: Ripening-specific gene expression and promoter analysis in transgenic tomato. *Plant Molecular Biology* 38:449–460.

Baldwin, E.A. and R. Pressey. 1988. Tomato polygalacturonase elicits ethylene production in tomato fruit. *Journal of the American Society for Horticultural Science* 113:92–95.

Balestrieri, C., D. Castaldo, A. Giovane, L. Quagliuolo, and L. Servillo. 1990. A glycoprotein inhibitor of pectin methylesterase in kiwi fruit (*Actinidia chinensis*). *European Journal of Biochemistry* 193:183–187.

Bartley, I.M. 1984. Lipid metabolism in ripening apples. *Phytochemistry* 24:2857–2859.

Beaulieu, J.C. and J.R. Gorny. 2001. Fresh-cut fruits. In *The Commercial Storage of Fruits, Vegetables, and Florist and Nursery Stocks*, ed. K. Gross. Agriculture Handbook Number 66, USDA, Washington DC.

Ben-Arie, R. and S. Lavee. 1971. Pectic changes occurring in Elberta peaches suffering from woolly breakdown. *Phytochemistry* 10:531–538.

Ben-Arie, R., N. Kislev, and C. Frankel. 1979. Ultrastructural changes in the cell walls of ripening apple and pear fruit. *Plant Physiology* 64:197–202.

Bishop, P.D., G. Pearce, J.E. Bryant, and C.A. Ryan. 1984. Isolation and characterization of the proteinase inhibitor-inducing factor from tomato leaves. *Journal of Biological Chemistry* 259:13172–13177.

Blume, B., and D. Grierson. 1997. Expression of ACC oxidase promoter-GUS fusions in tomato and *Nicotiana plumbaginifolia* regulated by developmental and environmental stimuli. *The Plant Journal* 12:731–746.

Bootten, T.J., P.J. Harris, L.D. Melton, and R.H. Newman. 2004. Solid-state ^{13}C-NMR spectroscopy shows that the xyloglucans in the primary cell walls of mung bean (*Vigna radiate* L.) occur in different domains: A new model for xyloglucan–cellulose interactions in the cell wall. *Journal of Experimental Botany* 55:571–583.

Bordenave, M. and R. Goldberg. 1994. Immobilized and free apoplastic pectinmethylesterases in mung bean hypocotyl. *Plant Physiology* 106:1151–1156.

Bourne, M.C. 1979. Fruit texture: Overview of trends and problems. *Journal of Texture Studies* 10:83–94.

Bourne, M.C. 2002. *Food Texture and Viscosity. Concept and Measurement.* London: Academic Press.

Brady, C.J., G. MacAlpine, W.B. McGlasson, and Y. Ueda. 1982. Polygalacturonase in tomato fruits and the induction of ripening in *Lycopersicon esculentum*. *Australian Journal of Plant Physiology* 9:171–178.

Brady, C.J., P.B.H. O'Connell, N.L Wade, and J. Smydzuk. 1970. Permeability, sugar accumulation, and respiration rate in ripening banana fruits. *Australian Journal of Biological Sciences* 23:1143–1152.

Brecht, J.K. and D.J. Huber. 1988. Products released from enzymically active cell wall stimulate ethylene production and ripening in preclimacteric tomato (*Lycopersicon esculentum* Mill.) fruit. *Plant Physiology* 88:1037–1041.

Brett, C. and K. Waldron. 1996. *Physiology and Biochemistry of Plant Cell Walls.* London: Chapman & Hall.

Brown, R.M., I.M. Saxena, and K. Kudlicka. 1996. Cellulose biosynthesis in higher plants. *Trends in Plant Science* 1:149–156.

Brummell, D.A. 2006. Cell wall disassembly in ripening fruit. *Functional Plant Biology* 33:103–119.

Brummell, D.A. and J.M. Labavitch. 1997. Effect of antisense suppression of endopolygalacturonase activity on polyuronide molecular weight in ripening tomato fruit and fruit homogenates. *Plant Physiology* 115:717–725.

Brummell, D.A. and M.H. Harpster. 2001. Cell wall metabolism in fruit softening and quality and its manipulation in transgenic plants. *Plant Molecular Biology* 47:311–340.

Brummell, D.A., M.H. Harpster, and P. Dunsmuir.1999a. Differential expression of expansin gene family members during growth and ripening of tomato fruit. *Plant Molecular Biology* 39:161–169.

Brummell, D.A., M.H. Harpster, P.M. Civello, J.M. Palys, A.B. Bennett, and P. Dunsmuir. 1999b. Modification of expansin protein abundance in tomato fruit alters softening and cell wall polymer metabolism during ripening. *Plant Cell* 11:2203–2216.

Brummell, D.A., V. Dal Cin, C.H. Crisosto, and J.M. Labavitch. 2004. Cell wall metabolism during maturation, ripening and senescence of peach fruit. *Journal of Experimental Botany* 55:2029–2039.

Brummell, D.A., W.J. Howie, C. Ma, and P. Dunsmuir. 2002. Postharvest fruit quality of transgenic tomatoes suppressed in expression of a ripening-related expansin. *Postharvest Biology and Technology* 25:209–220.

Bucheli, P., J. Lopez, E. Voirol, V. Petiard, and S.D. Tanksley. 1999. Definition of biochemical and molecular markers (quality trait loci) for tomato flavour as tools in breeding. *Acta Horticulturae* 487:301–306.

Bustamante, C.A., H.G. Rosli, M.C. Añón, P.M. Civello, and G.A. Martínez. 2006. β-xylosidase in strawberry fruit: Isolation of a full-length gene and analysis of its expression and enzymatic activity in cultivars with constrasting firmness. *Plant Science* 171:497–504.

Callahan, A.M., R. Scorza, C. Bassett, M. Nickerson, and F.B. Abeles. 2004. Deletions in an endopolygalacturonase gene cluster correlate with non-melting flesh texture in peach. *Functional Plant Biology* 31:159–168.

Campbell, A.D. and J.M. Labavitch. 1991. Induction and regulation of ethylene biosynthesis and ripening by pectic oligomers in tomato pericarp discs. *Plant Physiology* 97:706–713.

Campbell, A.D., M. Huysamer, H.U. Stotz, L.C. Greve, and J.M. Labavitch. 1990. Comparison of ripening processes in intact tomato fruit and excised pericarp discs. *Plant Physiology* 94:1582–1589.

Carey, A.T., D.L. Smith, E. Harrison, C.R. Bird, K.C. Gross, G.B. Seymour, and G.A. Tucker. 2001. Down-regulation of a ripening-related beta-galactosidase gene TBG1 in transgenic tomato fruits. *Journal of Experimental Botany* 52:663–668.

Carey, A.T., K. Holt, S. Picard, R. Wilde, G.A. Tucker, C.R. Bird, W. Schuch, and G.B. Seymour. 1995. Tomato exo-(1->4)-beta-D-galactanase. Isolation, changes during ripening in normal and mutant tomato fruit, and characterization of a related cDNA clone. *Plant Physiology* 108:1099–107.

Carpita, N. and M. McCann. 2000. The cell wall. In *Biochemistry and Molecular Biology of Plants*, eds. B.B. Buchanan, W. Gruissem, and R.L. Jones, 52–108. Rockville, Maryland: American Society of Plant Physiologists.

Carpita, N.C. and D.M. Gibeaut. 1993. Structural models of primary-cell walls in flowering plants: Consistency of molecular structure with the physical properties of the walls during growth. *The Plant Journal* 3:1–30.

Carrington, C.M.S. and R. Pressey. 1996. beta-Galactosidase II activity in relation to changes in cell wall galactosyl composition during tomato ripening. *Journal of the American Society for Horticultural Science* 121:132–136.

Causse, M., V. Saliba-Colombani, L. Lecomte, P. Duffe, P. Rousselle, and M. Buret. 2002. QTL analysis of fruit quality in fresh market to tomato: a few chromosome regions control the variation of sensory and instrumental traits. *Journal of Experimental Botany* 53:2089–2098.

Chaib, J., M.F. Devaux, M.G. Grotte, K. Robini, M. Causse, M. Lahaye, and I. Marty. 2007. Physiological relationships among physical, sensory, and morphological attributes of texture in tomato fruits. *Journal of Experimental Botany* 58:1915–1925.

Chen, N.J. and R.E. Paull. 2003. Endoxylanase expressed during papaya fruit ripening: Purification, cloning and characterization. *Functional Plant Biology* 30:433–441.

Chun, J.P. and D.J. Huber. 1998. Polygalacturonase-mediated solubilization and depolymerization of pectic polymers in tomato fruit cell walls. Regulation by pH and ionic conditions. *Plant Physiology* 117:1293–1299.

Chun, J.P. and D.J. Huber. 2000. Reduced levels of beta-subunit protein influence tomato fruit firmness, cell-wall ultrastructure, and PG2-mediated pectin hydrolysis in excised pericarp tissue. *Journal of Plant Physiology* 157:153–160.

Chung, T.T., G. West, and G.A.Tucker. 2006. Effect of wounding on cell wall hydrolase activity in tomato fruit. *Postharvest Biology and Technology* 40:250–255.

Clendennen, S.K., R. Lopez-Gomez, M. Gomez-Lim, C.J. Arntzen, and G.D. May. 1998. The abundant 31-kilodalton banana pulp protein is homologous to class-III acidic chitinases. *Phytochemistry* 47:613–619.

Coenen, G.J., E.J. Bakx, R.P. Verhoef, H.A. Schols, and A.G.J. Voragen. 2007. Identification of the connecting linkage between homo- or xylogalacturonan and rhamnogalacturonan type I. *Carbohydrate Research* 70:224–235.

Cohen, S.S. 1998. *A Guide to the Polyamines.* New York: Oxford University Press.

Cooley, M.B. and J.I. Yoder. 1998. Insertional inactivation of the tomato polygalacturonase gene. *Plant Molecular Biology* 38:521–530.

Cosgrove, D.J. 1986. Biophysical control of plant cell growth. *Annual Review of Plant Physiology* **37**:377–405.

Cosgrove, D.J. 2001. Wall structure and wall loosening. A look backwards and forwards. *Plant Physiology* 125:131–134.

Crookes, P.R. and D. Grierson.1983. Ultrastructure of tomato fruit ripening and the role of polygalacturonase isoenzymes in cell wall degradation of *Lycopersicon esculentum. Plant Physiology* 72:1088–1093.

Cutillas-Iturralde, A., D.C. Fulton, S.C. Fry, E.P. Lorences. 1998. Xyloglucan-derived oligosaccharides induce ethylene synthesis in persimmon (*Diospyros kaki* L.) fruit. *Journal of Experimental Botany* 49:701–706.

Da Silva, J., D. Arrowsmith, A. Hellyer, S. Whiteman, and S. Robinson. 1994. Xyloglucan endotransglycosylase and plant growth. *Journal of Experimental Botany* 45:1693–1701.

Dawson, D.M., L.D. Melton, and C.B. Watkins. 1992. Cell-wall changes in nectarines (*Prunus persica*): Solubilization and depolymerization of pectic and neutral polymers during ripening and in mealy fruit. *Plant Physiology* 100:1203–1210.

Deikman, J., R. Kline, and R.L. Fischer. 1992. Organization of ripening and ethylene regulatory regions in a fruit-specific promoter from tomato (*Lycopersicon esculentum*). *Plant Physiology* 100:2013–2017.

DellaPenna, D., C.C. Lashbrook, K. Toenjes, J.J. Giovannoni, R.L Fischer, and A.B. Bennett. 1990. Polygalacturonase isozymes and pectin depolymerization in transgenic rin tomato fruit. *Plant Physiology* 94:1882–1886.

DellaPenna, D., D.C. Alexander, and A.B Bennett. 1986. Molecular cloning of tomato fruit polygalacturonase: Analysis of polygalacturonase mRNA levels during ripening. *Proceedings of the National Academy of Sciences of the U.S.A.* 83: 6420–6424.

Demarty, M., C. Morvan, and M. Thellier. 1984. Calcium and the cell wall. *Plant, Cell and Environment* 7: 441–448.

Dirlewanger, E., A. Moing, C. Rothan, L. Svanella, V. Pronier, A. Guye, C. Plomion, and R. Monet. 1999. Mapping QTLs controlling fruit quality in peach (*Prunus persica* (L.) Batsch). *Theoretical and Applied Genetics* 98:18–31.

Doganlar, S., S.D. Tanksley, and M.A. Mutschler. 2000. Identification and molecular mapping of loci controlling fruit ripening time in tomato. *Theoretical and Applied Genetics* 100:249–255.

Dominguez-Puigjaner, E., I. Llop, M. Vendrell, and S. Prat. 1997. A cDNA clone highly expressed in ripe banana fruit shows homology to pectate lyases. *Plant Physiology* 114:1071–1076.

Doong, R.L., K. Liljebjelke, G. Fralish, A. Kumar, and D. Mohnen. 1995. Cell-free synthesis of pectin. Identification and partial characterization of a poly-galacturonate 4-α-galacturonosyltransferase and its products from membrane preparations of tobacco cell-suspension cultures. *Plant Physiology* 109:141–152.

Dotto, M.C., G.A. Martínez, and P.M. Civello. 2006. Expression of expansin genes in strawberry varieties with contrasting fruit firmness. *Plant Physiology and Biochemistry* 44:301–307.

Ducreux, L.J.M., W.L. Morris, I.M. Prosser, J.A. Morris, M.H. Beale, F. Wright, T. Shepherd, G.J. Bryan, P.E. Hedley, and M.A. Taylor. 2008. Expression profiling of potato germplasm differentiated in quality traits leads to the identification of candidate flavour and texture genes. *Journal of Experimental Botany* 59:4219–4231.

Dumville, J.C. and S.C. Fry. 2000. Uronic acid-containing oligosaccharins: their biosynthesis, degradation and signalling roles in non-diseased plant tissues. *Plant Physiology and Biochemistry* 38:125–140.

Fanutti, C., M.J. Gidley, and J.S.G. Reid. 1993. Action of a pure xyloglucan endo-transglycosylase (formerly called xyloglucan-specific endo-(1 → 4)-beta-D-glucanase) from the cotyledons of germinated nasturtium seeds. *The Plant Journal* 3:691–700.

Farkas, J.V., Z. Sulova, E. Stratilova, R. Hanna, and G. Maclachlan.1992. Cleavage of xyloglucan by nasturtium seed xlyoglucanase and transglycosylation to xyloglucan subunit oligosaccharides. *Archives of Biochemistry and Biophysics* 298:365–370.

Farmer, E.E., T.D. Moloshok, M.J. Saxton, and C.A. Ryan. 1991. Oligosaccharide signaling in plants. Specificity of oligouronide-enhanced plasma membrane protein phosphorylation. *Journal of Biological Chemistry* 266:3140–3145.

Fillion, L. and D. Kilcast. 2002. Consumer perception of crispness and crunchiness in fruits and vegetables. *Food Quality Preference* 1:23–29.

Fischer, R.L. and A.B. Bennett. 1991. Role of cell wall hydrolases in fruit ripening. *Annual Review of Plant Physiology and Plant Molecular Biology* 42: 675–703.

Fridman, E., F. Carrari, Y.S. Liu, A.R. Fernie, and D. Zamir. 2004. Zooming in on a quantitative trait for tomato yield using interspecific introgressions. *Science* 305:1786–1789.

Fry, S.C. 1986. Cross-linking of matrix polymers in the growing cell walls of angiosperms. *Annual Review of Plant Physiology* 37:165–186.

Fry, S.C. 1989. The structure and functions of xyloglucan. *Journal of Experimental Botany* 40:1–11.

Fry, S.C. 2004. Primary cell wall metabolism: tracking the careers of wall polymers in living plant cells. *New Phytologist* 161:641–675.

Fry, S.C., R.C. Smith, K.F. Renwick, D.J. Martin, S.K. Hodge, and K.J. Matthews. 1992. Xyloglucan endotransglycosylase, a new wall-loosening enzyme activity from plants. *Biochemistry Journal* 282:821–828.

Gaffe, J., D.M. Tieman, and A.K. Handa. 1994. Pectin methylesterase isoforms in tomato (*Lycopersicon esculentum*) tissues. Effects of expression of a pectin methylesterase antisense gene. *Plant Physiology* 105:199–203.

Giovannoni, J. 2001. Molecular biology of fruit maturation and ripening. *Annual Review of Plant Physiology and Plant Molecular Biology* 52:725–749.

Giovannoni, J., D. DellaPenna, A.B. Bennett, and R.L. Fisher. 1989. Expression of a chimeric polygalacturonase gene in transgenic rin (ripening inhibitor) tomato fruit results in polyuronide degradation but not fruit softening. *Plant Cell* 1:53–63.

Giovannoni, J., H. Yen, B. Shelton, S. Miller, J. Vrebalov, P. Kannan, D. Tieman, R. Hackett, D. Grierson, and H. Klee. 1999. Genetic mapping of ripening and ethylene-related loci in tomato. *Theoretical and Applied Genetics* 98:1005–1013.

Glenn, G.M., and B.W. Poovaiah. 1990. Calcium-mediated postharvest changes in texture and cell wall structure and composition in Golden Delicious apples. *Journal of the American Society for Horticultural Science* 115: 962–968.

González-Agüero, M., L. Pavez, F. Ibáñez, I. Pacheco, R. Campos-Vargas, L.A. Meisel, A. Orellana, J. Retamales, H. Silva, M. González, and V. Cambiazo. 2008. Identification of woolliness response genes in peach fruit after postharvest treatments. *Journal of Experimental Botany* 59:1973–1986.

Gonzalez-Bosch, C., D.A. Brummell, and A.B. Bennett. 1996. Differential expression of two endo-1,4-β-glucanase genes in pericarp and locules of wild-type and mutant tomato fruit. *Plant Physiology* 111:1313–1319.

Goulao, L.F., and C.M. Oliveira. 2008. Cell wall modifications during ripening: when a fruit is not the fruit. *Trends in Food Science and Technology* 19:4–25.

Goulão, L.F., Cosgrove, D.J., and C.M. Oliveira. 2008. Cloning, characterisation and expression analyses of cDNA clones encoding cell wall-modifying enzymes isolated from ripe apples. *Postharvest Biology and Technology* 48:37–51.

Grandillo, S., H.M. Ku, and S.D Tanksley.1999. Identifying the loci responsible for natural variation in fruit size and shape in tomato. *Theoretical and Applied Genetics* 99:978–987.

Grant, G.T., E.R. Morris, D.A. Rees, P.J.C. Smith, and D. Thom. 1973. Biological interactions between polysaccharides and divalent cations: the egg-box model. *FEBS Letters* 32:195–198.

Grierson, D. and G.A. Tucker. 1983. Timing of ethylene and polygalacturonase synthesis in relation to the control of tomato fruit ripening (*Lycopersicon esculentum*). *Planta* 83: 6420–6424.

Gross, K.C. and C.E. Sams. 1984. Changes in cell wall neutral sugar composition during fruit ripening: A species survey. *Phytochemistry* 23:2457–2461.

Guillon, F., S. Philippe, B. Bouchet, M.F. Devaux, P. Frasse, B. Jones, M. Bouzayen, and M. Lahaye. 2008. Down-regulation of an Auxin Response Factor in the tomato induces modification of fine pectin structure and tissue architecture. *Journal of Experimental Botany* 59:273–288.

Ha, M.A., D.C. Apperley, and M.C., Jarvis 1997. Molecular rigidity in dry and hydrated onion cell walls. *Plant Physiology* 115:593–598.

Hadfield, K.A. and A.B. Bennett. 1998. Polygalacturonases: Many genes in search of a function. *Plant Physiology* 117:337–343.

Hadfield, K.A., T. Dang, M. Guis, J.C. Pech, M. Bouzayen, and A.B. Bennett. 2000. Characterization of ripening-regulated cDNAs and their expression in ethylene-suppressed charentais melon fruit. *Plant Physiology* 122:977–983.

Hajirezaei, M.R., F. Börnke, M. Peisker, Y. Takahata, J. Lerchi, A. Kirakosyan, and U. Sonnewald. 2003. Decreased sucrose content triggers starch breakdown and respiration in stored potato tubers (*Solanum tuberosum*). *Journal of Experimental Botany* 54:477–488.

Hall, L.N., G.A. Tucker, C.J.S. Smith, C.F Watson, G.B Seymour, Y. Bundick, J.M. Boniwell, J.D. Fletcher, J.A. Ray, and W. Schuch. 1993. Antisense inhibition of pectin esterase gene expression in transgenic tomatoes. *Plant Journal* 3:121–129.

Hallett, I.C., E.A. MacRaeand, and T.F Wegrzyn. 1992. Changes in kiwifruit cell wall ultrastructure and cell packing during postharvest ripening. *International Journal of Plant Sciences* 153:49–60.

Harker, F.R. and I.C. Hallett. 1992. Physiological changes associated with development of mealiness of apple fruit during cool storage. *HortScience* 27:1291–1294.

Harker, F.R. and I.C. Hallett. 1994. Physiological and mechanical properties of kiwi-fruit tissue associated with texture change during cool storage. *HortScience* 27:1291–1294.

Harker, F.R., R.J. Redgwell, I.C. Hallett, S.H. Murray, and G. Carter. 1997a. Texture of fresh fruit. *Horticultural Reviews* 20:121–224.

Harker, F.R., M.G.H. Stec, I.C. Hallett, and C.L. Bennett. 1997b. Texture of paren-chymatous plant tissue: A comparison between tensile and other instrumen-tal and sensory measurements of tissue strength and juiciness. *Postharvest Biology and Technology* 11:63–72.

Harker, F.R., R. Volz, J.W. Johnston, I.C. Hallett, and N. DeBelie. 2000. What makes fruit firm and how to keep it that way. In: *16th Annual Postharvest Conference*, Yakima, WA.

Harpster, M.H., D.A. Brummell, and P. Dunsmuir. 2002a. Suppression of a ripen-ing-related endo-1,4-beta-glucanase in transgenic pepper fruit does not pre-vent depolymerization of cell wall polysaccharides during ripening. *Plant Molecular Biology* 50:345–355.

Harpster, M.H., D.M Dawson, D.J. Nevins, P. Dunsmuir, and D.A. Brummell. 2002b. Constitutive overexpression of a ripening-related pepper endo-1,4-beta-glucanase in transgenic tomato fruit does not increase xyloglucan depo-lymerization or fruit softening. *Plant Molecular Biology* 50:357–369.

Harriman, R.W., D.M. Tieman, and A.K. Handa. 1991. Molecular cloning of tomato pectin methylesterase gene and its expression in Rutgers, ripening inhibitor, nonripening and Never Ripe tomato fruits. *Plant Physiology* 97:80–87.

Hayama, H., A. Ito, T. Moriguchi, and Y. Kashimura. 2003. Identification of a new expansin gene closely associated with peach fruit softening. *Postharvest Biology and Technology* 29:1–10.

Hayashi, T. 1989. Xyloglucans in the primary cell wall. *Annual Review of Plant Physiology and Plant Molecular Biology* 40:139–168.

Hinton, D.M. and R.Pressey. 1974. Cellulase activity in peaches during ripening. *Journal of Food Science* 39:783–785.

Huber, D.J. 1983. The role of cell wall hydrolysis in fruit softening. *Horticultural Reviews* 5:169–219.

Huber, D.J. and E.M. O'Donoghue. 1993. Polyuronides in avocado (*Persea ameri-cana*) and tomato (*Lycopersicon esculentum*) fruits exhibit markedly different patterns of molecular weight downshifts during ripening. *Plant Physiology* 102:473–480.

Iannetta, P.P.M., C. Jones, D. Stewart, M.A. Taylor, R.J. McNicol, and H.V. Davies. 1998. Multidisciplinary approaches and the improvement of fruit quality in red raspberry (*Rubus idaeus* L.). *SCRI Annual Report*, 99–103.

Iannetta, P.P.M., J. van den Berg, R.E. Wheatley, R.J. McNicol, and H.V. Davies. 1999. The role of ethylene and cell wall modifying enzymes in raspberry (*Rubus idaeus*) fruit ripening. *Physiologia Plantarum*105:337–346.

Ishii, T., and T. Matsunaga. 1996. Isolation and characterization of a boron-rham-nogalacturonan-II complex from cell walls of sugar beet pulp. *Carbohydrate Research* 284:1–9.

Ishii, T., T. Matsunaga, P. Pellerin, M.A. O'Neill, A. Darvill, and P. Albersheim 1999. The plant cell wall polysaccharide rhamnogalacturonan II self-assembles into a covalently cross-linked dimer. *Journal of Biological Chemistry* 274:13098–13104.

Itai, A., K. Tanabe, F. Tamura, and T. Tanaka. 2000. Isolation of cDNA clones cor-responding to genes expressed during fruit ripening in Japanese pear (*Pyrus pyrifolia* Nakai): involvement of the ethylene signal transduction pathway in their expression. *Journal of Experimental Botany* 51:1163–1166.

Jarvis, M.C. 1984. Structure and properties of pectin gels in plant cell walls. Plant, *Cell and Environment* 7:153–164.

Jarvis, M.C., S.P.H. Briggs, and J.P. Knox. 2003. Intercellular adhesion and cell sep-aration in plants. *Plant, Cell and Environment* 26:977–989.

Jiménez-Bermúdez, S., J. Redondo-Nevado, J. Muñoz-Blanco, J.L. Caballero, J.M. López-Aranda, V. Valpuesta, F. Pliego-Alfaro, M.A. Quesada, and J.A. Mercado. 2002. Manipulation of strawberry fruit softening by antisense expression of a pectate lyase gene. *Plant Physiology* 128:751–759.

Johnston, J.W., K. Gunaseelan, P. Pidakala, M. Wang, and R. Schaffer, 2009. Co-ordination of early and late ripening events in apples is regulated through differential sensitivities to ethylene. *Journal of Experimental Botany* 60:2689–2699.

Jones, B., P. Frasse, E. Olmos, H. Zegzouti, Z.G. Li, A. Latché, J.C. Pech, and M. Bouzayen. 2002. Down regulation of AS-DR12, an auxin-response-factor homolog, in the tomato results in a pleiotropic phenotype including dark green and blotchy ripening fruit. *The Plant Journal* 32:603–613.

Karakurt, Y. and D.J. Huber. 2003. Activities of several membrane and cell-wall hydrolases, ethylene biosynthetic enzymes, and cell wall polyuronide degra-dation during low-temperature storage of intact and fresh-cut papaya (*Carica papaya*) fruit. *Postharvest Biology and Technology* 28:219–229.

Keegstra, K., K.W. Talmadge, W.D. Bauer, and P. Albersheim. 1973. The structure of plant cell walls. III. A model of the walls of suspension-cultured sycamore cells based on the interconnections of the macromolecular components. *Plant Physiology* 51:188–196.

Kenis, K., J. Keulemans, and M.W. Davey. 2008. Identification and stability of QTLs for fruit quality traits in apple. *Tree Genetics and Genomes* 4:647–661.

King, A.D. and H.R. Bolin. 1989. Physiological and microbiological storage stability of minimally processed fruits and vegetables. *Food Technology* 43:132–135.

King, G.J., C. Maliepaard, J.R. Lynn, F.H. Alston, C.E. Durel, K.M. Evans, B. Griffon, F. Laurens, A.G.Manganaris, E. Schrevens, S. Tartarini, and J. Verhaegh. 2000. Quantitative genetic analysis and comparison of physical and sen-sory descriptors relating to fruit flesh firmness in apple (*Malus pumila* Mill.). *Theoretical and Applied Genetics* 100:1074–1084.

Knee, M. 1982. Fruit softening. III. Requirement for oxygen and pH effects. *Journal of Experimental Botany* 33:1263–1269.

Kobayashi, M., H. Nakagawa, T. Asaka, and T. Matoh. 1999. Borate-rhamnogalac-turonan II bonding reinforced by Ca²⁺ retains pectic polysaccharides in higher-plant cell walls. *Plant Physiology* 119:199–203.

Kobayashi, M., T. Matoh, and J. Azuma. 1996. Two chains of rhamnogalacturonan II are cross-linked by borate-diol ester bonds in higher plant cell walls. *Plant Physiology* 110:1017–1020.

Koch, J.L. and D.J. Nevins. 1989. Tomato fruit cell wall. I. Use of purified tomato polygalacturonase and pectinmethylesterase to identify developmental changes in pectins. *Plant Physiology* 91:816–822.

Kojima, K., N. Sakurai, and S. Kuraishi. 1994. Fruit softening in banana: correlation among stress-relaxation parameters, cell wall components and starch during ripening. *Physiologia Plantarum* 90:772–778.

Kramer, M., R. Sanders, K. Bolkan, C. Waters, R.E. Sheehy, and W.R. Hiatt. 1992. Postharvest evaluation of transgenic tomatoes with reduced levels of polyga-lacturonase: processing, firmness and disease resistance. *Postharvest Biology and Technology* 1:241–255.

Ku, H.M., T. Vision, J. Liu, and S.D. Tanksley. 2000. Comparing sequenced seg-ments of the tomato and Arabidopsis genomes: Large-scale duplication fol-lowed by selective gene loss creates a network of synteny. *Proceedings of the National Academy of Sciences of the U.S.A.* 97:9121–9126.

Langley, K.R., A. Martin, R. Stenning, A.J. Murray, G.E. Hobson, W.W. Schuch, and C.R. Bird. 1994. Mechanical and optical assessment of the ripening of tomato fruit with reduced polygalacturonase activity. *Journal of the Science of Food and Agriculture* 66:547–554.

Lashbrook, C.C., C. Gonzalez-Bosch, and A.B. Bennett. 1994. Two divergent endo-1,4-β-glucanase genes exhibit overlapping expression in ripening fruit and abscission flowers. *Plant Cell* 6:1485–1493.

Lashbrook, C.C., J.J. Giovannoni, B.D. Hall, R.L. Fisher, and A.B. Bennett. 1998. Transgenic analysis of tomato endo-β-1,4-glucanase gene function. Role of Cel1 in floral abscission. *The Plant Journal* 13:303–310.

Lefever, G., M. Vieuille, N. Delage, A. d'Harlingue, J. de Monteclerc, and G. Bompeix. 2004. Characterization of cell wall enzyme activities, pectin com-position, and technological criteria of strawberry cultivars (*Fragaria* x *anan-assa* Duch). *Journal of Food Science* 69:221–226.

Lelièvre, J.M., A. Latché, B. Jones, M. Bouzayen, and J.C. Pech. 1997. Ethylene and fruit ripening. *Physiologia Plantarum* 101:727–739.

Lester, D.R., J. Speirs, G. Orr, and C.J. Brady. 1994. Peach (*Prunus persica*) endopoly-galacturonase cDNA isolation and mRNA analysis in melting and nonmelt-ing peach cultivars. *Plant Physiology* 105:225–231.

Li, Z.C., D.M. Durachko, and D.J. Cosgrove. 1993. An oat coleoptile wall protein that induces extension *in vitro* and that is antigenically related to a similar protein from cucumber hypocotyls. *Planta* 191:349–356.

Lorences, E.P., and S.C. Fry, 1993. Xyloglucan oligosaccharides with at least two alpha-D-xylose residues act as acceptor substrates for xyloglucan endotrans-glycosylase and promote the depolymerisation of xyloglucan. *Physiologia Plantarum* 88:105–112.

Lu, C., Z. Zainal, G.A. Tucker, and G.W. Lycett. 2001. Developmental abnormalities and reduced fruit softening in tomato plants expressing an antisense Rab11 GTPase gene. *Plant Cell* 13:1819–1833.

Maclachlan, G. and C. Brady. 1992. Multiple forms of 1,4-beta-glucanase in ripening tomato fruits include a xyloglucanase activatable by xyloglucan oligosaccharides. *Australian Journal of Plant Physiology* 19:137–146.

Maclachlan, G. and C. Brady. 1994. Endo-1,4-β-glucanase, xyloglucanase, and xyloglucan endo-transglycosylase activities versus potential substrates in ripening tomatoes. *Plant Physiology* 105:965–974.

Majewska-Sawka, A., and E.A. Nothnagel. 2000. The multiple roles of arabinogalactan proteins in plant development. *Plant Physiology* 122:3–10.

Manganaris, G.A., M. Vasilakakis, G. Diamantidis, and I. Mignani. 2006. Diverse metabolism of cell-wall components of melting and non-melting peach genotypes during ripening after harvest or cold storage. *Journal of the Science of Food and Agriculture* 86:243–250.

Mathieu,Y., A. Kurkdjian, H. Xia, J. Guern, A. Koller, M.D. Spiro, M. O'Neill, P. Albersheim, and A. Darvill. 1991. Membrane responses induced by oligogalacturonides in suspension-cultured tobacco cells. *The Plant Journal* 1:333–343.

Maurel, C. 1997. Aquaporins and water permeability of plant membranes. *Annual Review of Plant Physiology and Plant Molecular Biology* 48:399–429.

Maurel, C. 2007. Plant aquaporins: Novel functions and regulation properties. *FEBS Letters* 581: 2227–2236.

McNeil, M., A.G. Darvill, S.C. Fry, and P. Albersheim. 1984. Structure and function of the primary cell walls of plants. *Annual Review of Biochemistry* 53: 652–683.

McQueen-Mason, S., D.M. Durachko, and D.J. Cosgrove. 1992. Two endogenous proteins that induce cell wall extension in plants. *Plant Cell* 4: 1425–1433.

Medina-Escobar, N., J. Cardenas, E. Moyano, J.L. Caballero, and J. Munoz-Blanco. 1997. Cloning, molecular characterization and expression pattern of a strawberry ripening-specific cDNA with sequence homology to pectate lyase from higher plants. *Plant Molecular Biology* 34:867–877.

Mehta, R.A., T. Cassol, N. Li, N. Ali, A.K Handa, and A.K. Mattoo. 2002. Engineered polyamine accumulation in tomato enhances phytonutrient content, juice quality, and vine life. *Nature Biotechnology* 20:613–618.

Melotto, E., L.C. Greve, and J.M. Labavitch. 1994. Cell wall metabolism in ripening fruit. VII. Biologically active pectin oligomers in ripening tomato (*Lycopersicon esculentum* Mill.) fruits. *Plant Physiology* 106:575–581.

Mignani, I., L.C. Greve, R. Ben-Arie, H.U. Stotz, C. Li, K. Shackel, and J. Labavitch. 1995. The effects of GA$_3$ and divalent cations on aspects of pectin metabolism and tissue softening in ripening tomato pericarp. *Physiologia Plantarum* 93:108–115.

Moctezuma, E., D.L. Smith, and K.C. Gross. 2003. Antisense suppression of a beta-galactosidase gene (TBG6) in tomato increases fruit cracking. *Journal of Experimental Botany* 54:2025–2033.

Montgomery, J., S. Goldman, J. Deikman, L. Margossian, R.L. Fischer. 1993. Identification of an ethylene-responsive region in the promoter of a fruit ripening gene. *Proceedings of the National Academy of Sciences of the U.S.A.* 90: 5939–5943.

Moreno, E., J. Obando, N. Dos-Santos, J.P. Fernández-Trujillo, A.J. Monforte, and J. García-Mas. 2008. Candidate genes and QTLs for fruit ripening and softening in melon. *Theoretical and Applied Genetics* 116:589–602.

Murayama, H., M. Arikawa, Y, Sasaki, V. D. Cin, W. Mitsuhashi, and T. Toyomasu. 2009. Effect of ethylene treatment on expression of polyuronide-modifying genes and solubilization of polyuronides during ripening in two peach cultivars having different softening characteristics. *Postharvest Biology and Technology* 52:196–201.

Nicholass, F.J., C.J.S. Smith, W. Schuch, C.R. Bird, and D. Grierson. 1995. High levels of ripening-specific reporter gene expression directed by tomato fruit polygalacturonase gene-flanking regions. *Plant Molecular Biology* 28:423–435.

Nishitani, K. and R. Tominaga. 1991. *In vitro* molecular weight increase in xyloglucan by an apoplastic enzyme preparation from epicotyls of *Vigna angularis*. *Physiologia Plantarum* 82:490–497.

Nishitani, K., and R. Tominaga. 1992. Endo-xyloglucan transferase, a novel class of glycosyltransferase that catalyzes transfer of a segment of xyloglucan molecule to another xyloglucan molecule. *Journal of Biological Chemistry* 267:21058–21064.

Norman, C., S.Vidal, and E.T. Palva. 1999. Oligogalacturonide-mediated induction of a gene involved in jasmonic acid synthesis in response to the cell-wall degrading enzymes of the plant pathogen *Erwinia carotovora*. *Molecular Plant–Microbe Interactions* 12: 640–644.

O'Neill, M.A., D. Warrenfeltz, K. Kates, P. Pellerin, T. Doco, A.G. Darvill, and P. Albersheim. 1996. Rhamnogalacturonan II, a pectic polysaccharide in the walls of growing plant cells, forms a dimer that is cross-linked by a borate ester: *In vitro* conditions for the formation and hydrolysis of the dimer. *Journal of Biological Chemistry* 271:22923–22930.

O'Neill, M.A., P. Albersheim, and A. Darvill. 1990. The pectin polysaccharides of primary cell walls. In *Methods in Plant Biochemisty*, ed. P.M. Dey, 415–441. New York: Academic Press.

O'Neill, M.A., T. Ishii, P. Albersheim, and A.G. Darvill. 2004. Rhamnogalacturonan II: Structure and function of a borate cross-linked cell wall pectic polysaccharide. *Annual Review of Plant Biology* 55:109–139.

Oomen, R.J., C.H. Doeswijk-Voragen, M.S. Bush, J.P. Vincken, B. Borkhardt, L.A., van den Broek J. Corsar, P. Ulvskov, A.G. Voragen, M.C. McCann, and R.G. Visser. 2002. In muro fragmentation of the rhamnogalacturonan I backbone in potato (*Solanum tuberosum* L.) results in a reduction and altered location of the galactan and arabinan side-chains and abnormal periderm development. *The Plant Journal* 30:403–413.

Orfila, C., M.M. Huisman, W.G. Willats, G.J. van Alebeek, H.A. Schols, G.B. Seymour, and J.P. Knox. 2002. Altered cell wall disassembly during ripening of Cnr tomato fruit: implications for cell adhesion and fruit softening. *Planta* 215:440–447.

Orr, G. and C.J. Brady. 1993. Relationship of endopolygalacturonase activity to fruit softening in a freestone peach. *Postharvest Biology and Technology* 3:121–130.

Osorio, S., C. Castillejo, M.A. Quesada, N. Medina-Escobar, G.J. Brownsey, R. Suau, A. Heredia, M.A. Botella, and V. Valpuesta. 2008. Partial demethylation of oligogalacturonides by pectin methyl esterase 1 is required for eliciting defence responses in wild strawberry (*Fragaria vesca*). *The Plant Journal* 9:43–55.

Ozga, J.A. and D.M. Reinecke. 2003. Hormonal interactions in fruit development. *Journal of Plant Growth Regulation* 22:73–81

Palomer, X., I. Llops-Tous, M. Vendrell, F.A. Krens, J.G. Schaart, M.J. Boone, H. van der Valk, and E.M.J. Salentijn. 2006. Antisense down-regulation of strawberry *endo*-β-(1,4)-glucanase gene does not prevent fruit softening during ripening. *Plant Science* 171:640–646.

Paterson, A.H., S. Damon, J.H. Hewitt, D. Zamir, H.D. Rabinowitch, S.E. Lincoln, E.S. Lander, and S.D. Tanksley. 1991. Mendelian factors underlying quantitative traits in tomato: comparison across species, generations, and environments. *Genetics* 127:181–197.

Peumans, W.J., P. Proost, R.L. Swennen, and E.J. Van Damme. 2002. The abundant class III chitinase homolog in young developing banana fruits behaves as a transient vegetative storage protein and most probably serves as an important supply of amino acids for the synthesis of ripening-associated proteins. *Plant Physiology* 130:1063–1072.

Phan, T.D., W. Bo, G. West, G.W. Lycett, and G.A. Tucker. 2007. Silencing of the major salt-dependent isoform of pectinesterase in tomato alters fruit softening. *Plant Physiology* 144:1960–1967.

Pilling, E. and H. Hofte. 2003. Feedback from the wall. *Current Opinion in Plant Biology* 6: 611–616.

Pinheiro, S.C.F. and D.P.F. Almeida. 2008. Modulation of tomato pericarp firmness through pH and calcium: Implications for the texture of fresh-cut fruit. *Postharvest Biology and Technology* 47:119–125.

Platt-Aloia, K.A. and W.W. Thomson. 1981. Ultrastructure of the mesocarp of mature avocado fruit and changes associated with ripening. *Annals of Botany* 48: 451–465.

Poovaiah, B.W., Glenn, G.M. and A.S.N. Reddy. 1988. Calcium and fruit softening: Physiology and biochemistry. *Horticultural Reviews* 10:107–152.

Popper, Z.A. and S.C. Fry. 2005. Widespread occurrence of a covalent linkage between xyloglucan and acidic polysaccharides in suspension-cultured angiosperm cells. *Annals of Botany* 96:91–99.

Prasanna, V., T.N. Prabha, and R.N. Tharanathan. 2007. Fruit ripening phenomena: An overview. *Critical Reviews in Food Science and Nutrition* 47:1–19.

Pressey, R. 1983. β-galactosidase in ripening tomatoes. *Plant Physiology* 71: 132–135.

Pressey, R. and J.K. Avants. 1978. Difference in polygalacturonase composition of clingstone and freestone peaches. *Journal of Food Science* 43:1415–1417.

Pressey, R. and J.K. Avants. 1973. Two forms of polygalacturonase in tomatoes. *Biochimica et Biophysica Acta* 309:363–369.

Pressey, R. and J.K. Avants. 1982. Solubilization of cell walls by tomato polygalacturonases: effects of pectinesterases. *Journal of Food Biochemistry* 6:57–74.

Purugganan, M.M., J. Braam, and S.C. Fry. 1997. The arabidopsis Tch4 xyloglucan endotransglycosylase–substrate specificity, pH optimum, and cold tolerance. *Plant Physiology* 115:181–190.

Qin, Q., C.W. Bergmann, J.K. Rose, M. Saladie, V.S. Kolli, P. Albersheim, A.G. Darvill, and W.S. York, 2003. Characterization of a tomato protein that inhibits a xyloglucan-specific endoglucanase. *The Plant Journal* 34:327–338.

Quesada, M.A., R. Blanco-Portales, S. Posé, J.A. García-Gago, S. Jiménez-Bermúdez, A. Muñoz-Serrano, J.L. Caballero, F. Pliego-Alfaro, J.A. Mercado, and J. Muñoz-Blanco, 2009. Antisense down-regulaiton of the FaPG1 gene reveals an unexpected central role for polygalacturonase in strawberry fruit softening. *Plant Physiology* 150:1022–1032.

Ramakrishna, W., Z. Deng, C.K. Ding, A.K. Handa, and R.H.Jr. Ozminkowski. 2003. A novel small heat shock protein gene, vis1, contributes to pectin depolymerization and juice viscosity in tomato fruit. *Plant Physiology* 131:725–735.

Read, S.M. and A. Bacic. 1996. Cell wall porosity and its determination. In *Plant Cell Wall Analysis*, eds. H.F. Linskens, F.J. Jackson, 63–80. Berlin: Springer-Verlag.

Redgwell, R.J. and A.E. Percy. 1992. Cell wall changes during on-vine softening of kiwifruit. *New Zealand Journal of Crop and Horticultural Science* 20:453–456.

Redgwell, R.J. and R. Harker. 1995. Softening of kiwifruit discs: Effect of inhibition of galactose loss from cell walls. *Phytochemistry* 39:1319–1323.

Redgwell, R.J., E. MacRae, I. Hallett, M. Fisher, J. Perry, and R. Harker. 1997. *In vivo* and *in vitro* swelling of cell walls during fruit ripening. *Planta* 203:162–173.

Redgwell, R.J., L.D. Melton, and D.J. Brasch. 1992. Cell wall dissolution in ripening kiwifruit (*Actinidia deliciosa*). Solubilization of the pectic polymers. *Plant Physiology* 98:71–81.

Redondo-Nevado, J., E. Moyano, N. Medina-Escobar, J.L. Caballero, and J. Munoz-Blanco. 2001. A fruit-specific and developmentally regulated endopolygalacturonase gene from strawberry (*Fragaria* x *ananassa* cv. Chandler). *Journal of Experimental Botany* 52:1941–1945.

Reeve, R.M., H. Timm, and M.L. Weaver. 1973a. Parenchyma cell growth in potato tubers I. Different tuber regions. *American Journal of Potato Research* 50:49–57.

Reeve, R.M., H. Timm, and M.L. Weaver. 1973b. Parenchyma cell growth in potato tubers II. Cell divisions vs. cell enlargement. *American Journal of Potato Research* 50:71–78.

Ricard, J., G. Noat, M. Crasnier, and D. Job. 1981. Ionic control of immobilized enzymes. Kinetics of acid phosphatase bound to plant cell walls. *Biochemical Journal* 195:357–367.

Ridley, B.L., M.A. O'Neill, and D. Mohnen. 2001. Pectins: Structure, biosynthesis, and oligogalacturonide-related signaling. *Phytochemistry* 57:929–967.

Roberts, K., 1990. Structure at the plant cell surface. *Current Opinion in Cell Biology* 2:920–928.

Rodríguez, R., A. Jiménez, R. Guillén, A. Heredia, and J. Fernández-Bolaños. 1999. Postharvest changes in white asparagus cell wall during refrigerated storage. *Journal of the Science of Food and Agriculture* 47:3551–3557.

Rose, J.K.C., H.H. Lee, and A.B. Bennett. 1997. Expression of a divergent expansin gene is fruit-specific and ripening-regulated. *Proceedings of the National Academy of Sciences of the U.S.A.* 94:5955–5960.

Rose, J.K.C., J. Braam, S.C. Fry, and K. Nishitani. 2002. The XTH family of enzymes involved in xyloglucan endotransglucosylation and endohydrolysis: Current perspectives and a new unifying nomenclature. *Plant and Cell Physiology* 43:1421–1435.

Rose, J.K.C., K.A. Hadfield, J.M. Labavitch, and A.B. Bennett. 1998. Temporal sequence of cell wall disassembly in rapidly ripening melon fruit. *Plant Physiology* 117:345–361.

Rosli, H.G., P.M. Civello, and G.A. Martínez. 2009. Alpha-l-Arabinofuranosidase from strawberry fruit: Cloning of three cDNAs, characterization of their expression and analysis of enzymatic activity in cultivars with contrasting firmness. *Plant Physiology and Biochemistry* 47:272–281.

Ross, G.S., R.J. Redgwell, and E.A. MacRae. 1993. Kiwifruit beta-galactosidase: Isolation and activity against specific fruit cell-wall polysaccharides. *Planta* 189:499–506.

Ross, G.S., T. Wegrzyn, E.A. MacRae, and R. Redgwell. 1994. Apple β-galactosidase. Activity against cell wall polysaccharides and characterization of a related cDNA clone. *Plant Physiology* 106:521–528.

Rousseaux, M.C., C.M. Jones, D. Adams, R. Chetelat, A. Bennett, and A. Powell. 2005. QTL analysis of fruit antioxidants in tomato using *Lycopersicon pennellii* introgression lines. *Theoretical and Applied Genetics* 111:1396–1408.

Rushing, J.W. and D.J. Huber. 1984. *In vitro* characterization of tomato fruit softening. The use of enzymically active cell walls. *Plant Physiology* 75:891–894.

Sacher, J.A. 1962. Relations between changes in membrane permeability and the climacteric in banana and avocado. *Nature* 195:577–578.

Saftner, R.A., W.S. Conway, and C.E. Sams.1998. Effect of postharvest calcium chloride treatments on tissue water relations, cell wall calcium levels and postharvest life of Golden Delicious apples. *Journal of the American Society for Horticultural Science* 123:893–897.

Saladié, M., A.J. Matas, T. Isaacson, M.A. Jenks, S.M. Goodwin, K.J. Niklas, R. Xiaolin, J.M. Labavitch, K.A. Shackel, A.R. Fernie A. Lytovchenko, M.A. O'Neill, C.B.Watkins, and J.K.C. Rose. 2007. A reevaluation of the key factors that influence tomato fruit softening and integrity. *Plant Physiology* 144:1012–1028.

Salentijn, E.M.J., A. Aharoni, J.G. Achaart, M.J. Boone, and F.A. Frenks. 2003. Differential *Physiologia Plantarum* 118:571–578.

Santiago-Doménech, N., S. Jiménez-Bermúdez, A.J. Matas, J.K.C. Rose, J. Muñoz-Blanco, J.A. Mercado, and M.A. Quesada, 2008. Antisense inhibition of a pectate lyase gene supports a role for pectin depolymerization in strawberry fruit softening. *Journal of Experimental Botany* 59:2769–2779.

Schols, H.A., E.J. Bakx, D. Schipper, and A.G.J. Voragen. 1995. A xylogalacturonan subunit present in the modified hairy regions of apple pectin. *Carbohydrate Research* 279:265–279.

Serrano, M., D. Martinez-Romero, F. Guillén, S. Castillo, and D. Valero. 2006. Maintenance of broccoli quality and functional properties during cold storage as affected by modified atmosphere packaging. *Postharvest Biology and Technology* 39:61–68.

Seymour, G.B., I.J. Colquhoun, M.S. Dupont, K.R. Parsley, and R.R. Selvendran. 1990. Composition and structural features of cell wall polysaccharides from tomato fruits. *Phytochemisty* 29:725–731.

Seymour, G.B., K. Manning, E.M. Eriksson, A.H. Popovich, and G.J. King. 2002. Genetic identification and genomic organization of factors affecting fruit texture. *Journal of Experimental Botany* 53:2065–2071.

Seymour, G.B., M. Poole, K. Manning, and G.J. King, 2008. Genetics and epigenetics of fruit development and ripening. *Current Opinion in Plant Biology* 11:58–63.

Shackel, K.A., C. Greve, M.J. Labavitch, and H. Ahmadi. 1991. Cell turgor changes associated with ripening in tomato pericarp tissue. *Plant Physiology* 97:814–816.

Sheehy, R.E., M. Kramer, and W.R. Hiatt.1988. Reduction of polygalacturonase activity in tomato fruit by antisense RNA. *Proceedings of the National Academy of Sciences of the U.S.A.* 85:8805–8809.

Sheng, J., J. Ye, L. Shen, and Y. Luo. 2003. Effect of lipoxygenase and jasmonic acid on ethylene biosynthesis during tomato fruit ripening. *Acta Horticulturae* 620:119–125

Showalter, A.M. 1993. Structure and function of plant cell wall proteins. *Plant Cell* 5:9–23.

Simón, A., E. González-Fandos, and D. Rodríguez. 2008. Effect of film and temperature on the sensory, microbiological and nutritional quality of minimally processed cauliflower. *International Journal of Food Science and Technology* 43:1628–1636.

Simpson, S.D., D.A. Ashford, D.J. Harvey, and D.J. Bowles.1998. Short chain oligogalacturonides induce ethylene production and expression of the gene encoding aminocyclopropane 1-carboxylic acid oxidase in tomato plants. *Glycobiology* 8:579–583.

Smith, C.J.S., C.F. Watson, J. Ray, C.R. Bird, P.C. Morris, W. Schuch, and D. Grierson. 1988. Antisense RNA inhibition of polygalacturonase gene expression in transgenic tomatoes. *Nature* 334:724–726.

Smith, C.J.S., C.F. Watson, P.C. Morris, C.R. Bird, G.B. Seymour, J.E. Gray, C. Arnold, G.A.Tucker, W. Schuch, and S. Harding. 1990. Inheritance and effect on ripening of antisense polygalacturonase genes in transgenic tomatoes. *Plant Molecular Biology* 14:369–379.

Smith, D.L., D.A. Starrett, and K.C. Gross. 1998. A gene coding for tomato fruit β-galactosidase II is expressed during fruit ripening. *Plant Physiology* 117:417–423.

Smith, D.L. and K.C. Gross. 2000. A family of at least seven beta-galactosidase genes is expressed during tomato fruit development. *Plant Physiology* 123:1173–1183.

Smith, D.L., J.A. Abbott, and K.C. Gross. 2002. Down-regulation of tomato β-galactosidase 4 results in decreased fruit softening. *Plant Physiology* 129:1755–1762.

Staehelin, L.A. and I. Moore. 2005. The plant Golgi apparatus: structure, functional organization and trafficking mechanisms. *Annual Review in Plant Physiology and Plant Molecular Biology* 46:261–288.

Steele, N.M., M.C. McCann, and K. Roberts. 1997. Pectin modification in cell walls of ripening tomatoes occurs in distinct domains. *Plant Physiology* 114: 373–381.

Sung, S.K. and G. An. 1997. Molecular cloning and characterization of a MADS-box cDNA clone of the Fuji apple. *Plant and Cell Physiology* 38:484–489.

Sung, S.K., G.H. Yu, J. Nam, D.H. Jeong, and G. An. 2000. Developmentally regulated expression of two MADS-box genes, MdMADS3 and MdMADS4, in the morphogenesis of flower buds and fruits in apple. *Planta* 210:519–528.

Taiz, L. and E. Zieger. 1998. Cell walls: Structure, biogenesis and expansion. In *Plant Physiology*, 409–443. USA: Sinauer Associates.

Talbott, L.D. and P.M. Ray. 1992. Molecular size and separability features of pea cell wall polysaccharides. Implications for models of primary wall structure. *Plant Physiology* 98:357–368.

Taylor, M.A., G.J. McDougall, and D. Stewart. 2007. Potato flavour and texture. In *Potato Biology and Biotechnology: Advances and Perspectives*, ed. D. Vreugdenhil, 525–540. Amsterdam: Elsevier.

Thakur, B.R., R.K. Singh, and A.K. Handa. 1996. Effect of an antisense pectin methylesterase gene on the chemistry of pectin in tomato (*Lycopersicon esculentum*) juice. *Journal of Agricultural and Food Chemistry* 44:628–630.

Thomas, T.R., K.A. Shackel, and M.A. Matthews. 2008. Mesocarp cell turgor in *Vitis vinifera* L. berries throughout development and its relation to firmness, growth, and the onset of ripening. *Planta* 228:1067–1076.

Thompson, A.J., M. Tor, C.S. Barry, J. Verbalov, C. Orfila, M.C. Jarvis, J.J. Giovannoni, D. Grierson, and G.B. Seymour 1999. Molecular and genetic characterisation of a novel pleiotropic tomato-ripening mutant. *Plant Physiology* 120:383–389.

Thompson, J.E and S.C. Fry. 2000. Evidence for covalent linkage between xyloglucan and acidic pectins in suspension-cultured rose cells. *Planta* 211:275–286.

Tieman, D.M. and A.K. Handa. 1994. Reduction in pectin methylesterase activity modifies tissue integrity and cation levels in ripening tomato (*Lycopersicon esculentum* Mill.) fruits. *Plant Physiology* 106:429–436.

Tieman, D.M., H.D. Kausch, D.M. Serra, and A.K. Handa. 1995. Field performance of transgenic tomato with reduced pectin methylesterase activity. *Journal of the American Society for Horticultural Science* 120:765–770.

Tieman, D.M., R.W. Harriman, G. Ramamohan, and A.K. Handa. 1992. An antisense pectin methylesterase gene alters pectin chemistry and soluble solids in tomato fruit. *Plant Cell* 4:667–679.

Toivonen, P.M.A. and D.A. Brummell. 2008. Biochemical bases of appearance and texture changes in fresh-cut fruit and vegetables. *Postharvest Biology and Technology* 48:1–14.

Tong, C., D. Krueger, Z. Vickers, D. Bedford, J. Luby, A. El-Shiekh, K. Schackel, and H. Ahmadi. 1999. Comparison of softening-related changes during storage of "Honeycrisp" apple, its parents, and "Delicious." *Journal of the American Society for Horticultural Science* 124:407–415.

Tong, C.B., J.M. Labavitch, and S.F. Yang. 1986. The induction of ethylene production from pear cell culture by cell wall fragments. *Plant Physiology* 81:929–930.

Trainotti, L., D. Zanin, and G. Casadoro. 2003. A cell wall-oriented genomic approach reveals a new and unexpected complexity of the softening in peaches. *Journal of Experimental Botany* 54:1821–1832.

Ugalde, T.D., P.H. Jerie, and D.J. Chalmers. 1988. Intercellular pH of peach and apricot mesocarp. *Australian Journal of Plant Physiology* 15:505–517.

Ulvskov, P., H. Wium, D. Bruce, B. Jørgensen, K. Bruun Quist, M. Skjøt, D. Hepworth, and S. Sørensen. 2005. Biophysical consequences of remodeling the neutral side chains of rhamnogalacturonan I in tubers of transgenic potatoes. *Planta* 220:609–620.

Van Haaren, M.J.J. and C.M. Houck. 1993. A functional map of the fruit-specific promoter of the tomato 2A11 gene. *Plant Molecular Biology* 21:625–640.

Vicente, A.R., C. Greve, and J.M. Labavitch. 2006. Recent findings in plant cell wall structure and metabolism: future challenges and potential implications for softening. *Stewart Postharvest Review* 2(2):1–8.

Vicente, A.R., M. Saladié, J.K.C. Rose, and M.J. Labavitch. 2007a. The linkage between cell wall metabolism and fruit softening: looking to the future. *Journal of the Science of Food and Agriculture* 87:1435–1448.

Vicente, A.R., C. Ortungo, A.L.T. Powell, L.C. Greve, and J.H. Labavitch. 2007b. Temporal sequence of cell wall disassembly events in developing fruits. 1. Analysis of raspberry (*Rubus idaeus*). *Journal of Agricultural and Food Chemistry* 55:4119–4124.

Vicente, A.R., C. Ortungo, H. Rosli, A.L.T. Powell, L.C. Greve, and J.H. Labavitch. 2007c. Temporal sequence of cell wall disassembly events in developing fruits. 2. Analysis of blueberry (*Vaccinium* sp.). *Journal of Agricultural and Food Chemistry* 55:4125–4130.

Vickery, R.S. and J. Bruinsma. 1973. Compartments and permeability for potassium in developing fruits of tomato (*Lycopersicon esculentum* Mill.). *Journal of Experimental Botany* 24:1261–1270.

Villarreal, N.M., H.G. Rosli, G.A. Martínez, and P.M. Civello. 2008. Polygalacturonase activity and expression of related genes during ripening of strawberry cultivars with contrasting fruit firmness. *Postharvest Biology and Technology* 47:141–150.

Viña, S.Z. and A.R. Chaves. 2003. Texture changes in fresh cut celery during refrigerated storage. *Journal of the Science of Food and Agriculture* 83:1308–1314.

Vincken, J.P., H.A. Schols, R.J.F.J. Oomen, M.C. McCann, P. Ulvskov, A.G.J. Voragen and R.G.F. Visser. 2003. If homogalacturonan were a side chain of rhamnogalacturonan I. Implications for cell wall architecture. *Plant Physiology* 132:1781–1789.

Voragen, A.G.J., W. Pilnik, J.-F. Thibault, M.A.V. Axelos, and C.M.G.C. Renard. 1995. Pectins. In *Food Polysaccharides and Their Applications*, ed. A.M. Stephen, 287–339. New York: Marcel Dekker.

Vrebalov, J., D. Ruezinsky, V. Padmanabhan, R. White, D. Medrano, R. Drake, W. Schuch, and J. Giovannoni. 2002. A MADS-box gene necessary for fruit ripening at the tomato ripening-inhibitor (*Rin*) locus. *Science* 296:343–346.

Wada, H., K. Shackel, and M. Matthews. 2008. Fruit ripening in *Vitis vinifera*: Apoplastic solute accumulation accounts for pre-veraison turgor loss in berries. *Planta* 227:1351–1361.

Wakasa, Y., H. Kudo, R. Ishikawa, S. Akada, M. Senda, M. Niizeki, and T. Harada. 2006. Low expression of an endopolygalacturonase gene in apple fruit with long-term storage potential. *Postharvest Biology and Technology* 39:193–198.

Wang, T.W., C.G. Zhang, W. Wu, L.M. Nowack, E. Madey, and J.E. Thompson. 2005. Antisense suppression of deoxyhypusine synthase in tomato delays fruit softening and alters growth and development. *Plant Physiology* 138:1372–1382.

Warrilow, A.G.S., R.J. Turner, and M.G. Jones. 1994. A novel form of pectinesterase in tomato. *Phytochemistry* 35:863–868.

Watada, A.E., K. Abe, and N. Yamauchi. 1990. Physiological activities of partially processed fruits and vegetables. *Food Technology* 44:116–122.

Watada, A.E. and L. Qi. 1999. Quality of fresh-cut produce. *Postharvest Biology and Technology* 15:201–205.

Watson, C.F., L. Zheng, and D. DellaPenna. 1994. Reduction of tomato polygalacturonase beta subunit expression affects pectin solubilization and degradation during fruit ripening. *Plant Cell* 6:1623–1634.

Willats, W.G.T., L. McCartney, W. Mackie, and J.P. Knox. 2001. Pectin: cell biology and prospects for functional analysis. *Plant Molecular Biology* 47:9–27.

Woolley, L.C., D.J. James, and K. Manning, 2001. Purification and properties of an endo-β-1,4-glucanase from strawberry and down-regulation of the corresponding gene, cel1. *Planta* 214:11–21.

Yao, J.L., Y.H. Dong, A. Kvarnheden, and B. Morris. 1999. Seven MADS-box genes in apple are expressed in different parts of the fruit. *Journal of the American Society for Horticultural Science* 124:8–13.

Zainal, Z., G.A. Tucker, and G.W. Lycett. 1996. A rab11-like gene is developmentally regulated in ripening mango (*Mangifera indica* L.) fruit. *Biochimica et Biophysica Acta* 1314:187–190.

chapter five

Selection of the indicator enzyme for blanching of vegetables

Vural Gökmen

Contents

5.1 Introduction

Blanching is one of the most important unit operations preceding other processing techniques, such as canning, freezing, and dehydration. Enzyme activities in raw vegetables are responsible for the undesirable colors and flavors that may develop during processing and storage. Usually the raw vegetables cannot be stored for longer periods, even under frozen conditions. Most vegetables require a short-term heat treatment to inactivate naturally occurring enzymes and stabilize sensorial and nutritional quality.

The term *blanching* was originally used to designate heat-treatment operations in the processing of plant foods for frozen storage, which prevent deteriorative changes. The blanching process is associated with the inactiva-tion of enzymes, and so, is very important to prevent quality changes during prolonged storage. It involves exposing plant tissue to steam or hot water for a prescribed time at a specified temperature. Hot water blanching is usually carried out between 75 and 95°C for between 1 and 10 minutes, depending on the size of the individual vegetable pieces. Typical times for water blanch-ing at 95°C are 2–3 minutes for green beans and broccoli, 4–5 minutes for Brussels sprouts, and 1–2 minutes for peas (Holdsworth, 1983).

The blanching process is utilized in different industries for slightly different purposes. Blanching of vegetables has several advantages as well as a number of disadvantages. Depending on the final way of pre-serving the products, blanching can fulfill one or several of the following purposes (Poulsen, 1986):

1. Inactivation of enzymes prevents discoloration and development of unpleasant taste during storage. Colors caused by the presence of chloro-phylls or carotenoids are also protected from enzymatic degradation.
2. Blanching modifies the structure of macromolecules in vegetables. Proteins are forced to coagulate and shrink under liberation of water. Also, starch that could otherwise cause a cloudy appearance can be removed.
3. Blanching forms a cooked flavor to a certain extent in vegetables.
4. Air that is confined to plant tissues is expelled and the product is eas-ier to can or pack. Oxidation risk during frozen storage is reduced.
5. Blanching improves visual quality of vegetables. Many vegetables obtain a clearer color after the blanching process.
6. Defective parts of vegetables become more visible so the product can be sorted more effectively.
7. The microbial status is improved because vegetative cells, yeasts, and molds are partially killed during the blanching process.
8. Cooking time of the finished product is shortened.

Besides the above-mentioned advantages, blanching may result in some loss of soluble solids (especially in water blanching), and may have adverse environmental impacts due to requirements for large amounts of water and energy (Williams et al., 1986).

As a pre-freezing operation, blanching is the primary means of inactivat-ing undesirable enzymes present in vegetables (Barrett and Theerakulkait, 1995). Enzymes catalyze most of the quality changes that occur during the storage of frozen vegetables. Optimization of the blanching process involves measuring the rate of enzyme destruction, so the blanching time is just long enough to destroy the indicator enzyme. The selection of an enzyme as an indicator of the adequacy of a blanching process is critical for the success of the operation.

This chapter focuses on the selection of indicator enzymes for blanching of vegetables. The following sections discuss the enzymes responsible for the quality changes in vegetables and thermal stabilities of potential indicator enzymes. The effects of residual activity of selected indicator enzymes on possible quality changes in frozen vegetables during storage are also discussed in detail.

5.2   Blanching systems

Hot-water and steam blanching are two processes widely applied in the vegetable processing industry (Figure 5.1). A good blanching technique should fulfill the following demands (Poulsen, 1986):

1. A uniform heat distribution to the individual units of product
2. A uniform blanching time to all units of product
3. No damage to the product during the entire blanching and cooling process
4. A high product yield and quality
5. Low consumption of energy and water

In the freezing industry, blanching is the operation with the second largest energy consumption after freezing itself. The energy balance for a

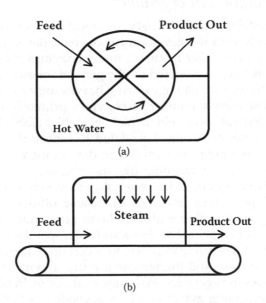

Feed   **Product Out**

Hot Water

(a)

Steam

Feed   **Product Out**

(b)

Figure 5.1 Schematics of (a) rotary hot-water and (b) pure-steam blanching systems.

steam blancher can be written as follows:

$$Q_H = m_F C_p \Delta T + Q_L \qquad (5.1)$$

$$m_S = \frac{Q_H}{\lambda} \qquad (5.2)$$

where Q_H is the heat supplied to the blancher, m_F is the mass feed rate of the product to the blancher, C_p is the heat capacity of the product, ΔT is the difference between the raw vegetables and the blanching temperature, and Q_L represents energy losses. m_S is the mass flow rate of steam and λ is the heat of vaporization of steam. In an ideal blancher, $Q_L = 0$; assuming that $C_p \approx 4.18$ kJ/kg·K and $\lambda = 2330$ kJ/kg, steam requirements would be 134 kg/ton vegetables (Bomben, 1979). Product retention time at a constant product feed rate (and therefore equipment size) is determined by the rate of heat transfer from the heating medium to the product. The rate of heat transfer depends on the thermal conductivity of the product, heat transfer coefficient, and temperature gradients between the heating medium and the product.

5.3 Enzymes responsible for quality deterioration in vegetables

Vegetables contain a wide variety of naturally occurring enzymes, which are involved in the development of color, flavor, aroma, texture, and nutrient quality. After maturity, many enzymes continue to act on remaining substrates, accelerated by the general senescence of the tissue and aided further by the damage during harvesting and storage (Velasco et al., 1989). There are a number of enzymes primarily responsible for the quality deterioration of unblanched vegetables (Table 5.1). In general, the quality deteriorations may be related to sensorial changes such as discoloration, browning, and off-flavor development, and nutritional changes such as loss of vitamins like ascorbic acid and thiamine and loss of bioactive compounds like phenolic compounds and carotenoids. Lipolytic and proteolytic enzymes can cause off-flavor development. Pectolyic enzymes together with cellulases are mainly responsible for textural changes. Polyphenol oxidases and chlorophyllase can cause color changes. Peroxidases can also cause, to a certain extent, color changes. Ascorbic acid oxidase and thimaniase are the enzymes responsible for nutritional losses in vegetables (Williams et al., 1986). In addition to their primary action, some enzymes may have secondary actions, which result in color and nutritional changes. In recent years, phenolic compounds

Table 5.1 Enzyme Responsible for Quality Deterioration in Unblanched Vegetables

Type of Deterioration		Responsible Enzymes
Sensorial	Off-flavor development	Lipoxygenases
		Proteases
		Lipases (secondary action)
	Textural changes	Pectinases
		Cellulase
	Color changes	Polyphenol oxidases
		Chlorophyllase
		Peroxidases (lesser extend)
		Lipoxygenases (secondary action)
Nutritional		Ascorbic acid oxidase
		Thiaminase
		Polyphenol oxidases
		Lipoxygenases (secondary action)

Source: Adapted from Williams et al. (1986) and Barrett and Theerakulkait (1995).

have gained importance as plant originated natural antioxidants, thus their oxidation by the action of polyphenol oxidases is also considered a nutritional loss (Altunkaya and Gökmen, 2008). Carotenoids are also another group of plant-originated lipophilic antioxidants. During the oxidation of polyunsaturated fatty acids having *cis,cis*-1,4-pentadien moiety by the action of lipoxygenases, carotenoids act as the inhibitor of lipoxygenases, which means their co-oxidation results in a nutritional loss in addition to off-flavor development and color changes (Serpen and Gökmen, 2006).

5.4   Thermal inactivation of enzymes by blanching

5.4.1   Selection of blanching indicator enzyme

It has been known for a long time that heat treatment could prolong the high-quality storage life of vegetables, especially at low temperatures, preferably below freezing (Kochman, 1936). Prolonged frozen storage results from the inactivation of deteriorative enzymes. Complete inactivation of all enzymes is easily achieved by heating. Heating is also associated with some losses in color, texture, flavor, aroma, and nutritional quality. In a blanching treatment, the need is clearly for sufficient heat treatment to stabilize the product against quality deterioration, but at the same time, to minimize quality loss. This need led to the use of an endogenous enzyme as an indicator of adequate heat treatment. The criteria on selecting an indicator enzyme are considered as follows (Velasco et al., 1989):

1. The loss of enzyme activity should be correlated with quality retention of vegetables during storage.
2. The activity of enzymes should be easily measured in the processing plant.
3. Inactivation of enzymes should be irreversible; the activity should not be regained during consequential processes.
4. It is advantageous if the same enzyme could be used as an indicator for other vegetables.

   Since catalase and peroxidase are known to be relatively resistant to heat, these two enzymes have been widely used as the indicators of blanching adequacy. As early as 1932, catalase has been used to monitor the adequacy of the blanching process of English green peas (Diehl, 1932). After extensive research by others, the loss of peroxidase activity has been shown to correlate with off-flavor development more closely than the loss of catalase activity (Joslyn, 1949). There has been a period of 20–30 years in which catalase has been used as the indicator enzyme for English green peas and some other vegetables (Sapers and Nickerson, 1962) and peroxidase has been used as the indicator enzyme for other vegetables. Catalase in most plant materials is inactivated in about 50–70% of the time required to inactivate peroxidase at the same temperature (Velasco et al., 1989). In 1975, the U.S. Department of Agriculture recommended that catalase inactivation is not a satisfactory indicator of adequate blanching for the majority of vegetables, and that inactivation of peroxidase is necessary to minimize the possibility of future deterioration of quality (USDA, 1975).

   Enzymes other than peroxidase and catalase have been used less frequently to monitor adequacy of heat treatment of vegetables. These include polyphenol oxidase for browning development, polygalacturonase for loss of consistency, and lipoxygenase and lipase for off-flavor development (Williams et al., 1986; Velasco et al., 1989).

   In industrial applications, the majority of vegetables have been blanching to the point of complete peroxidase inactivation, because peroxidase appears to be one of the most heat stable enzymes in plants. It has been generally accepted that if peroxidase is completely destroyed then it is quite unlikely that other enzymes will survive (Schwimmer, 1981). However, various studies have indicated that the quality of a blanched, frozen stored product is improved if some peroxidase activity remains at the end of the blanching process (Winter, 1969; Delincée and Schaefer,

1975). Experimental findings have suggested that a complete inactivation of peroxidase is not necessary for quality preservation during frozen storage of vegetables (Böttcher, 1975).

The use of peroxidase as an indicator enzyme is not without problems because peroxidase can regain activity under certain conditions. Peroxidase, to a lesser extent, can cause color changes and is not directly responsible for quality deterioration during frozen storage of vegetables. In general, it is considered that peroxidase activity is not directly associated with quality deterioration in vegetables (Velasco et al., 1989). Heating vegetables to a complete inactivation of peroxidase may lead to too severe blanching, which ultimately impairs the quality of the frozen product and wastes energy (Williams et al., 1986).

One of the problems in completely inactivating peroxidase is the presence of 1–10% of more heat stable isoenzymes of peroxidase in most vegetables (Winter, 1969; Böttcher, 1975; Delincée and Schaeffer, 1975; Güneş and Bayındırlı, 1993; Yemenicioglu et al., 1998; Morales-Blancas et al., 2002). The relative amounts of isoenzymes vary from vegetable to vegetable and even in the same vegetable can vary with variety, age, and environmental factors (Williams et al., 1986). Another problem in the use of peroxidase as the indicator for adequate blanching is that the peroxidases from different vegetables have different thermal stabilities.

5.4.2   Thermal stability of indicator enzymes

Thermal inactivation of enzymes responsible for quality deterioration in vegetables during storage requires careful determination of temperature stabilities.

Enzymes of different vegetables may have different thermal stabilities. With this respect, it is necessary to determine the thermal stability of indicator enzymes for different vegetables prior to the blanching operation. In general, peroxidases in low acid foods are more resistant to heat treatment than are those in acid foods (Williams et al., 1986). The variation in thermal stability of indicator enzymes in different vegetables requires the processor to determine the time needed for each vegetable with the blanching equipment and conditions used.

The presence of isoenzymes can cause problems in the blanching treatment of most vegetables. The relative amounts of isoenzymes vary from vegetable to vegetable, and even in the same vegetable can vary with variety, age, and environmental factors. The problems with the heat inactivation of different enzymes and different isoenzymes of the same enzyme occur because the inactivation begins at different temperatures and may proceed at different rates (Williams et al., 1986).

Thermal resistance of enzymes is traditionally expressed in terms of
D-values and z-values. D-value is the time at a specified temperature for
the enzyme activity to decrease by one log cycle (90%). z-value is the change
in temperature needed to alter the D-value by one log cycle. The purpose
of mathematical modeling of enzyme inactivation in heated foods is to
assess the effect of different heat treatments on residual enzyme activity
without performing numerous trial runs (Adams, 1991). Thermal inactivation of enzymes is usually described as classical loglinear (monophasic)
approach or a biphasic model.

Residual enzyme activity in heat-treated food is expressed as a fraction of initial activity (A_o);

$$\text{Residual Activity} = \frac{A}{A_o} \qquad (5.3)$$

Table 5.2 gives comparative temperature stability of some enzymes in different plant materials.

Table 5.2 Comparative Thermal Stability of Some Enzymes in Fruits and Vegetables

Enzyme	Food	z-Value (°C)	Reference
Catalase	Vegetables	16	Sapers and Nickerson, 1962
Peroxidase	Peas	12-27	Gökmen et al., 2005
			Williams et al., 1986
	Green bean	27-47	Bahçeci et al., 2005
			Williams et al., 1986
	Asparagus	31-89	Williams et al., 1986
	Corn	39	Vetter et al., 1959
	Tomato	10	Williams et al., 1986
	Spinach	33-45	Williams et al., 1986
	Carrot	18	
Lipoxygenase	Peas	9	Gökmen et al., 2005
			Williams et al., 1986
	Green bean	20	Bahçeci et al., 2005
Chlorophyllase	Spinach	12	Resende et al., 1969
Pectin esterase	Citrus juice	8	Williams et al., 1986
Ascorbate oxidase	Peach and vegetables	33	Williams et al., 1986
Polygalacturonase	Citrus juice	9	Williams et al., 1986
	Papaya	11	Aylward and Haisman, 1969
Galactolipase	Spinach	8	Kim et al., 2001
	Carrot	9	Kim et al., 2001
Phospholipase	Spinach	9	Kim et al., 2001
	Carrot	16	Kim et al., 2001

5.4.2.1   Loglinear (monophasic) model

Thermal inactivation of an enzyme may be considered, theoretically, as a first-order decay process. Figure 5.2 exemplifies a first order thermal inactivation kinetics of pea lipoxygenase at different temperatures.

The first-order kinetic model is based on the assumption that the disruption of a single bond or structure is sufficient to inactivate the enzyme. Considering the complexity of the structure of an enzyme and the variety of different phenomena involved in the inactivation, this explanation seems to be exceedingly simple. The following processes have been found to be involved in thermal denaturation of enzymes (Vámos Vigyázó, 1981):

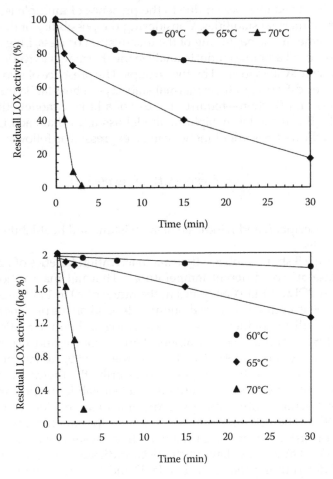

Figure 5.2 Thermal inactivation kinetics of green pea lipoxygenase at different temperatures (a) raw data, (b) monophasic behavior. (Adapted from Bahçeci, 2003.)

1. Disassociation of the prosthetic group from the holoenzyme
2. A conformation change in the apoenzyme
3. Modification or degradation of the prosthetic group

A first-order model has been used to describe the enzyme inactivation in different vegetables, such as carrot, potato, green bean, and pumpkin. (Anthon and Barrett, 2002; Anthon et al., 2002; Bifani et al., 2002; Gonçalves et al., 2007).

5.4.2.2   Biphasic model

Thermal inactivation of enzymes has long been observed as biphasic, the two phases with different rate constants. The deviation from first-order kinetics has been interpreted due to the presence of multiple isoenzymes of different thermal stabilities. Considering the possibility of the presence of isoenzymes at the beginning of the inactivation process, Ling and Lund (1978) proposed a simple model to analyze the thermal inactivation kinetics of an enzyme system formed by two groups. The presence of two groups of isoenzymes differing in their thermal stability—a heat-labile fraction and a heat-resistant fraction—requires the use of a kinetic model other than a simple loglinear model. A biphasic model assumes that each fraction of enzyme follows first-order kinetics can be expressed as follows:

$$\frac{A}{A_o} = f_L \exp(-k_L t) + f_R \exp(-k_R t) \tag{5.4}$$

where subscripts R and L indicate heat-resistant and heat-labile fractions, respectively.

Figure 5.3 shows biphasic thermal inactivation kinetics of green bean peroxidase at two different temperatures. Thermal inactivation kinetic studies in POD and LOX enzymes in the range of 70 to 100°C have clearly shown biphasic curves that are thought to depend on the presence of isoenzymes with different thermal stabilities (Wang and Luh, 1983; Powers and others, 1984; Ganthavorn and others, 1991; Sarikaya and Özilgen, 1991; Günes and Bayindirli, 1993; Forsyth and others, 1999; Agüero et al., 2008). A biphasic model has been proposed to describe the thermal inactivation kinetics of an enzyme system formed by a heat-labile fraction and a heat-resistant fraction, both with first-order inactivation kinetics (Ling and Lund, 1978). The differences between kinetic parameters for heat-labile and heat-resistant isoenzymes fractions from several sources (Ling and Lund, 1978; Günes and Bayindirlı, 1993) indicate the need and importance of determining the kinetics of POD and LOX in different vegetable extracts. The latter is important because the residual enzyme activity is exponentially related to the activation energy (E_a) and to the inactivation

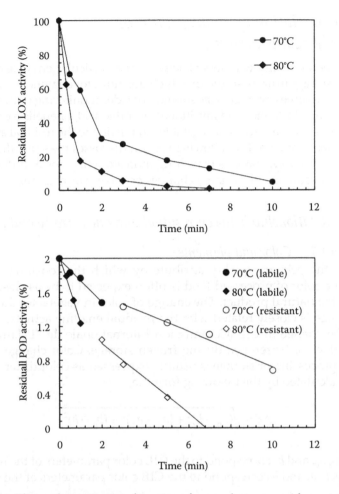

Figure 5.3 Thermal inactivation kinetics of green bean peroxidase at different temperatures (a) raw data, (b) biphasic behavior. (Adapted from Bahçeci, 2003.)

rate constant (k). Thus, small errors in the calculations of these parameters or inappropriate values can have a big impact on residual enzyme activity predictions (Arabshahi and Lund, 1985).

The design of efficient blanching treatments requires knowledge of critical factors such as enzymatic distribution within the tissue, inactivation kinetic parameters, and relative proportions of heat-labile and heat-resistant fractions (Adams, 1991). This type of information usually is not available in the literature and is unique to each vegetable, species, cultivar, and environmental condition, among other factors (Vámos-Vigyázó, 1981; Kushad et al., 1999).

5.5 Correlation of quality with loss of enzyme activity

Correlation of quality with loss of activity of a particular enzyme generally involves storage (frozen) studies in which the indicator enzyme is inactivated to various degrees. Storage tests should consider monitoring both sensorial (color, texture, taste, etc.) and nutritional (ascorbic acid, phenolic compounds, carotenoids, etc) properties of vegetables in relation to the residual activity of the indicator enzyme. The following section discusses the suitability of peroxidase and lipoxygenase as blanching indicator enzymes in different vegetables as exemplified by quality changes occurring during frozen storage.

5.5.1 Relationship between residual enzyme activity and quality

5.5.1.1 Color and pigments

Color is the primary quality attribute by which the consumer assesses food. The color of processed food is often expected to be as close as possible to the natural product. The change of color in vegetables during frozen storage is closely related with the residual enzyme activity. Table 5.3 gives the change in CIE (Commission Internationale de l'Eclairage) Lab color values of frozen pea during frozen storage. Color change in foods during processing or storage is usually expressed as color difference (ΔE) that is calculated by the following formula;

$$\Delta E = \sqrt{(L_0 - L)^2 + (a_0 - a)^2 + (b_0 - b)^2} \tag{5.5}$$

where L_0, a_0, and b_0 correspond to the CIE color parameters of the reference, whereas L, a, and b correspond to the CIE color parameters of the sample.

Table 5.3 Change of Color in Some Blanched and Unblanched Green Peas during Frozen Storage at –18°C

	Before Storage	After Storage (12 Months)		
		UB	B_{LOX}[a]	B_{POD}[b]
L	67.56	64.34	55.64	57.26
a	−16.10	−12.28	−12.51	−15.63
b	+30.91	+24.71	+22.80	+32.57
ΔE		15.09	14.86	10.44

[a] Hot water blanching condition to inactivate initial LOX activity ≥90% : 70°C × 4 min.
[b] Hot water blanching condition to inactivate initial POD activity ≥90% : 80°C × 2 min.
Source: Data from Bahçeci (2003).

Unblanched (UB) peas are susceptible to significant color changes during frozen storage. Blanching treatment, which takes lipoxygenase as an indicator enzyme (B_{LOX}), does not prevent loss of natural color of peas during frozen storage. However, color can be protected better when peas are blanched prior to frozen storage, taking peroxidase as an indicator enzyme (B_{POD}). The effect of blanching on color changes during frozen storage may vary for different vegetables. Therefore, the effect of blanching on color changes should be determined for each vegetable separately. For vegetable soybeans, the color scores have been found similar for the samples using either lipoxygenase or peroxidase as the blanching index, but blanched samples appeared greener than unblanched samples after 160 days of frozen storage (Sheu and Chen, 1991). For pumpkins, blanching treatment has been found to influence the color significantly. Blanched pumpkins become darker and lose redness, yellowness, and vivid characteristics (Gonçalves et al., 2007).

The most important color modifications in vegetables are related to three biochemical mechanisms:

1. Changes in the natural pigments of vegetable tissues (chlorophylls, anthocyanins, carotenoids)
2. Development of enzymatic browning
3. Breakdown of the cellular chloroplasts and chromoplasts

Chlorophylls are mainly responsible for the color in green vegetables. The change from bright green to olive green in unblanched frozen vegetables during prolonged storage results from the conversion of chlorophylls a and b to the corresponding pheophytins. Chlorophyll content of vegetables is affected by blanching conditions. The degree of heat treatment is proportional to the conversion of chlorophylls to pheophytins (Walker, 1964).

The loss of chlorophyll during frozen storage is also proportional to the residual enzyme activity. Blanching conditions to inactivate potential indicator enzymes in vegetables may have direct influence on the rate of chlorophyll loss during frozen storage. Figure 5.4 shows the change of chlorophylls in peas and green beans during frozen storage as affected by different blanching conditions. It is clear that chlorophyll content of vegetables tends to decrease during frozen storage for both unblanched and blanched vegetables. For peas, a blanching treatment that takes lipoxygenase as the indicator enzyme (LOX) increases the rate of chlorophyll loss during frozen storage. The same is true for green beans. However, blanching treatment that takes peroxidase as the indicator enzyme (POD) may help protect chlorophylls in green beans (Bahçeci et al., 2005; Gökmen et al., 2005).

Carotenoids are generally stable to the heat treatments involved in blanching. However, they are rapidly lost during storage due to oxidation. The conjugated double bond system makes carotenoids especially

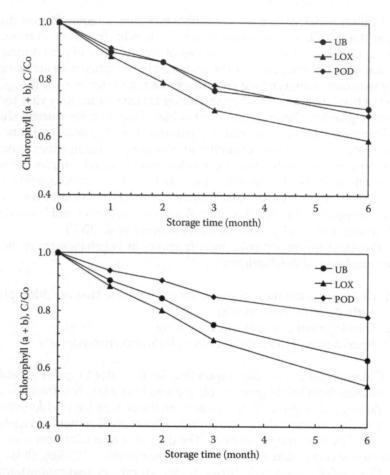

Figure 5.4 Change of chlorophyll concentration of (a) green pea (UB: Unblanched, LOX: Blanched to inactivate lipoxygenase ≥90% in hot water, 70°C × 4 min, POD: Blanched to inactivate peroxidase ≥90% in hot water, 80°C × 2 min), and (b) green bean (UB: Unblanched, LOX: Blanched to inactivate lipoxygenase ≥90% in hot water, 70°C × 2 min, POD: Blanched to inactivate peroxidase ≥90% in hot water, 90°C × 3 min) during frozen storage at –18°C. (Adapted from Serpen, 2003.)

susceptible to oxidative changes, which usually lead to discoloration or bleaching (Serpen and Gökmen, 2006; Serpen and Gökmen, 2007). The rate of the degradation of carotenoids during frozen storage is closely associated with the residual enzyme activity. Figure 5.5 clearly shows that a blanching treatment taking peroxidase as the indicator enzyme (POD) significantly protects β-carotene against oxidation during frozen storage of carrot.

Since the choice of indicator enzyme for blanching directly affects the degradation of naturally occurring pigments in vegetables during frozen storage, it is worth to see their percentage loss for a defined storage period.

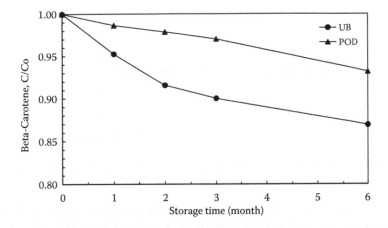

Figure 5.5 Change of β-caroten concentration of carrot during frozen storage at –18°C (UB: Unblanched, POD: Blanched to inactivate peroxidase ≥90% in hot water, 80°C × 1.5 min). (Adapted from Serpen, 2003.)

The shelf-life of frozen vegetables is usually limited to one year. So, the loss of chlorophylls and carotenoids within a storage period of one year may help us understand the effect of residual enzyme activity on product quality. Chlorophylls (a and b), lutein, lutein-5,6-epoxide, and neoxhantin are the major pigments naturally occurring in most green vegetables. Table 5.4 gives the percentage loss of these pigments in pea blanched under different conditions prior to frozen storage. It is clear from the Table 5.4 that a significant loss in pea pigments occurs without blanching (UB) within 12 months of frozen storage at –18°C. When peas are blanched choosing lipoxygenase as the indicator enzyme (B_{LOX}), the loss of carotenoids can be slightly prevented. However, blanching treatment applied to inactivate lipoxygenase

Table 5.4 Loss of Chlorophylls and Carotenoids in Green Pea Blanched under Different Conditions to Inactivate Lipoxygenase (B_{LOX}) and Peroxidase (B_{POD})

Pigment	Loss at –18°C in 12 Months (%)		
	UB	B_{LOX} [a]	B_{POD} [b]
Chlorophyll a	53	61	49
Chlorophyll b	42	49	38
Neoxhantin	60	58	45
Lutein	51	39	32
Lutein-5,6-epoxide	32	22	20

[a] Hot water blanching condition to inactivate initial LOX activity ≥90% : 70°C × 4 min.
[b] Hot water blanching condition to inactivate initial POD activity ≥90% : 80°C × 2 min.
Source: Data from Serpen (2003).

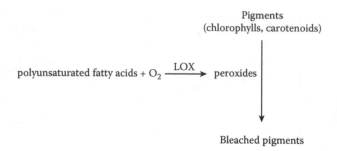

Figure 5.6 Reaction mechanism for bleaching of chlorophylls and carotenoids by the action of lipoxygenase.

increases the loss of chlorophylls in pea during frozen storage. If the blanching conditions are adjusted to inactive peroxidase (B_{POD}) instead of lipoxygenase, both chlorophylls and carotenoids can be better protected against oxidation during frozen storage (Gökmen et al., 2005). It is thought that a mild heat treatment applied to inactivate heat-labile lipoxygenase modifies the pea microstructure and makes pigments more susceptible to oxidation. It should be noted here that the situation might be different for different green vegetables. Therefore, results from one vegetable cannot be directly extrapolated to another vegetable without an experimental approval.

Lipoxygenase catalyzes the incorporation of molecular oxygen into polyunsaturated fatty acids (hydroperoxidation) containing *cis,cis*-1,4-pentadiene moieties (Eskin et al., 1977). From the technological point of view, the enzyme lipoxygenase is worth special attention because of its possible impact on the final color of vegetables. A medium in which hydroperoxide is formed enzymatically causes the destruction of naturally occurring pigments including chlorophylls and carotenoids (Weber et al., 1974). This type of bleaching is also known as co-oxidation reaction (Figure 5.6). Many researchers relate the off-flavor development with the oxidation reactions of polyunsaturated fatty acids catalyzed by lipoxygenase (Heimann and Franzen, 1978).

5.5.1.2   Vitamins

Ascorbic acid is the most abundant vitamin in vegetables. Since it is one of the most labile nutrients, it is often used as a nutritional quality indicator of food processing treatments. Ascorbic acid is readily oxidized by ascorbic acid oxidase or under strong oxidation conditions (Serpen and Gökmen, 2007; Serpen et al., 2007). The following reaction scheme depicts the reversible equilibrium that is well known to occur between ascorbic acid and dehydroascorbic acid in which dehydroascorbic acid irreversibly hydrolyzes to diketogluconic acid (Serpen and Gökmen, 2007; Serpen et al., 2007).

$$\text{Ascorbic acid} \underset{-2H^+}{\overset{+2H^+}{\longleftrightarrow}} \text{Dehydroascorbic acid} \xrightarrow{\text{H}_2\text{O}} \text{Diketogluconic acid}$$

A significant loss of ascorbic acid occurs in unblanched vegetables during frozen storage. The degree of ascorbic acid lost is closely related to the oxidation-reduction conditions and the residual enzyme activity. Figure 5.7 shows the change of ascorbic acid concentrations in frozen-stored pea and green bean as influenced by different blanching conditions targeted to inactivate lipoxygenase and peroxidase. Blanching significantly protects ascorbic acid in peas during frozen storage. The blanching conditions are enough to inactivate either lipoxygenase or peroxidase

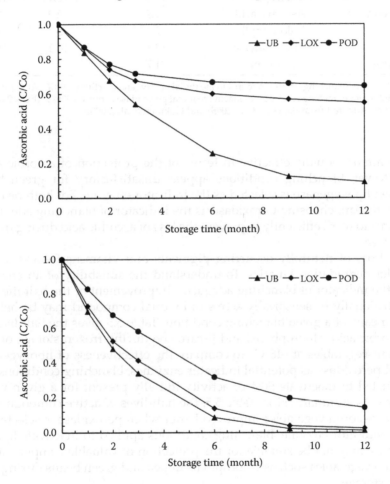

Figure 5.7 Change of ascorbic acid concentration during frozen storage at –18°C; (a) green pea (UB: Unblanched, LOX: Blanched to inactivate lipoxygenase ≥90% in hot water, 70°C × 4 min, POD: Blanched to inactivate peroxidase ≥90% in hot water, 80°C × 2 min), (b) green bean (UB: Unblanched, LOX: Blanched to inactivate lipoxygenase ≥90% in hot water, 70°C × 2 min, POD: Blanched to inactivate peroxidase ≥90% in hot water, 90°C × 3 min). (Adapted from Bahçeci, 2003.)

Table 5.5 Half-Life of Ascorbic Acid in Green Pea during Frozen Storage at –18°C

| | | Half-Life, Month | | |
| | | | Blanched | |
Vegetable	Constituent	Unblanched	B_{LOX}[a]	B_{POD}[b]
Green pea	Ascorbic acid	3.3	12.8	19.0
	Chlorophyll a	14.0	8.4	17.4
	Chlorophyll b	16.8	12.1	17.5
Green bean	Ascorbic acid	1.9	2.2	3.5
	Chlorophyll a	7.3	5.1	8.3
	Chlorophyll b	13.1	10.1	16.7
Carrot	β-Carotene	31.7		96.9

[a] Hot water blanching condition to inactivate initial lipoxygenase activity ≥90% : 70°C × 4 min.
[b] Hot water blanching condition to inactivate initial peroxidase activity ≥90% : 80°C × 2 min.
Source: Adapted from Bahçeci et al. (2005) and Gökmen et al. (2005).

are almost equally effective in terms of the protection of ascorbic acid. However, blanching conditions appear unsatisfactory for green bean when lipoxygenase is selected as the indicator enzyme. For both pea and green bean, choosing peroxidase as the indicator of blanching adequacy seems to work efficiently to prevent the loss of ascorbic acid during frozen storage.

Loss of naturally occurring pigments and vitamins follows a first order degradation kinetics. To understand the suitability of an enzyme as the indicator of blanching adequacy, improvement of the half-life of a nutritionally or sensorially active individual compound may be helpful by means of a given blanching condition. Table 5.5 gives the half-lives of ascorbic acid, chlorophylls, and β-carotene during frozen storage of different vegetables at –18°C. To compare the effectiveness of lipoxygenase and peroxidase as potential indicator enzymes, blanching conditions are adjusted to inactivate 90% of activity initially present for a given vegetable. As summarized in Table 5.5, the half-lives of active compounds in frozen-stored vegetables improved better when peroxidase is selected as the indicator enzyme. Blanching conditions applied to inactivate lipoxygenase may not be suitable for the protection of valuable components of green vegetables such as chlorophylls in pea and green beans during frozen storage.

The development of volatile aroma compounds in vegetables during frozen storage may also be related to residual enzyme activity. For leeks, formation of aldehydes such as hexanal has been found higher in unblanched samples than in blanched ones (Nielsen et al., 2004).

5.6 Conclusion

There are various enzymes naturally occurring in vegetables. For a better preservation of vegetables, blanching inactivates enzymes, which are responsible for the loss of quality. For a single vegetable, determination of the thermal inactivation conditions of deteriorative enzymes is a crucial step to design the blanching process. However, this step should be combined with a complementary storage test in which the stabilities of individual quality parameters are determined. In most cases, thermal inactivation of peroxidase at a ratio of 90% stabilizes the nutritional and sensorial quality of vegetables during a prolonged frozen storage. However, blanching to inactivate peroxidase commonly requires applying more thermal energy to the given vegetable. Some vegetables are apparently very sensitive to thermal energy, and their stability decreases during storage as the thermal load increases during blanching.

Abbreviations

A: enzyme activity
B: blanched sample
C: concentration
LOX: lipoxygenase
POD: peroxidase
UB: unblanched sample

References

Adams, J.B. 1991. Review: Enzyme inactivation during heat processing of food stuffs. *International Journal of Food Science and Technology* 26(1), 1–20.

Agüero, M.V., M.R. Ansorena, S.I. Roura, and C.E. del Valle. 2008. Thermal inactivation of peroxidase during blanching of butternut squash. *LWT-Food Science and Technology* 41, 401–407.

Altunkaya, A. and V. Gökmen. 2008. Effect of various inhibitors on enzymatic browning, antioxidant activity and total phenol content of fresh lettuce (Lactuca sativa). *Food Chemistry* 107, 1173–1179.

Anthon, G. E. and D.M. Barrett. 2002. Kinetic parameters for the thermal inactivation of quality-related enzymes in carrots and potatoes. *Journal of Agricultural and Food Chemistry* 50(14), 4119–4125.

Anthon, G.E., Y. Sekine, N. Watanabe, and D.M. Barrett. 2002. Thermal inactivation of pectin methylesterase, polygalacturonase, and peroxidase in tomato juice. *Journal of Agricultural and Food Chemistry* 50(21), 6153–6159.

Arabshahi, A. and D.B. Lund. 1985. Considerations in calculating kinetic parameters from experimental data. *Journal of Food Process Engineering* 7(4), 239–251.

Aylward, F. and D.R. Haisman. 1969. Oxidation systems in fruits and vegetables: Their relation to the quality of preserved products. *Advances in Food Research* 17, 1–76.

Bahçeci, K.S., A. Serpen, V. Gökmen, and J. Acar. 2005. Study of lipoxygenase and peroxidase as indicator enzymes in green beans: change of enzyme activity, ascorbic acid and chlorophylls during frozen storage. *Journal of Food Engineering* 66, 187–192.

Bahçeci, K.S. 2003. Determination of thermal inactivation and reactivation of peroxidase and lipoxygenase in some frozen vegetables. Master's thesis, Hacettepe University, Ankara, Turkey.

Barrett, D.M. and C. Theerakulkait. 1995. Quality indicators in blanched, frozen stored vegetables. *Food Technology* 49(1), 62, 64–65.

Bifani, V., J. Inostroza, M.J. Cabezas, and M. Ihl. 2002. Determination of kinetic parameter of peroxidase and chlorophyll in green beans (*Phaseolus vulgaris* cv. Win) and their stability when frozen. *Revista de Quimica Teorica y Aplicada, Febrero,* LIX (467), 57–64.

Bomben, J.L. 1979. Waste reduction and energy use in blanching. In *Food Processing Waste Management*; Green, J.H., Kramer, A., Eds., AVI Publishing: CT, 151–173.

Bottcher, H. 1975. Enzyme activity and quality of frozen vegetables. I. Remaining residual activity of peroxidase. *Nahrung* 19, 173.

Delincée, H. and W. Schaefer. 1975. Influence of heat treatments of spinach at temperatures up to 100°C on important constituents. Heat inactivation of peroxidase. *LWT-Food Science and Technology* 8, 217–221.

Diehl, H.C. 1932. A physiological view of freezing preservation. *Industrial Engineering and Chemistry* 24, 661.

Eskin, N.A.M., S. Grossman, and A. Pinsky. 1977. Biochemistry of lipoxygenase in relation to food quality, *Critical Reviews in Food Science and Nutrition* 204, 1–40.

Forsyth J.L., R.K.O. Apenten, and D.S. Robinson. 1999. The thermostability of purified isoperoxidases from *Brassica oleraceae* var. *gemmifera*. *Food Chemistry* 65(1), 99–109.

Ganthavorn, C., C.W. Nagel, and J.R. Powers. 1991. Thermal inactivation of asparagus lipoxygenase and peroxidase. *Journal of Food Science* 56(1), 47–49.

Gökmen, V., K.S. Bahçeci, A. Serpen, and J. Acar. 2005. Study of lipoxygenase and peroxidase as blanching indicator enzymes in peas: Change of enzyme activity, ascorbic acid and chlorophylls during frozen storage. *LWT-Food Science and Technology* 38, 903–908.

Gonçalves, E.M., J. Pinheiro, M. Abreu, T.R.S. Brandau, and C.L.M. Silva. 2007. Modelling the kinetics of peroxidase inactivation, colour and texture changes of pumpkin (*Cucurbita maxima* L.) during blanching. *Journal of Food Engineering* 81, 693–701.

Günes, B. and A. Bayındırlı. 1993. Peroxidase and lipoxygenase inactivation during blanching of green beans, green peas and carrots. *LWT-Food Science and Technology* 26, 406–410.

Heimann, W. and K.H. Franzen. 1978. Development of volatile compounds from radicals arising during intermediate steps of lipoxygenase-reaction, *Zeitschrift für Lebensmittel Untersuchung und Forschung* 167, 78–81.

Holdsworth, S.D. 1983. *The Preservation of Fruits and Vegetable Food Products.* Macmillan Press, London

Joslyn, M.A. 1949. Enzyme activity in frozen vegetable tissues. Advances in Enzymology 9, 613–652.

Kim, M.J., J.M. Oh, S.H. Cheon, T.K. Cheong, S.H. Lee, E.O. Choi, H.G. Lee, C.S. Park, and K.H. Park. 2001. Thermal inactivation kinetics and application of phospho- and galactolipid-degrading enzymes for evaluation of quality changes in frozen vegetables. *Journal of Agricultural and Food Chemistry* 49, 2241–2248.

Kohman, E.F. 1936. Enzymes and the storage of perishables. *Food Industries* 8, 287–288.

Kushad, M.M., M. Guidera, and D. Bratsch. 1999. Distribution of horseradish peroxidase activity in horseradish plants. *HortScience* 34(1), 127–129.

Ling, A. and D. Lund. 1978. Determining kinetic parameter for thermal inactivation of heat-resistant and heat-labile isozymes from thermal destruction curves. *Journal Food Science* 43(4), 1307–1310.

Lu, A.T. and J.R. Whitaker. 1974. Some factors affecting rates of heat inactivation and reactivation of horseradish peroxidase. *Journal of Food Science* 39, 1173–1178.

Morales-Blancas, E.F., V.E. Chandia, and L. Cisneros-Zevallos. 2002. Thermal inactivation kinetics of peroxidase and lipoxygenase from broccoli, green asparagus and carrots. *Journal of Food Science* 67(1), 146–154.

Nielsen, G.S., L.M. Larsen, and L. Poll. 2004. Impact of blanching and packaging atmosphere on the formation of aroma compounds during long-term frozen storage of leek (*Allium ampeloprasum* Var. Bulga) slices. *Journal of Agricultural and Food Chemistry* 52, 4844–4852.

Poulsen, K.P. 1986. Optimization of vegetable blanching. *Food Technology* 40(6) 122–129.

Powers, J.R., M.J. Costello, and H.K. Leung. 1984. Peroxidase fractions from asparagus of varying heat stabilities. *Journal of Food Science* 49(6), 1618–1619.

Resende, R., F.J. Francis, and C.R. Stumbo. 1969. Thermal destruction and regeneration of enzymes in green bean and spinach puree. *Food Technology* 23, 63–66.

Sapers, G.M. and J.T.R. Nickerson. 1962. Stability of spinach catalase. I. Purification and stability during storage. *Journal of Food Science* 27(3), 277.

Sarikaya, A. and M. Özilgen. 1991. Kinetics of peroxidase inactivation during thermal processing of whole potatoes. *LWT-Food Science and Technology* 24(2), 159–163.

Schwimmer, S. 1981. *Source Book of Food Enzymology.* AVI Publishers, Westport, CT.

Serpen, A. and V. Gökmen. 2007. Reversible degradation kinetics of ascorbic acid under reducing and oxidizing conditions. *Food Chemistry* 104, 721–725.

Serpen, A., V. Gökmen, K.S. Bahçeci, and J. Acar. 2007. Reversible degradation kinetics of vitamin C in peas during frozen storage. *European Food Research and Technology* 224, 749–753.

Serpen, A. 2003. Determination of the effects of peroxidase and lipoxygenase on pigment co-oxdation in some frozen vegetables. Master's thesis, Hacettepe University, Ankara, Turkey.

Sheu, S. C. and A.O. Chen. 1991. Lipoxygenase as blanching index for frozen vegetable soybeans. *Journal of Food Science* 56(2), 448–451.

USDA. 1975. Enzyme Inactivation Tests Frozen Vegetables, Technical Inspection Procedures for the Use of USDA Inspectors, U.S. Dept. Agric. Agri. Mktg. Serv., Washington, DC.

Vámos Vigyázó, L. 1981. Polyphenol oxidase and peroxidase in fruits and vegetables. *Critical Reviews in Food Science and Nutrition* 15(1), 49–127.

Velasco, P.J., M.H. Lim, R.M. Pangborn, and J.R. Whitaker. 1989. Enzymes responsible for off-flavor and off-aroma in blanched and frozen-stored vegetables. *Biotechnology and Applied Biochemistry* 11, 118–127.

Walker, G.C. 1964. Color deterioration of frozen green beans (*Phaseolus vulgaris*). *Journal of Food Science*, 29(4), 383.

Wang, Z. and B.S. Luh. 1983. Characterization of soluble and bound peroxidases in green asparagus. *Journal of Food Science* 48(5), 1412–1417, 1421.

Weber, F., G. Laskawy, and W. Grosch. 1974. Co-oxidation of carotene and crocin by soybean lipoxygenase isoenzymes, *Zeitschrift für Lebensmittel Untersuchung und Forschung* 155(3), 142–150.

Williams, D.C., M.H. Lim, A.O. Chen, R.M. Pangborn, and J.R. Whitaker. 1986. Blanching of vegetables for freezing: Which indicator enzyme to choose. *Food Technology* 40(6), 130–140.

Winter, E. 1969. Behavior of peroxidase during blanching of vegetables. *Zeitschrift für Lebensmittel Untersuchung und Forschung* A 141, 201.

Winter, E. 1969. Behaviour of peroxidase during blanching of vegetables. *Zeitschrift Lebensmittel Untersuchung und Forshung* 141 201–208.

Yemenicioğlu, A., M. Özkan, S. Velioğlu, and B. Cemeroğlu. 1998. Thermal inactivation kinetics of peroxidase and lipoxygenase from fresh pinto beans (*Phaseolus vulgaris*). *Zeitschrift für Lebensmittel Untersuchung und -Forschung* A 206, 294–296.

chapter six

Enzymatic peeling of citrus fruits

Maria Teresa Pretel, Paloma Sánchez-Bel,
Isabel Egea, and Felix Romojaro

Contents

6.1 Introduction

The conventional industrial methods for the peeling of citrus fruits include the manual or mechanical separation of the rind and the later chemical degradation of the rest of the albedo and segment membranes. These methods imply a high cost of labor, require high volumes of water for washing, and cause significant environmental problems due to the use of peeling caustics. Enzymatic peeling is an alternative method for removing the rind of the fruits with lower water consumption and causing less contamination than the traditional methods. The principle of the enzymatic peeling is based on the digestion, by an enzymatic preparation, of pectic substances found in cell walls of

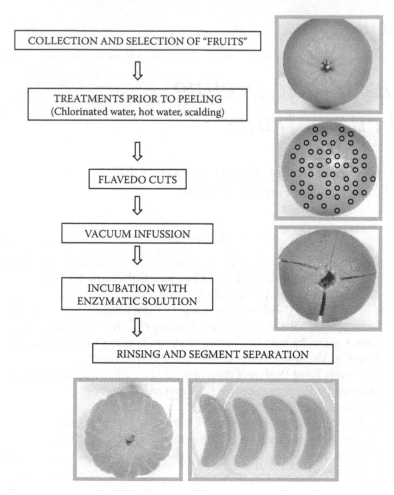

Figure 6.1 Enzymatic peeling of citrus fruits.

plants (Bruemmer et al., 1978; Berry et al., 1988). The enzymatic peeling of citrus fruits is presented in Figure 6.1. The operation starts with a selection of fruits followed by washing with chlorinated water and, in certain cases, by a later treatment with hot water or scalding. Then, the incisions are made on the flavedo to allow the penetration of the enzymatic solution to the albedo. Fruits are later dipped into the enzymatic solution in a tank and vacuum is applied. Finally, fruits are washed with pressurized water and the finished product is obtained (whole fruit or segments).

From the molecular point of view, pectin, cellulose, and hemicellulose are responsible for the adherence of the skin to the fruit (Whitaker, 1984). Therefore, both pectinases and cellulases are needed for the enzymatic peeling. The cellulases are probably needed for the release of the pectins in the albedo, and the pectinases contribute to the hydrolysis of the polysaccharides of the cell wall (Ben-Shalom et al., 1986; Coll, 1996). Bruemmer et al. (1978) were the first authors employing this enzymatic method in the peeling of grapefruits by vacuum infusion of commercial pectolytic preparations, showing that the obtained sections maintained their original taste and texture with higher efficiency and quality than those obtained by conventional peeling processes. They also observed how the commercial pectinases considerably differ on their peeling efficiency. Likewise, Berry et al. (1988) showed that the loss of juice in both enzymatically peeled entire grapefruits and segments by the method developed by Bruemmer et al. (1978), was lower than in fruits obtained by conventional chemical or manual methods. To obtain a better knowledge of the enzymatic degradation, Ben Shalom et al. (1986) stated the importance of evaluating the effect of commercial enzymes on the substrates to be degraded. This fact was later confirmed by other authors (Rouhana and Mannheim, 1994; Soffer and Mannheim, 1994; Baker and Wicker, 1996; Pretel et al., 2005; Pinnavaia et al., 2006). However, not only the enzymatic preparation is critical for obtaining a good peeling efficiency; there are many other determining parameters. For instance, the adherence of the peel to the fruit and its thickness are different according to the species or the citrus varieties, and the design of the cuts, the vacuum conditions, temperature, and pH also affect the peeling (Berry et al., 1988; Adams and Kirk, 1991; McArdle and Culver, 1994; Pretel et al., 1997; Pretel et al., 1998a,b; Prakash et al., 2001; Suutarinen et al., 2003; Liu et al., 2004; Pagán et al., 2005; Pretel et al., 2007a,b). On the other side, the reuse of the enzymatic solution has been studied by different authors since, from the economical point of view, this is one of the important factors when using the operation at industrial level (Pretel et al., 1997; Rouhana and Mannheim, 1994; Pagán et al., 2006; Pretel et al., 2007b).

The enzymatic peeling could be a significant alternative for the food industry because it can be applied to different types of fruits and vegetables such as grapefruit (Roe and Bruemmer, 1976; Bruemmer et al., 1978); grapefruit and orange (Berry et al., 1988); mandarin (Coll, 1996); orange (Pretel et al., 1997); apricots, nectarines, and peaches (Toker and Bayindirli, 2003); or potatoes, carrots, and Swedish turnips (Suutarinen et al., 2003). The use of the enzymatic peeling and the possibilities of the commercialization of these products could be increased to face the demand of innovation from the world markets for products that have been minimally processed or for replacing the traditional processes.

6.2   Effects of morphological and physiological characteristics of citrus fruits on enzymatic peeling

All cultivated citrus fruits show the same anatomical structure (Figure 6.2a), although the elements of this structure are different depending on the species and variety. The colored portion of the peel is the epicarp and it is also known as the flavedo. In the flavedo, there are cells containing carotenoids that give the characteristic color to the different citrus fruits such as orange, mandarin or tangerine, grapefruit and lemon. The flavedo oil glands are the raised structures on the skin or contain the essential oils characteristic of each citrus cultivar. The mesocarp or albedo is typically a thick, white, spongy layer. However, in some varieties of orange and mandarins the albedo is very thin and difficult to separate from the flavedo, a fact affecting the enzymatic peeling. The albedo consists of large parenchymatous cells rich in pectic substances and hemicelluloses (Ting and Rouseff, 1986). It completely envelopes the endocarp that is the edible portion of citrus fruits.

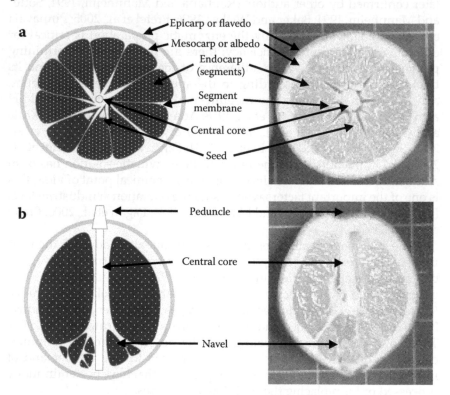

Figure 6.2 Transverse cut (a) and longitudinal cut (b) of a navel orange.

The combined form of albedo and flavedo is called the pericarp, commonly known as the rind or peel. The endocarp is made up of a set of vesicles containing juice and grouped in the segments that are found in a number between 5 and 19, depending on the species. Oranges usually have 9–11 segments, while lemons and grapefruits have 8–11 and 12–15 segments, respectively. The number of seeds varies depending on species, variety, and pollination conditions. However, the varieties most suitable for enzymatic peeling are those without seeds since they are the most appreciated by the consumers (Pretel et al., 2008). Some orange varieties develop in the base of the fruit a secondary small and atrophied orange that resembles a navel (Figure 6.2b). The "navel" and some morphological characteristics like the adhesion between albedo and segments, the adhesion between segments and, above all, the homogeneity of the segment membrane are characteristic of each species and variety (Pretel et al., 2007a) and they directly affect enzymatic peeling. The presence of the navel hampers the diffusion of the enzymatic solution into the tissue and the irregularity of the segments decreases the quality of the final product, as could be verified in the Thomson variety (Pretel et al., 2007a). The strong adhesion between the albedo and the segments and the close union between the segments increase the levels of vacuum needed for the process because the diffusion of the enzymatic solution in the tissues would be hampered (Pretel et al., 1997). The presence of fractures in the carpelar membrane and, therefore, the lack of homogeneity of the skin, make it difficult to obtain segments since the enzymatic solution penetrates even with low levels of vacuum (Pretel et al., 1997).

It is important from an economical point of view to know the thickness of the albedo, because the higher thickness of albedo means the higher quantity of enzymatic solution needed for the process (McArdle and Culver, 1994). On the other side, a too much thin albedo could not absorb enzymatic solution enough for an adequate fruit peeling. Although the thickness of the albedo can give an idea about the quantity of enzymatic solution that each species or variety can absorb. Its degree of compaction is one of the parameters affecting enzymatic peeling (Pretel et al., 1997) because it directly influences the penetration of the pectolytic enzyme solution when vacuum is applied. The enzymatic solution better diffuses when the tissue is porous, that is, when the parenchymatous tissue of the albedo shows wide intercellular spaces and, therefore, the volume of the albedo is higher than its weight (Baker and Bruemmer, 1989). Then, when vacuum is applied, the air in the intercellular spaces is more easily replaced by the enzymatic solution. The species and varieties with a less porous albedo need more severe vacuum conditions for the enzymatic solution to penetrate in the albedo because the compaction of the tissue hampers the distribution of the enzymatic solution among the cells (Pretel et al., 1997).

Finally, some physiological, nutritional, and functional properties of the fruits, like the respiratory intensity, the sugar content, and the antioxidant capacity, play an indirect role in the enzymatic peeling because they affect the possibilities of storage and the quality of the finished product. The respiratory intensity of the citrus fruits that becomes apparent with the consumption of O_2 and the release of CO_2 tends to decrease during ripening and it does not suffer many changes after harvesting because these are non-climacteric fruits (Pretel et al., 1997). In some varieties of orange (Pretel et al., 2008), the respiratory intensity varies between 8 and 26 mg of CO_2 $kg^{-1}h^{-1}$. This low respiratory intensity allows the fruits to remain in the tree or to be stored for longer periods of time. Pretel et al. (2008) determined the influence of the ripening state of fruits on the enzymatic degradation of the albedo and the carpelar membrane of citrus fruits, and they conclude that within the ripeness limits studied (8.5 to 15.5° Brix), the ripening degree of fruits affects the enzymatic peeling since variations of this parameter will probably modify both the concentrations of enzymatic preparation necessary for the peeling and the optimum vacuum conditions. Ismail et al. (2005) also verified that the peeling efficiency was variable over the season with fruit harvested in March and May being most easily peeled (over 50% efficiency) and fruit harvested in February, April, and June being peeled less efficiently (11% in February, 30% in April, and 45% in June). These authors also stated that the storage of intact Valencia oranges and Ruby Red grapefruit did not affect peeling efficiency for up to 12 weeks, but peeling efficiency declined after 15 weeks of storage.

6.3   Citrus species used in enzymatic peeling studies

Most studies about enzymatic peeling of citrus fruits have been carried out using different varieties of oranges and grapefruits, although other species like mandarin, lemon, and *Citrus maxima* Burm. Merrill variety Cimboa have been also studied.

The orange (*Citrus sinensis*) is the more significant species of the genus *Citrus* because it is the most consumed one. The fruits differ in form and color according to the varieties, and this allows the classification of oranges into four groups (Figure 6.3): navel, white, blood, and sweet (Loussert, 1992). Navel oranges (Figure 6.3a) show a small and primitive fruit, called the *navel*, in the base of the fruit. These navel oranges are a group of oranges without seeds, with a very early ripening, and with an excellent organoleptic quality. However, the presence of the navel makes them inadequate for obtaining segments by enzymatic peeling because it hampers the diffusion of the enzymatic solution. The segments have a great organoleptic quality due to the thin skin, although this characteristic,

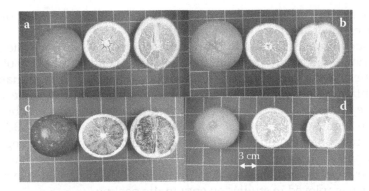

Figure 6.3 Photographs of fruits from the four groups of oranges (*Citrus sinensis* L. Osbeck): navel: Navelina (a); white or blond oranges: Salustiana (b); blood: Sangrina (c); and sweet oranges: Grano de Oro (d).

together with the irregularities, hampers the enzymatic peeling. However, whole peeled navel oranges can be obtained, as it happens with the variety Thomson (Pretel et al., 2007a). The variety Navelina, one of the most cultivated oranges in spite of the fact that these oranges have a navel, has been employed in a study focused on determining the changes that happen in the peel albedo (Pagan et al., 2006).

Among the group of white oranges, Fine ones are included, and they have been selected according to the fruit quality, the production, and the harvesting season (Figure 6.3b). White oranges have neither seeds nor a navel, so they can be employed for obtaining segments by enzymatic peeling. This is the case of the variety Salustiana that has been studied by Pretel et al. (1997) for optimizing different parameters that are essential in the enzymatic peeling like the influence of the cuts in the flavedo, the vacuum effect, or the peeling temperature. This variety was also successfully employed for storage studies of orange segments and whole peeled oranges as "ready to eat" products (Pretel et al., 1998a). The variety Valencia, one of the most cultivated varieties in the world, was employed by Soffer and Mannheim (1994) for studying the efficiency of different enzymatic preparations and the effect of scalding, as a previous step to peeling, on the incubation time and the final product quality. Other authors (Ismail et al., 2005) studied changes in peeling efficiency of Valencia oranges according to the harvesting season and the quality of the final product. Pinnavaia et al. (2006) compared the fresh segments and slices from this orange variety infused under vacuum with different enzyme solutions along with post treatment acid dips and temperature conditioning to show the effect on quality, residual enzyme activity related juice leakage, shelf life, and microbial stability. Pao and Petracek (1997) studied this variety to identify microorganisms that are responsible for the spoilage of oranges peeled with water or citric acid solution infusion and to show the effect of citric acid treatment at different storage conditions.

Blood oranges (Figure 6.3c) are different from white ones due to the presence of red anthocyanins in the flavedo and endocarp. Many of these varieties have a very compact (little porous) albedo, a fact that hampers the obtaining of segments by enzymatic peeling, although some varieties like *Citrus sinensis* (L.) Osbeck cv. Sangrina are suitable for obtaining orange segments (Pretel et al., 2007b). Sweet oranges (Figure 6.3d) have good morphological qualities (Pretel et al., 2005) for obtaining segments by enzymatic peeling, but due to their characteristic acid-free flavor and the presence of seeds, they are not appreciated by consumers, so they have not been employed in studies on enzymatic peeling.

Mandarins are an easily peeled citrus fruit due to the weak adherence between epicarp and endocarp. These fruits are widely cultivated in the world, especially for their fresh consumption, although some varieties like Satsuma are commercialized advertised in syrup. Mandarins are employed in several studies. Coll (1996) used mandarins to study the enzymatic degradation of the carpelar membrane. Pretel et al. (1998b) employed *Citrus unshiu* Marc. Satsuma for studying the modeling design of cuts for enzymatic peeling with the optimization of the parameters of peeling. Liu et al. (2004) also determined the optimum conditions for enzyme infusion peeling of a local variety of mandarins or limau madu (*Citrus reticulata* B.) from Malaysia. The grapefruit (*Citrus x paradisi* Macfad) have been employed for a long time in studies about enzymatic peeling because it is a widely used fruit as a breakfast fruit in sections or as fruit juice due to its refreshing taste, rich nutritional composition, mild bitterness claimed, and tonic effects. Bruemmer et al. (1978) were the first researchers to develop a process for the obtaining of peeled segments of citrus fruits by vacuum infusion of commercial pectolytic preparations. For this study, they used grapefruits (*Citrus paradisi* Macfad cv. Duncan). Ben-Shalom et al. (1986) carried out studies on the membrane composition and the characteristics of commercial pectolytic enzymes using segments of the variety Marsh Seedless. Soffer and Mannheim (1994) verified in grapefruit the importance of the pH of the enzymatic solution and the peeling cycles for the effectiveness of the enzymatic preparation. Rouhana and Mannheim (1994) carried out a study on the optimization of enzymatic peeling of grapefruit and they could verify the effectiveness of different commercial preparations. On the other side, the toughness of its peel and its close adherence to the inner fruit sections make grapefruit difficult to peel, so Prakash et al. (2001) studied different previous scalding treatments, vacuum levels, and incubation time to improve the enzymatic peeling of grapefruits. Ismail et al. (2005) used grapefruit var. Ruby Red for studying the efficacy of enzymatic peeling depending on storage time. Other citrus fruits like fruits of *Citrus maxima* Burm Merrill

variety Cimboa have been employed in studies about enzymatic degradation of albedo and segment membrane (Pretel et al., 2005) due to their high
proportion of albedo in fruit. Finally, lemon (*Citrus limon*) has been used
for studies on extraction, characterization, and enzymatic degradation of
lemon pectins (Ros et al., 1996).

It must be considered that enzymatic peeling is also used for removing the skins of other species like apricots (*Prunus armeniaca*), nectarines
(*P. persica* var. nucipersica schneid) and peaches (*P. persica*), and that the
main advantage of this new technology in comparison to mechanical or
chemical peelings is the quality of the final product, as well as the reduced
requirement of heat treatment and industrial waste (Toker and Bayirdirli,
2003). The suitability of commercial cellulase and pectinase preparations
for the hydrolysis of isolated peels and whole vegetables such as potatoes
(*Solanum tuberosum* cv. Asterix), carrots (*Daucus carota* L.), Swedish turnips (*Brassica napus* L.), and onions (*Allium cepa* L.) was also investigated
(Suutarinen et al., 2003). Although the enzymatic pretreatment enhanced
the degradation of the peels of carrots and onions, they did not obtain
good results for potatoes and Swedish turnips because the high content
of cutin/suberin in the skin made the enzymatic degradation difficult.
Therefore, the authors propose further research to improve the enzyme-
aided peeling method.

6.4   Treatments prior to enzymatic peeling of citrus fruits

The effect of scalding or other treatments with hot water prior to enzymatic peeling has been studied (Table 6.1). One of the main works where
a treatment with hot water before enzymatic peeling was employed for
improving the process was that of Bruemmer et al. (1978) in grapefruit.
The authors immersed the fruits in a water bath at 60°C for 30 min. After
this time, the temperatures of the albedo and the central core were 60°C
and 35°C, respectively. The fruit was then removed from the bath and
the peeling process went on. At the end, high quality grapefruit segments were obtained. Later, Rouhana and Mannheim (1994) examined
the effect of scalding on grapefruits, and Soffer and Mannheim (1994)
on Valencia oranges and grapefruits. Both groups reported that scalding
at 100°C from 2 to 4 min, depending on the thickness of the fruits' skin,
is a necessary step prior to enzymatic peeling of grapefruits. Increasing
the scalding time improved the efficiency of enzymatic peeling and
decreased the peeling time. The probable explanations for this result
are that heat treatment decreased the viscosity of pectin, changed the
crystalline structure of cellulose to an amorphic structure (Alberts et al.,
1989), and improved the ability of the peel to absorb the enzyme solution.

Table 6.1 Treatments Prior to Enzymatic Peeling

Fruit	Treatment	Time of Treatment	Result	Reference
Grapefruit	Water bath 60°C	30 min (until albedo reaches 50°C)	Good	Bruemmer et al. (1978)
Grapefruit	Scalding	2–4 min (depending on ripeness and peel thickness)	Increasing the time, improve the peeling	Rouhana and Mannheim (1994)
Orange Valencia	Scalding	0–4.5 min (various tests)	Increasing the time, worsen the peeling	Soffer and Mannheim (1994)
Mandarin Satsuma	Hot water (already 45°C)	Until albedo reaches 40°C	Good	Pretel et al. (1998)
Grapefruit	Scalding	1–4 min (various tests)	Increasing the time, improve the peeling	Prakash et al. (2001)
Orange Valencia	Hot water (already 45°C)	Not specified	Good	Pinnavaia et al. (2006)
Orange Thomson and Mollar	Hot water (already 45°C)	Until albedo reaches 35–40°C	Good	Pretel et al. (2007a)
Orange Sangrina	Hot water (already 45°C)	Until albedo reaches 35–40°C	Good	Pretel et al. (2007b)

As a result, the peel components were readily digested by the enzymes. On the other hand, too long a scalding time resulted in decreased quality of the final product. Rouhana and Mannheim (1994) found that the effect of scalding on Valencia oranges is very different to that observed in grapefruits because, for this variety of orange, the better results were obtained without scalding before the peeling process. The authors also checked that the scalding caused disgusting flavors in the oranges and decreased the quality of the final product. Prakash et al. (2001) observed that with 1 min of scalding of grapefruit, peeling was very difficult even after infusion with the enzymatic preparation (Table 6.1). Peeling was moderately easy when the scalding time was raised to 2 min, while above 2 min, peeling became extremely easy. Thus, an increase of the scalding time improved the efficiency of the enzymatic peeling process. However, a too long scalding time decreased the quality of the final product. Some authors (Table 6.1) applied a treatment of the fruits in a hot water bath, independent of the period of time, until the albedo reaches 35–40°C prior to peeling (Pretel et al., 1997, 1998a, 2007a, 2007b), while others start the peeling process directly, without heat treatment (Pao and Petracek, 1997; Liu et al., 2004; Pagán et al., 2005). The treatment of fruits by hot water dipping shows the best results to obtain a good enzymatic peeling. In most cases, fruits are placed into chlorinated water (300 ppm) before being transferred to the hot water bath to reduce possible microbial con- tamination of the finished product (Pretel et al., 1997, 1998b, 2007a, 2007b) or, as Pinnavaia et al. (2006) suggested, fruits are pre-treated with an acid solution (0.1 N HCl).

6.5   Effect of the pattern of flavedo cuts on peeling efficiency

The degradation of the flavedo by the enzymes to obtain peeled citrus fruits is very difficult and expensive since it is a barrier for the attack of degra- dative enzymes to the softest tissues of albedo and segment membranes. For the enzymatic solution to penetrate inside the albedo and among the segments, it is necessary to make cuts in the flavedo of the fruits before applying vacuum (Bruemmer et al., 1978). The same authors proposed hand-scored the peel of grapefruits in quadrants (Figure 6.4a), while Soffer and Mannheim (1994) and Prakash et al. (2001) made four radial lines, tak- ing care not to cut the fruit sections underneath (Figure 6.4b). Rouhana and Mannheim (1994), before the vacuum infusion of the enzymatic solution, also made from 4 to 6 radial cuts in grapefruit using a knife (Figure 6.4b). Pretel et al. (1997) found that cuts recommended above did not result in a good peeling process since areas of undegraded albedo remained and it was difficult to separate the segments. Presumably the furthest albedo part from the cuts had not been saturated with the enzymatic solution. To obtain whole Salustiana oranges, the best results were obtained with three transversal cuts made in the calycinal, peduncular, and equatorial zones of the fruit, and two longitudinal cuts (Figure 6.4d). Using this cut pattern, there was minimal difficulty in separating the remnants of the rind, as the enzyme solution spread easily through the albedo without penetrating the segments. However, when orange segments were needed, the best results were obtained when the transversal cuts near the calycinal and peduncular zones were substituted by removal of the peel in these zones (Figure 6.4e). Under these conditions, the peel remnants were eas- ily removed and the solution penetrated between several segments of the orange (Pretel et al., 1997).

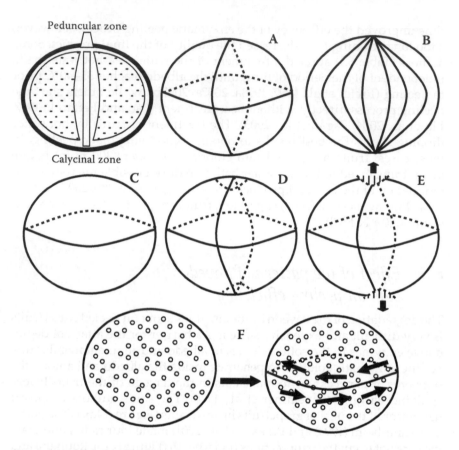

Figure 6.4 Design of the cuts made on the flavedo by different authors (a, b, c, d, e, f) for favoring the penetration of the enzymatic solution when vacuum is applied.

In subsequent studies, Pretel et al. (1998b) examined different cut patterns in the flavedo (Figure 6.4c, d, e) for the enzymatic peeling of mandarin by analyzing characteristics of enzymatic saturation at different pressures and times of vacuum. The authors demonstrate that the cuts significantly influenced the penetration rate and the distribution of the enzyme solution inside the albedo. With the design *c*, in which only an equatorial incision was made, it was observed that when enzymatic saturation was at its maximum, a large percentage of the albedo was not degraded by the enzyme. This was due to the softening of the skin of the segments in the area adjacent to the cuts, favoring the passing of liquid into the interior of the juice vesicles, thus impeding its homogeneous distribution throughout the rest of the albedo. By applying an inferior vacuum it would be possible to avoid the entrance of the enzymatic solution into the vesicular zone (Pretel et al., 1997), but this would suppose an increase in the percentage of albedo not attacked in the zones furthest from the cut. Thus, the presence of altered vesicles, as well as the difficulty of removing remains of residual peel and separating the segments, meant that this cut design was considered inadequate for enzymatic peeling of mandarin (Pretel et al., 1998b). The design *d*, consisting of making three transverse cuts in the calycinal, peduncular, and equatorial zones of the fruit and two longitudinal cuts, was proposed by Pretel et al. (1997) for obtaining whole peeled oranges Salustiana. The mandarins presented a low percentage of non-attacked albedo and, moreover, this design showed an advantage with respect to cut design *c* in that the greater number of incisions made to the fruit's skin permitted the uniform distribution of liquid in the interior of the albedo, which favored the easy separation of the remains of the residual peel. The vacuum conditions under which more than 95% enzymatic saturation was reached were not valid, as penetration of the enzymatic solution into the segments was noted. In short, it can be stated that the design of cut *d* proved acceptable, as much for obtaining segments as for obtaining whole fruits. In other work dealing with enzymatic peeling of Salustiana orange (Pretel et al., 1997), it was observed that this design was only valid for obtaining whole fruits. Probably, the reduced adhesion of the albedo to the skin of the segments allowed, in this case, to obtain segments with good peeling quality. The cut design *e* consisted of substituting the transverse cuts of the peduncular and calycinal zones with the removal of a portion of skin from these zones. With this design, the penetration of the enzymatic solution into the segments was reduced under all the conditions tested. This cut design with the application of vacuum conditions permitted between 85 and 95% of enzymatic saturation, and allowed for a homogeneous distribution of the enzymatic solution in the interior of the albedo, favoring easy separation of remains of the residual peel, as occurred with design *d*. In addition, removing a portion of the skin from the peduncular and calycinal zones allowed the entrance of the solution into the fruit, assisting the passage of enzymatic solution between the segments, with a subsequent increase in efficiency of segment separation (Pretel et al., 1998b).

Pagan et al. (2006), to enzymatically peel orange Navelina, previously scored the peel with a sharp knife so that the cut reached approximately half of the albedo thickness. Cuts were made following marked meridian lines at a distance of 1 cm between them (Figure 6.4b). This distance was considered after studying the width of the spot formed in the albedo impregnated with the enzymatic buffered solution after the penetrations through a cut. Moreover, the peel was perforated with a thumbtack, punching holes at 1 cm distance. The whole orange peel was removed very carefully from the fruit by hand by means of four deeper meridian cuts. Pinnavaia et al. (2006) employed a similar cut design in Valencia oranges; the peel was scored by hand with a citrus peeler

(Sunkist) parking six cuts from the stem to the blossom end to permit infusion of solution into the albedo. In following studies (Pretel et al., 2007a, 2007b), the entire fruit surface was homogeneously perforated by rolling the oranges over a 1 m² wood plate to which cylindrical metal projections 5 mm long and 1 mm in diameter were attached (Figure 6.4f). The distance between each cylindrical metal projection was 10 mm. With this system, 8 ± 3 perforations per cm² in the albedo fruit were produced. After the application of vacuum with the enzymatic solution and the needed incubation time, an equatorial cut was carried out and each half of skin was turned in the opposite direction. This method, which favored the penetration of the solution into the albedo, could more easily be adapted for industrial peeling than those used in earlier studies (Pretel et al., 1997; 1998a,b).

6.6   Effect of vacuum conditions (pressure and time) on citrus fruit peeling efficiency

The suitable application of vacuum after perforation of the flavedo is basic in the enzymatic peeling (Baker and Wicker, 1996; Pretel et al., 1997; Prakash et al., 2001; Pagán et al., 2006; Pretel et al., 2007a, 2007b). This allows the penetration of the enzymatic solution under the fruit skin when entire peeled fruits are required, and also the penetration of the enzymes within the segments to obtain peeled segments. However, for some porous tissues, such as citrus albedo, the sudden excessive pressure change can cause irreversible tissue collapse, preventing the entry of the enzyme solution (McArdle and Culver, 1994). In addition, low vacuum pressures may not be enough to obtain good peeling efficacy (Pretel et al., 1997). In spite of the importance of vacuum pressure for enzymatic peeling of citrus fruits at the industrial level, only a few studies have been carried out. Most authors do not measure the real vacuum pressure applied (Rouhana and Mannheim, 1994; Soffer and Manheim, 1994), while others (Prakash et al., 2001) apply 93 kPa for 0.5–4 min.

Furthermore, the effectiveness of vacuum pressure required for the entrance of the enzymatic solution toward the internal fruit tissues depends on the morphological characteristics of each variety (McArdle and Culver, 1994). In addition, the type of final product required needs to be taken into consideration. Thus, if the aim is to obtain peeled segments, the vacuum pressure–time combination should allow the saturation of the albedo by the enzyme solution, and moreover, its penetration between the segments (Baker and Bruemmer, 1989; Baker and Wicker, 1996). However, if the aim is to obtain a whole peeled fruit for fresh consumption or for canning, the enzymatic solution should not penetrate between the segments, because the residual enzyme could result in alterations such as undesired flavor and the destruction of juice vesicles just before and during the application of preservation techniques applied (Pretel et al., 1997).

Pretel et al. (2007a), to estimate the optimum quantity of enzymatic solution needed for the enzymatic peeling of different varieties of oranges, employed the Potential Enzymatic Saturation of Albedo (PESA), which was considered as the amount of enzymatic solution that the albedo fruit would be able to absorb. To evaluate PESA, fruits without any visible external damage were placed in a water bath at 40°C. The entire surface of fruits was homogeneously perforated. The fruits were placed in a vacuum tank containing 9 L of the Peelzym II solution (1 ml L⁻¹) with Chinese ink (1:10) at 40°C and submitted to different vacuum pressures and times. The PESA was measured using two parameters: percentage of fruit weight increase and percentage of the albedo surface dyed with ink. The authors carried out a study on two different orange varieties (Mollar and Thomson) to show the effect of vacuum application way, continuously or in pulses (during 6 minutes or in three pulses of 2 minutes, respectively) on PESA and the optimal distribution of enzyme between the empty spaces of albedo. As shown in Figure 6.5 (a and b), it was observed that the increment in fruit weight in both varieties was higher when the different vacuum pressures were applied in pulses than when applied in continuous mode, although in the Thomson variety there were no significant differences when the vacuum pressure applied was 67 kPa or higher. Considering the albedo surface dyed with the ink (Figures 6.5c and 6.5d), the application of a vacuum in three pulses of 2 minutes was also more effective than the continuous mode during 6 minutes in both varieties for pressures of 27, 40, and 53 kPa. The results showed that the distribution of enzymatic solution is independent of the method of vacuum applied at pressures higher than 67 kPa. Therefore, according to our results, it seems more adequate to apply low vacuum pressures in pulses than high vacuum pressures in continuum to obtain similar PESA. Adams and Kirk (1991) suggested that alternating pulses of pressure with decompression while the fruit was submerged in a pectinase solution allowed the greatest entry of the solution, possibly as a result of the alternative compression and relaxation of the albedo. In addition, high pressures could produce damage in the tissues and the possibility of liquids entering the vesicles would increase (Baker and Wicker, 1996).

The morphological characteristics of each variety could also affect the ideal vacuum system for enzymatic peeling and even the type of product that can be obtained, whole peeled orange fruits or segments (Pretel et al., 2001). In addition, other factors, such as fruit ripening stage (Kunz, 1995) or peel thickness (Toker and Bayindirli, 2003) affect PESA. According to Figure 6.6, the increase in fresh weight was significantly higher as the absolute value of vacuum pressure increased, indicating a higher PESA. However, an excess of enzyme solution in highly porous

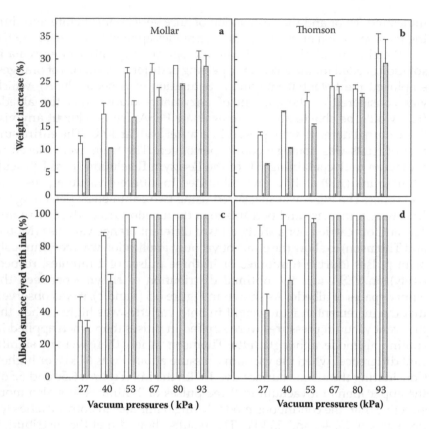

Figure 6.5 Potential enzymatic saturation of the albedo (PESA) of Mollar and Thomson orange varieties: increase in weight (a, b) and estimation of percentage of albedo surface dyed with ink (c, d), applying different vacuum pressures (27 kPa, 40 kPa, 53 kPa, 67 kPa, 80 kPa, and 93 kPa) for six minutes in three pulses (white bars: 2 + 2 + 2 min) and six minutes continuously (light gray bars: 6 min). (Pretel, M.T. et al. 2007. Optimization of vacuum infusion and incubation time for enzymatic peeling of Thomson and Mollar oranges. *LWT-Food Science and Technology* 40: 12–30. With permission.)

tissues like albedo could be a disadvantage for this process, in which the infiltrated material cannot be used again (Baker and Wicker, 1996). The results gained from the PESA studies are presented in Figures 6.5 b and c. To obtain segments from the Mollar variety, the best vacuum conditions are 80 kPa with three vacuum pulses. With these optimum conditions, a dying of the central core was obtained as result of the penetration of the ink solution between the segment membranes. The fruit weight increment of Thomson orange fruits with the application of different pressures and vacuum pulses is presented in Figures 6.6c and d. The obtaining of segments from the Thomson variety is difficult, though there is possible the

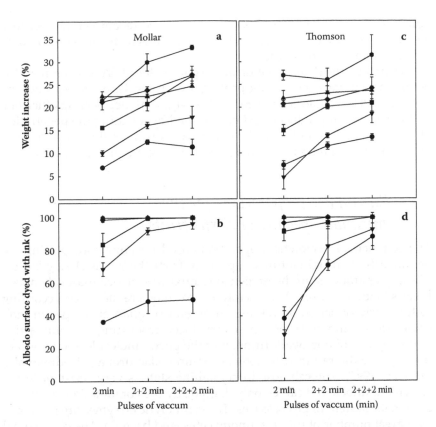

Figure 6.6 Increase in weight of fruits (a) and visual estimation of percentage of albedo surface dyed with ink (b) of fruits of the Mollar variety, and increase in weight of fruits (c) and visual estimation of percentage of albedo surface dyed with ink (d) of fruits of the Thomson variety after applying different vacuum pressures—27 kPa (●), 40 kPa (▼), 53 kPa (■), 67 kPa (♦), 80 kPa (▲), and 93 kPa (●)—for different times: two minutes (2 min), four minutes in two pulses (2 + 2 min), and six minutes in three pulses (2 + 2 + 2 min). (Pretel, M.T. et al. 2007. Optimization of vacuum infusion and incubation time for enzymatic peeling of Thomson and Mollar oranges. *LWT-Food Science and Technology* 40: 12–30. With permission.)

obtaining of entire peeled orange. The best vacuum conditions to obtain entire peeled orange were 53 kPa with two vacuum pulses.

Pretel et al. (2007b) employed this same parameter (PESA) for determining the ideal vacuum conditions for the obtaining of segments of orange Sangrina, and they conclude that to obtain segments from the Sangrina variety, the best vacuum conditions are 67 kPa with two vacuum pulses. With these conditions, the dying of central core and the penetration of the ink solution between the segment membranes were obtained. The use of 80 or 93 kPa would not be recommended because an excess of enzyme solution in tissue as porous as the albedo could be a disadvantage for this process.

Therefore, the determination of the parameter PESA is recommended before the starting of enzymatic peeling of each species or variety since the optimum PESA could be different. This way, a higher peeling efficiency and a better final product could be obtained by avoiding the penetration of the enzymatic solution in the endocarp as well as the appearance of albedo areas that have not been degraded by the enzymatic solution.

6.7   Enzymatic preparations for the enzymatic peeling of citrus fruits

The commercial pectolytic preparations used for enzymatic peeling are obtained from fungi cultures, especially from the genus *Aspergillus* sp. These preparations are heterogeneous mixtures of pectinases, hemicellulases, and cellulases. The pectolytic enzymes are classified according to their way of action on the region of galacturonan of the pectin molecule. Most commercial pectinases produced from fungi are only active in the regions of homogalacturonan of the pectin molecule and they cannot degrade the ramified regions of rhamnogalacturonan (Beldman and Voragen, 1993). The cellulose in crystalline state is very resistant to the enzymatic attack and can be only degrade by the combination of a multi-enzymatic system of glycosidases. The cellulolytic enzymes are produced by a great number of fungi, actinomycetes, and bacteria. The commercial cellulases come from fungi cultures from the genus *Trichoderma* sp. with a high activity on the crystalline cellulose.

Pectinases vary widely with regard to their efficacy for peel removal, with those containing high levels of endo-polygalacturonase activity (as measured by the reduction in viscosity of a standard pectin solution) tending to be the most effective (Baker and Wicker, 1996). However, the difference between the conditions found for optimum degradation of the membrane components and of the commercial pectin indicates that the enzyme has to be selected by studying the process conditions directly with the natural substrate (Ben-Shalom et al., 1986). *In vitro* studies for determining the most adequate preparations for the degradation of the albedo and the carpelar membrane of citrus fruits have been carried out, like the studies of commercial pectolytic enzymes with grapefruit membranes and citrus pectin (Ben-Shalom et al., 1986), study of albedo and carpelar membrane degradation for further application in enzymatic peeling of citrus fruits using *Citrus maxima* 'Cimboa' (Pretel et al., 2005) or a study about the extraction, characterization, and enzymatic degradation of lemon peel pectins (Ros et al., 1996).

Table 6.2 shows some enzymatic preparations for the enzymatic peeling of citrus fruits. It shows the name of the enzymatic preparation with its concentration, the main enzymatic activities in the preparation, the final product obtained, and the peeling time. Bruemmer et al. were the first ones, in 1978, to use enzymatic preparations for obtaining sections of grapefruit by vacuum infusion of commercial pectolytic preparations, and they stated that the efficiency of peeling was directly related to the increase of polygalacturonase activity. The concentration ratios for cellulase activity showed no relationship to effective peeling ratio but the ratios for pectinesterase and polygalacturonase activities showed some similarity.

Berry et al. (1988) assessed the effectiveness of some commercial enzymatic preparations for obtaining grapefruit and orange segments since the success of this process depends, among other factors, on the employed enzyme. These authors found that the most effective enzymes were those with the highest polygalacturonase activities, and Spark L HPG was one of the most adequate ones. Later, Rouhana and Mannheim (1994) also assessed the effectiveness for the enzymatic peeling of grapefruit of different enzymatic preparations provided by NOVO Nordisk ferment AG, Switzerland; Rohm, Germany; Miles, U.S.A., and Grindsted, Denmark. These authors found that the most adequate enzymatic preparations for obtaining segments of grapefruit were Pextinex ULTRA spl with Celluclast 1.5 L or Rohapect D5S with Rohament CT. Using these preparations, under optimal conditions, an incubation time of 35 min is required for complete peeling of grapefruit. Moreover, they found that lower enzyme concentrations increased the process time, while higher enzyme concentration did not decrease the process time significantly. In this work, they stated that both pectolases and cellulases were needed for a successful enzymatic peeling, though the best pectinases were found to be those that contained high concentration of polygalacturonase, pectin-transeliminase, and pectinesterase.

Soffer and Mannheim optimized in 1994 the enzymatic peeling of orange and grapefruit and obtained the best results with the combination of Pectinex Ultra spl and Celluclast 1.5 L. Using these preparations, an incubation time of 20–25 min was required for complete peel removal. Lower cellulose concentrations increased the process time, while higher concentrations caused damage to the fruit appearance. Lower or higher pectolytic enzyme concentrations did not affect peeling time but reduced the appearance of the peeled oranges. Cellulases were probably needed to liberate the pectin from the albedo. This is accomplished by hydrolysis of the polysaccharides, which hold the pectin to the cell wall (Ben-Shalom et al., 1986).

Pretel et al. (1997) optimized the obtaining of segments and whole oranges of the variety Salustiana with the glycohydrolase Rohament

Table 6.2 Some Effective Enzymatic Preparations for the Enzymatic Peeling of Citrus Fruits

Enzymatic Preparation and Concentration	Main Activities	Final Product Obtained	Incubation Time	Reference
Irgazyme A and B	P (+++) PG (++) C (+)	Grapefruit (segments)	Several minutes	Bruemmer et al. (1978)
Spark L HPG	P (+++) PG (++) C (+)	Grapefruit (segments)	30–60 min	Berry et al. (1988)
Pectinex ULTRA spl (2 g kg⁻¹) with Celluclast 1 g kg⁻¹	PG (+++) PE (+++) PME (+++) C (+)	Grapefruit (segments)	35 min	Rouhana and Mannheim (1994)
Rohapect D5S (2 g kg⁻¹) whit Celluclast 1.5 L (1 g kg⁻¹)	PG (+++) PE(+++) PME (+++) C (+)	Grapefruit (segments)	35 min	Rouhana and Mannheim (1994)
Pectinex ULTRA spl (2 g kg⁻¹) with Celluclast 1 g kg⁻¹	PG (+++) PE(+++) PME (+++) C (+)	Orange, Valencia	20–25 min	Soffer and Mannheim (1994)
Rohament PC (10 g L⁻¹)	PG (+++) C (+) PEC (+)	Orange, Salustiana (segments)	50 min	Pretel et al. (1997)
Rohament PC (10 g L⁻¹)	PG (+++) C (+) PEC (+)	Orange, Salustiana (whole fruit)	10 min	Pretel et al. (1998)
Brand A (0.1–5 ml L⁻¹)	PG (+++) PME (+++) C (+++)	Indian grapefruit	12 min	Prakash et al. (2001)
Brand B (1 ml L⁻¹)	PG (+++) PME (+++) C (+++)	Indian grapefruit	12 min	Prakash et al. (2001)
Peelzym II (0.4% v/w)	P (+++) PG (+++) C (+)	Local mandarins (segments)	20–30 min	Liu et al. (2004)
Peelzym II (5 mL/30 g peel)	P (+++) PG (+++) C (+)	Oranges, Navelina	—	Pagán et al. (2006)
Peelzym II (1 ml L⁻¹)	P (+++) PG (+++) C (+)	Orange, Mollar	30 min	Pretel et al. (2007a)
Peelzym II (1 ml L⁻¹)	P (+++) PG (+++) C (+)	Orange, Thomson (whole fruit)	40 min	Pretel et al. (2007a)
Peelzym II (1 ml L⁻¹)	P (+++) PG (+++) C (+)	Segment of orange, Sangrina	30 min	Pretel et al. (2007b)

Note: Pectinase (P), Polygalacturonase (PG), PE (Pectinesterase), PME (Pectinmethylesterase), Cellulase (C), High (+++), Moderate (++), Low (+).

PC (10 g L^{-1}) from Rohm GMBH (Germany) because it was one of the enzymatic preparations employed in the fruit maceration. Later, Pretel et al. (1998a) employed Rohament PC (10 g L^{-1}) for obtaining whole orange fruits Salustiana and their storage as a "ready to eat" product. Then, Prakash et al. (2001) employed enzymes coded *Brand A* and *Brand B*, obtained from a commercial source, for the peeling of Indian grapefruit. Both enzymatic preparations have high polygalacturonase, polymethylgalacturonase, and pectinmethylesterase activity. They only differ in the fact that Brand A does not have hemicellulase and shows a low xylanase activity. Similar results were obtained for both of them and, therefore, they conclude that the constituent enzymes in the peeling enzyme preparations may be acting in unison rather than as single constituents. Also, the similarities in activities of the complexes may be responsible for the similar results obtained.

Pretel et al. (2005) studied the enzymatic activities of four commercial enzymatic preparations provided by the company Novo Nordisk Ferment Ltd.®: Peelzym I, II, and III produced by *Aspergillus niger*, and Peelzym IV produced by *Aspergillus niger* and *Trichoderma reesi*. The Peelzym II showed the highest activity on citrus pectin and polygalacturonic acid but presented low activity cellulase. According to most authors (Bruemmer et al., 1978; Bruemmer, 1981; Berry et al., 1988; Coll, 1996), Peelzym II would be the most appropriate for albedo and carpelar membrane degradation since it presented the highest polygalacturonase activity. Due to the high efficiency showed by the enzymatic preparation, several authors used Peelzym II to determine the optimum conditions for peeling different citrus. Liu et al. (2004) employed this preparation to peel local mandarins; Pretel et al. (2007a), for the enzymatic peeling of orange Thomson and Mollar; Pretel et al. (2007b), for obtaining segments of orange Sangrina; and Pagán et al. (2006), for peeling orange Navelina. An inverse relationship was observed between the enzyme concentration and duration (overall incubation time) needed for complete peel removal. An increase in enzyme concentration showed a decrease in the overall incubation time needed for complete peel removal (Liu et al., 2004). On the other side, Pinnavaia et al. (2006), following the recommendations of previous works (Pretel et al., 1997; Rouhana and Mannheim, 1994), employed two preparations with higher activities of pectinase and cellulase for the enzymatic peeling of Valencia oranges. The two preparations were 0.1% Ultrazym 100G (Novozymes, Dittingen, Switzerland) and 0.1% Rohapect PTE (AB Enzymes, Darmstadt). After a treatment with HCl (0.1 N) and a water washing for removing the rest of the enzyme, they research the storage period at 2°C. Of the two commercial enzymes, Ultrazym provided the minimal juice leakage, softening of slices and microbial contamination stored up to two weeks, despite the infusion process applied to the fruits.

Some pectolytic preparations employed for the enzymatic peeling of citrus fruits (Peelzym I, II, III, and IV) have been also employed for the enzymatic peeling of apricots, nectarines, and peaches (Toker and Bayindirli, 2003), and it has been found that the most adequate preparation, as well as the rest of conditions like pH and temperature, are different for each fruit since the skin differs in the composition of pectin, cellulose and hemicellulose according to the fruit (Toker and Bayindirli, 2003). Although most authors, as we have above mentioned, consider the enzymatic preparation and its concentration as essential factors, Pao and Petracek (1997) obtained peeled oranges Valencia by introducing fruits in a vacuum chamber containing deionized water or citric solution without using enzymatic solution.

On the other side, apart from the composition of the enzymatic preparation, the concentration is also critical to obtain a good peeling effectiveness (Bruemmer, 1981; Baker and Bruemmer, 1989; Soffer and Mannheim, 1994). For instance, Pretel et al. (1997) indicated that to obtain a good peeling efficacy of Salustiana oranges it was necessary to use 10 g L^{-1} of Rohament C, while later works (Pretel et al., 2007b) using a different enzymatic preparation (Peelzym II) demonstrated that 1 ml L^{-1} is the most adequate concentration for obtaining a good peeling efficiency.

On the other hand, and due to the high price of enzymatic solutions, it is possible to reduce the amount of enzymatic solution by increasing incubation time (Prakash et al., 2001), thus reducing production costs (Bruemmer et al., 1978; Pretel et al., 1997; Prakash et al., 2001; Toker and Bayindirli, 2003). To find the optimum incubation time with the enzymatic solution is an important challenge for obtaining a good finished product after the vacuum application (Soffer and Mannheim, 1994; Pretel et al., 1998a; Prakash et al., 2001). An incubation time above the optimum could cause the degradation of the juice vesicles from the surface and the softening of the segments (Pretel et al., 1997), while an incubation time below the optimum one causes the appearance of areas of albedo that have not been degraded by the enzyme linked to the membranes of the segments (Pretel et al., 2007a,b). Table 6.2 shows the most adequate incubation times for obtaining different products from enzymatically peeled citrus fruits. For considering a citrus fruit to have a good peeling quality, the percentage of fruit surface not attacked by the enzymatic solution, the ease of skin removing (albedo + flavedo) and the fruit firmness are usually assessed. If the segment obtaining is intended, the ease of segment separation and the percentage of segments without defects (Pretel et al., 1997; Pretel et al., 2007a,b; Prakash et al., 2001) is also assessed. Table 6.2 shows that the most adequate incubation times range from 10 to 80 minutes depending on the employed enzyme and the final product. Although, in most cases, the incubation time for obtaining citrus fruits segments ranges from 30 to 40 minutes.

6.8 Influence of temperature and pH on enzymatic peeling

The action of enzymes on the degradation of cell walls is specially affected by temperature and pH and, therefore, they directly affect the process of enzymatic peeling.

Pretel et al. (1997) studied the influence of temperature on the enzymatic peeling of oranges between 10 and 60°C, and they observed that at low temperatures a proportional increase in time was needed for the whole fruit to be peeled adequately. For temperatures above 50°C, the orange peel became soft since, while passage of the enzyme solution into the segments was favored, there was a poor distribution through the albedo. The epicuticular wax, which holds the juice sacs of the segments, melts at temperatures above 45°C, causing the segments to become soft and to disintegrate. At temperatures below 20°C, an excessively long incubation time (more than 2.5 h) was needed to allow easy peeling and the separation of the albedo from the segments and the division of individual segments themselves was poor. Additionally, the longer times led to deterioration of the vesicular structure of the fruit. The best peeling results were observed at a temperature between 30 and 45°C. Similar results were obtained by Pagán et al. (2005) in a study about the effect of temperature on enzymatic peeling process of oranges since with temperatures below 30°C or above 50°C no changes in the peeling efficiency were observed. Related to these results, Rouhana and Mannheim (1994) found out that 40°C was the best temperature for enzymatic peeling of grapefruit, since lowering the temperature extended the process time for the peeling and working at higher temperatures caused a decrease in fruit integrity. The epicuticular wax, which holds the juice sacs of the segments, melts at temperatures above 45°C, causing the segments to soften and disintegrate (Soffer and Mannheim, 1994; Rouhana and Mannheim, 1994). However, for the enzymatic peeling of the Salustiana orange, the best temperature was 35°C (Pretel et al., 1997). This range of temperature between 35 and 40°C, besides being within the optimum range of peeling, would become economically profitable for the industry since a minimum addition of energy is required, the experiments thus getting the closest to the industrial peeling requirements. In a study about enzymatic peeling of apricots, nectarines and peaches, Toker and Bayindirli (2003) observed that 20°C was too low to obtain successful enzymatic peeling and that an excessively long incubation time (more than 2 h) was needed to allow peeling. They also observed that long peeling times resulted in decreased quality of the final product and that, at relatively high temperatures, such as 50°C, the texture changes due to fruit softening. On their part, Ben-Shalom et al. (1986) found out that the optimum temperature for degradation of segment membranes by pectinase C-80 was 55°C and also the optimum temperature for degradation of commercial pectin was 50°C, thus concluding that the optimum conditions for the action of the enzymatic preparations depends on the degraded substrate. Most of the authors employ temperatures between 30 and 40°C for enzymatic peeling of citrus fruits, 30°C for grapefruit (Bruemmer et al., 1978), 40°C for Salustiana oranges (Pretel et al., 1998a), 40°C for mandarin (Pretel et al., 1998b), 30±2° C for Indian grapefruit (Prakash et al., 2001), 35°C for Valencia (Pinnavaia et al., 2006), Thomson and Mollar (Pretel et al., 2007a), and Sangrina oranges (Pretel et al., 2007b). In the enzymatic peeling of potatoes, carrots, Swedish turnips, and onions, the employed temperature was also 40°C (Suutarinen et al., 2003).

Other of the more important attributes for the enzymes is their dependence on pH (Ben-Shalom et al., 1986). These authors, when studying the pH dependence of pectinase C-80, proved that the pH ranged between 4 and 5 to get the maximal degradation in the membranes of the segments of grapefruit. Soffer and Mannheim (1994) found out that the optimum pH for pectinase and cellulase action on citrus albedo and membranes was established between 3.5 and 3.8. Rouhana and Mannheim (1994) also established that the optimal pH for the enzymatic digestion lies between 4 and 5 since when the citrus are placed into an enzymatic solution, the pH decreases to 3.5 or 3.8, probably due to the dissolution of the acids. Usually, the enzymatic peeling solution is stabilized with a buffer solution of sodium citrate/citric acid; thus, the incubation time to get a quality product considerably diminishes (Rouhana and Mannheim, 1994; Pretel et al., 1997). However, the optimum pH for the enzymatic peeling is established according to both the level of activity shown by the enzyme preparation and its stability in the operational conditions. For this reason, the stability of the preparation was ascertained by studying the evolution with time of its cellulase and pectinase activities at different pH, using specific substrates (citrus pectin and CM-cellulose, respectively) as standard substrate, finding that in both overall enzymatic activities, the half-life decreased with increased acidity of the medium. Minimum deactivation was obtained for both at pH 4–4.5. Cellulase activity showed greater pH stability, while pectinase activity was reduced to 36% under the same conditions. Since peeling efficiency is established at pH 4, this value was chosen as the best for the peeling process (Pretel et al., 1997). Therefore, it could be concluded that, although the most suitable pH for the enzymatic degradation of the albedo is 3.5, probably for the enzymatic peeling the pH range could be wider, between 3.5 and 4.5. Most works on enzymatic peeling carry out the enzymatic digestion within this temperature range (Ben-Shalom et al., 1986; Liu et al., 2004; Rouhana and Mannheim, 1994; Pretel et al., 2007a; 2007b; Pretel et al., 1998a,b; Pretel et al., 2005; Pretel et al., 2007a,b).

6.9   Reuse of the enzyme preparation in an industrial peeling process

Although this is one of the biggest problems associated with the enzymatic peeling of citrus fruits, very few studies deal with operational stability at the industrial level. Rouhana and Mannheim (1994) simulated the industrial peeling process with three consecutive peelings of grapefruit and they observed a loss of activity polygalacturonase of 70% and a higher incubation time needed for obtaining a high quality product. These authors associated the loss of activity to the accumulation of degradation products, i.e., pectin and cellulytic substances, during the peeling process. In a later study, Pretel et al. (1997) simulated a process by reusing the enzyme solution (Rohament PC) for 8 days in continuous periods of 8 h followed by periods of storage (16 h) in a cold room. Every 8 h of continuous operation, the enzyme solution was microfiltered. These authors analyzed the evolution of the activity pectinase and cellulase according to the operation time, and they found that the enzymatic solution could be employed for 22 peeling cycles without any losses of activity or peeling efficiency. Pagán et al. (2006) also checked that the ultrafiltration process resulted in the renewal of the reducing sugars and a slight loss of enzyme; 11% for polygalacturonase and 14% for cellulase when tested at the same enzymatic concentration as the initial enzyme preparation. Later, Pretel et al. (2007b) tried to take the enzymatic peeling assays to the industrial process by studying the loss of cellulase and pectinase activity after four peeling cycles and during the cold storage of the solution. For this purpose, four peeling processes were carried out with the same enzymatic solution. After that, the solution was kept at 4°C during 10 weeks after a decanting but without the ultrafiltration of the solution to reach the industrial level of the processes and to reduce costs. The enzymatic activity was calculated by determining the fluidity of the carboximethylcellulose and citric pectin solutions before the peeling process, after each one of the four peeling processes and weekly during 74 days. Figure 6.7 shows that fluidity increased with reaction time and the carboximethylcellulase activity was maintained at 100%, with no significant differences after four peeling cycles and even after 14 days under cool conditions (4°C). After 74 days under cool conditions the carboximethylcellulase activity was 75% of the initial activity, and therefore it can be considered that the enzymatic solution keeps a high percentage of activity that would not decrease the peeling efficacy (Pretel et al., 1997). However, as has been mentioned above, some authors (Bruemmer et al., 1978; Berry et al., 1988; Coll, 1996; Pretel et al., 2005) indicate that the pectinase activity is more important than carboximethylcellulase activity for the enzymatic peeling of citrus. Figure 6.8 shows than pectinase activity was reduced by 30% after four peeling cycles and 50% after 74 days of conservation, which was higher than that observed in carboximethylcellulase

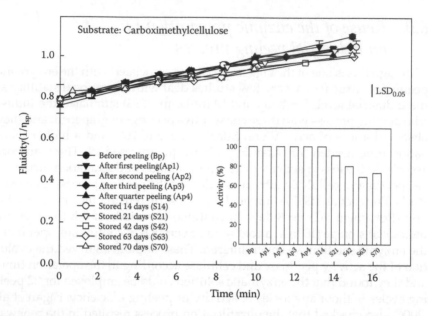

Figure 6.7 Evolution of carboximethylcellulose solution fluidity ($1/\eta_{sp}$) reciprocal of specific viscosity using Peelzym II at 5 g ml^{-1} after reusing this enzyme preparation in continuous four peeling cycles and after different days of storage at 4°C. Inserted relative enzymatic activity of Peelzym II after reusing in continuous four peeling cycles and after different days of storage at 4°C. Carboximethylcellulase activity of Peelzym II: 2172 ± 154 U ml^{-1}. $1/\eta_{sp}$: reciprocal of specific viscosity. (Pretel, M.T. et al. 2007. Obtaining fruit segments from a traditional orange variety (*Citrus sinensis* (L.). Osbeck cv. Sangrina) by enzymatic peeling. *European Food Research and Technology* 225: 783–788. With permission.)

activity. However, the highest loss of pectinase activity occurred in the second peeling process (24.6%), while from the second peeling process to the fourth only a loss of 10% occurred. In addition, after 42 days of storage under cool conditions, pectinase activity decreased only 8%, reaching a 17.7% loss after 74 days of storage. Thus, the storage under cool conditions affected less the pectinase activity than the subsequent peeling processes. These results coincide with those obtained by Soffer and Manheim (1994) in that, after three peeling cycles the residual enzymatic activities diminished by 20%, probably due to the accumulation of degradation products during the peeling process (Kimball, 1999). This problem would be partially solved by microfiltration of the enzyme solution, which has been shown to maintain the residual enzymatic activity (Pretel et al., 1997). An additional problem of the enzymatic peeling process that has been already referred by other authors (Pagán et al., 2006) is the enzymatic solution that remains absorbed in the fruit albedo when vacuum is applied and therefore the enzyme solution is lost when the skin is removed.

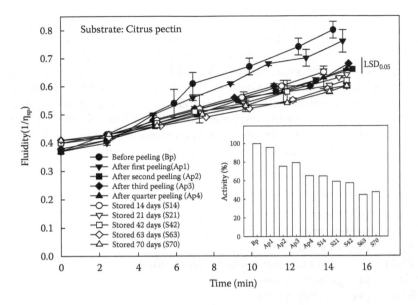

Figure 6.8 Evolution of pectinase solution fluidity ($1/\eta_{sp}$) using Peelzym II at 5 g ml^{-1} after reusing this enzyme preparation in continuous four peeling cycles and after different days of storage at 4°C. Inserted relative enzymatic activity of Peelzym II after reusing in continuous four peeling cycles and after different days of storage at 4°C. Pectinase activity of Peelzym II: 5263 ± 165 U ml^{-1}. $1/\eta_{sp}$: the reciprocal of specific viscosity. $1/\eta_{sp}$: reciprocal of specific viscosity. (Pretel, M.T. et al. 2007. Obtaining fruit segments from a traditional orange variety (*Citrus sinensis* (L.). Osbeck cv. Sangrina) by enzymatic peeling. *European Food Research and Technology* 225: 783–788. With permission.)

6.10   Conclusions

Enzymatic peeling is an alternative method for removing the rind of the fruits with lower water consumption and causing less contamination than traditional processes. Most studies about enzymatic peeling of citrus fruits have been carried out using different varieties of oranges and grapefruits, although other species like mandarin, lemon, and cimboba have been also studied. There are many parameters to take into account to obtain good peeling effectiveness, such as type of enzymatic preparation and its concentration, the design of the cuts of the flavedo, and the vacuum conditions to allow the entrance of enzymatic solution towards the interior of the mesocarp. In addition, the morphologic characteristics of fruits, such as skin adherence and its thickness, and the degree of union between the segments, are also important factors that determine enzymatic peeling efficiency. Finally, the effect of scalding and the incubation time and temperatures on peeled fruit quality may also be important. Here we showed that most of the parameters mentioned above are well studied and the optimum conditions are known in citrus fruits. Therefore, the determination of the parameter PESA (Potential Enzymatic Saturation of Albedo) is recommended before the starting of enzymatic peeling of each species or variety since the optimum PESA could be different. One of the biggest problems associated with the enzymatic peeling of citrus fruits is the recovering of the enzymatic solution after the process and the minimization of the decrease of the enzymatic activity of the solution. However, very few studies deal with operational enzymatic recovery and stability at the industrial level and further studies are required.

Abbreviations

C:	cellulase
P:	pectinase
PE:	pectinesterase
PESA:	potential enzymatic saturation of albedo
PG:	polygalacturonase
PME:	pectinmethylesterase

References

Adams, B. and W. Kirk. 1991. Process for enzyme peeling of fresh citrus fruit. U.S. patent. No 5000967.

Alberts, B., D. Bray, J. Lewis, M. Raff, K. Roberts, and J.D. Watson. 1989. *Molecular biology of the Cell,* 2nd ed., London: Garland.

Baker, R.A. and J.H. Bruemmer. 1989. Quality and stability of enzymatically peeled and sectioned citrus fruit. In: *Quality Factors of Fruits and Vegetables,* American Chemical Society, Ed. J.J. Jen, 140–148. Washington, DC.

Baker, R.A., and L. Wicker. 1996. Current and potencial applications of enzyme infusion in the food industry. *Trends in Food Science and Technology* 7: 279–284.

Ben-Shalom, N., A. Levi, and R. Pinto. 1986. Pectolytic enzyme studies for peeling of grapefruit segment membrane. *Journal of Food Science* 51 (2): 421–423.

Berry, R.E., R.A. Baker and J.H. Bruemmer. 1988. Enzyme separated sections: A new lighly proccessed citrus product. In: *Proceedings of the Sixth International Citrus Congress,* Ed. R. Goren and K. Hendel, 1711–1716. Tel-Aviv, Israel, Philadelphia, PA: Balaban.

Bruemmer, J.H., A.W. Griffin, and O. Onayami. 1978. Sectionizing grapefruit by enzyme digestion. *Proceedings of the Florida State Horticultural Society* 91:112–114.

Bruemmer, J.H. 1981. Method of preparing citrus fruit. Sections with fresh fruit flavor and appereance. US Patent 4.284.651.

Coll, L. 1996. Polisacáridos estructurales y degradación enzimática de la membrana carpelar de mandarina Satsuma (Citrus unshiu Marc.). Pelado enzimático de los segmentos. Doctoral thesis. Universidad de Murcia, Murcia.

Ismail, M.A., H. Chen, E.A. Baldwin, and A. Plotto. 2005. Changes in enzyme-assisted peeling efficiency and quality of fresh Valencia orange and of stored Valencia orange and Ruby red grapefruit. *Proceedings of the Florida State Horticultural Society* 118: 403–405.

Kimball, D.A. 1999. *Citrus Processing: A Complete Guide.* Gaithersburg, MD: Spen.

Liu, F., A. Osman, F.S. Yuso, and H.M. Ghazali. 2004. Effects of enzyme-aided peeling on the quality of local mandarin (*Citrus reticulate* B.) segments. *Journal of Food Processing and Preservation* 28: 336–347.

Loussert, R. 1992. *Los Agrios.* Madrid: Mundiprensa.

McArdle, R.N. and C.A. Culver. 1994. Enzyme infusión: A developing technology. *Food Technology* 11: 85–89.

Pagán, A., A. Ibarz, and J. Pagán. 2005. Kinetics of the digestion products and effect of temperature on the enzymatic peeling process of oranges. *Journal of Food Engineering* 71: 361–365.

Pagán, A., J. Conde, A. Ibarz, and J. Pagán. 2006. Orange peel degradation and enzyme recovery in the enzymatic peeling process. *International Journal of Food Science and Technology* 41: 113–120.

Pao, S. and P.D. Petracek. 1997. Shelf life extension of peeled oranges by citric acid treatment. *Food Microbiology* 14: 485–491.

Pinnavaia, S., E.A. Baldwin, A. Plotto, J. Narciso, and E. Senesi. 2006. Enzyme-peeling of Valencia oranges for fresh-cut slices. *Proceedings of the Florida State Horticultural Society* 119: 335–339.

Prakash, S., R.S. Singhal, and P.R. Kulkarni. 2001. Enzymic peeling of Indian grapefruit (*Citrus paradisi*). *Journal of the Science of Food and Agriculture* 81: 1440–1442.

Pretel, M.T., P. Lozano, F. Riquelme, and F. Romojaro. 1997. Pectic enzymes in fresh fruit processing: optimization of enzymic peeling of oranges. *Process Biochemistry* 32 (1): 43–49.

Pretel, M.T., P.S. Fernández, A. Martínez, and F. Romojaro. 1998a. The effect of modified atmosphere packaging on "ready-to-eat" oranges. *Lebensmittel-Wissenschaft und Technology* 31: 322–328.

Pretel, M.T., P.S. Fernández, A. Martínez, and F. Romojaro. 1998b. Modelling design of cuts for enzymatic peeling of mandarin and optimization of different parameters of the process. *Z Lebensmittel Unterssuchung und Forschung* 207: 322–327.

Pretel, M.T., F. Romojaro, M. Serrano, A. Amorós, M.A. Botella, and C. Obón. 2001. New commercial uses for traditional varieties of citrus fruits of the Spanish southeast. *Levante agrícola* 357: 320–325.

Pretel, M.T., P. Sanchez-Bel, I. Egea, and F. Romojaro. 2008. Enzymatic peeling of citrus fruits: factors affecting degradation of the albedo. In: *Tree and Forestry Science and Biotechnology*, Ed. Teixeira da Silva, J.A. Islework, UK: Global Science Books.

Pretel, M.T., A. Amorós, M.A. Botella, M. Serrano and F. Romojaro. 2005. Study of albedo and carpelar membrane degradation for further application in enzymatic peeling of citrus fruits. *Journal of the Science of Food and Agriculture* 84: 86–90.

Pretel, M.T., M.A. Botella, A. Amorós, P.J. Zapata, and M. Serrano. 2007a. Optimization of vacuum infusion and incubation time for enzymatic peeling of Thomson and Mollar oranges. *LWT-Food Science and Technology* 40: 12–30.

Pretel, M.T., M.A. Botella, A. Amorós, M. Serrano, I. Egea, and F. Romojaro. 2007b. Obtaining fruit segments from a traditional orange variety (*Citrus sinensis* (L.). Osbeck cv. Sangrina) by enzymatic peeling. *European Food Research and Technology* 225: 783–788.

Roe, B. and J.H. Bruemmer. 1976. New grapefruit product: Debitterizing albedo. *Proceedings of the Florida State Horticultural Society* 89: 191–194.

Ros, J.M., A. Henk, H.A. Schols, and G.J. Voragen. 1996. Extraction, characterisation, and enzymatic degradation of lemon peel pectins. *Carbohydrate Research* 282: 271–284.

Rouhana, A and C.H. Mannheim. 1994. Optimization of enzymatic peeling of grapefruit. *Lebensmittel Wissenschaft und Technology* 27: 103–107.

Soffer, T., and C.H. Mannheim. 1994. Optimization of enzymatic peeling of oranges and pomelo. *Lebensmittel Wissenschaft und Technology* 27: 245–248.

Suutarinen, M., A. Mustranta, K. Autio, M. Salmenkallio-Marttila, R. Ahvenainen, and J. Buchert. 2003. The potential of enzymatic peeling of vegetables. *Journal of the Science of Food and Agriculture* 83, 1556–1564.

Toker, I. and A. Bayındırlı. 2003 Enzymatic peeling of apricots, nectarines and peaches. *Lebensmittel-Wissenschaft und Technologie* 36 (2): 215–221.

Whitaker, J.R. 1984 Pectic substances, pectic enzymes and haze formation in fruit juices. *Enzyme and Microbial Technology* 6: 341–349.

chapter seven

Use of enzymes for non-citrus fruit juice production

Liliana N. Ceci and Jorge E. Lozano

Contents

7.1   Introduction

Commercialized non-citrus fruit juices include, among others, apple, pear, grape, peach, melon, and berry juices, ready to drink, frozen or concentrated. By far the most important non-citrus fruit commodities are apples, which are mostly processed in the form of juices. The fruit juice industry is a very important enzyme consumer. Commercial sources of fungal pectic enzymes have been used in fruit juice processing since the 1930s for clarifying fruit juices and disintegrating plant pulps to increase juice yields. Industrially used enzymes are similar to the naturally occurring pectinases, cellulases, and hemicellulases found in fruit during ripening. Most enzymes are marketed on the basis that they are generally recognized as safe (GRAS) for their intended use in the juice process.

In this chapter, after reviewing the non-citrus fruit juice processing, focus is put on the use of enzymes in this particular industry. Maceration and clarification of apple juice through mixture of enzymes (pectinases, cellulases, etc.) and the application of other specific enzymes in juice production, as amylases, are extensively covered. Particular items as enzyme activity determination and pH dependence are treated in some detail. Finally, the use of immobilized enzymes and other miscellaneous applications are also summarized.

7.1.1   Non-citrus fruit juice production: World market

Fruit and vegetable juice world trade averaged near US$4,000 million last decade (FAOSTAT, 2005). By far the most important non-citrus fruit commodities are apples, which are mostly processed in the form of juices. In 2006, apple world production was 1.2 million tons. There are, however, many other products elaborated from fruits, as canned, dried, and frozen fruit, pulps, purees, and marmalades. In addition, developments in aseptic processing have brought new dimension and markets to the juice industry. In the period 1997–2006 the world production of concentrated apple juice increased 50%, following a growing trend. This increase is explained by the extraordinary progress of China, which elaborates half of the world total and grows at an average annual rate of 30%, although with a more moderate trend at present. On the other hand, big consumers such as the United States present a decreasing production trend, with an average annual decrease of 2%. The lower price of imported juice, among other factors, explains this trend. Global apple juice trade was expected to have another record year in 2007–2008. In 2007, the value of juice imports grew to $1.5 billion, up from $1 billion in 2006 (USDA, 2008).

It is well known that fungal pectic enzymes have been used in commercial applications for fruit juice processing since the 1930s, such as clarification and disintegration of plant pulps to increase juice yields. Commercial enzymes are similar to the naturally occurring pectinases, cellulases, and hemicellulases found in fruit during ripening. Most enzymes are marketed on the basis that they are GRAS for their intended use in the juice process.

Enzymes are used in the juice industry to aid in the separation of juice from the fruit cells and to clarify the juice by the removal of pectin and naturally occurring starches that contribute to undesired viscosity, poor filtration, and a cloudy appearance. A considerable number of food enzymes are on the market.

7.2   Non-citrus fruit juice processing

Fruits are mainly water (75–90%), most located in vacuole causing turgor to the fruit tissue, and fruit juice is prepared by mechanically squeezing or macerating fresh fruits without the application of heat or solvents. In many countries, the term *fruit juice* can only legally be used to describe a product which is 100% fruit juice (FDA, 2001; Food England, 2003). Fruit cell wall consists of crystalline cellulose microfibrils embedded in an amorphous matrix of pectin and hemicelluloses. The definition of a mature fruit varies with each type. Typically, sugar and organic acid levels, and their ratio indicate maturity stage. Juice is composed of water; soluble solids (sugars and organic acids); aroma and flavor compounds; vitamins and minerals; pectic substances; pigments; and, to a very small degree, proteins and fats. During ripening, fruits decrease in acidity and starch and increase in sugars. Juices are products for direct consumption and are obtained by the extraction of cellular juice from fruit; this operation can be done by pressing or by diffusion. Fruit juices were categorized in juices without pulp (*clarified* or *cloudy*); and juices with pulp (*pulps*, *purees*, and *nectars*). Juices obtained by removal of a major part of their water content by vacuum evaporation or fractional freezing will be defined as *concentrated juices*.

Non-citrus fruit processing plants can vary from a simple facility for single juice extraction and canning, to a complex manufacturing facility including ultrafiltration (UF) and reverse osmosis equipments, cold storage and waste treatment plant. A simplified characteristic flow diagram of a non-citrus juice processing line is shown in Figure 7.1. Processed products can be either as single-strength or bulk concentrate, both in clarified or cloudy juice. Basic steps in the production of fruit juices can be divided into four principal stages: Front-end operations; juice extraction; juice clarification and fining; juice pasteurization and concentration.

7.2.1   Extraction methods

The way to extract juice depends on the fruit variety. In general there are few problems in reducing the size of pome fruits, as apples or pears. As Figure 7.1 shows, after washing pome fruits are milled. For the disintegration fixed positioned or rotating grinding knives may be used. Fruit mills

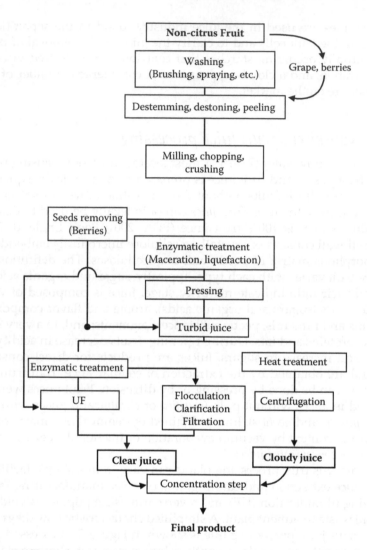

Figure 7.1 Typical non-citrus fruit juices (clear or cloudy) processing line steps. (Lozano, J.E. 2006. Fruit manufacturing: Scientific basis, engineering properties and deteriorative reactions of technological importance. In *Food Engineering Series*; Springer, USA, pp. 1–72. With permission.)

generally used are rotating disc mills, rasp or grater mill, and fixed blade hammer mills, in which the rotor with fixed blades rotates within a perforated screen. Milled particles should be about the same size and bigger enough to facilitate pressing. Grater mills are found more efficient with firm fruit while hammer mills are more efficient than graters for mature or softer fruit, provided speed is properly adjusted (Lozano, 2006).

Apple juice produced with enzymes typically follows one of two primary methods of extraction; traditional pressing or decanter extraction (Nagy et al., 1993). In the traditional pressing operation, whole apples are milled and treated with enzymes prior to pressing to loosen cell walls and promote free run juice. In a number of pressing operations, pressing aids such as rice hulls are mixed with the apple mash prior to pressing. Most systems for extracting juices from apple and similar fruit pulps use some method of pressing juice through cloths of various thicknesses, in which pomace is retained. These filter presses include (Lozano, 2003) rack and cloth presses; horizontal pack press; and continuous belt press. In a *rack and cloth press* the milled fruit pulp is placed in a cloth (plastic fabric) forming a "cheese" of pulp separated by racks made of hardwood or plastic, which are stacked up to 1 m or more in height, and pressed hydraulically. After pressing, the juice is transferred to clarification tanks where additional enzyme is added to the juice to degrade pectin and hydrolyze starch, if required, prior to filtration. The enzymatic process will be later described in detail.

Other extraction techniques include centrifugation, diffusion extraction, and ultrafiltration. First method includes horizontal decanters, which are nowadays especially used for juice clarification. Diffusion extraction was adapted from the sugar industry. Extraction is a typical counter current type process. The diffusion extraction process is influenced by a number of variables, including temperature, thickness, water, and fruit variety. Slices from extractors pass through a conventional press system and the very dilute juice may be returned to the extractors. It comes out that the extra juice yield from diffusion extraction compensates for the extra energy cost involved for concentration. Membrane clarification will be specifically considered further on.

Using decanter extraction pectinase preparations, and sometimes cellulase and hemicellulase preparations, are used to further reduce viscosity. Because the mash is heated and stirred in the presence of enzyme prior to centrifugation, some additional juice is recovered that may not be extracted during traditional pressing operation. The decanter extraction process is sometimes described as liquefaction, or whole fruit liquefaction. However, the whole fruit is not completely liquefied.

7.2.2   Clarification and fining

As previously described, the conventional route to concentrate is to strip aroma, then depectinize juice with enzymes; centrifuge to remove heavy sediment; and filter through pressure precoat filters and polish filters. The juice is then usually concentrated through multi stage vacuum evaporator. This process involves a slight concentration of juice during the stripping step in which up to 10% volume is removed to eliminate methanol released during depectinization with pectin methyl esterase. However, viscosity increased with concentration, which may slow flocculation and filtration. If a cloudy product is required, the juice is pasteurized immediately after pressing to denature any residual enzymes. Centrifugation then removes large pieces of debris, leaving most of the small particles in suspension.

Fruit juices, both clarified and opalescent, may be concentrated to 4 folds ($\cong 50°$Brix) with little problem with respect to natural pectin gelling. At this point in the concentration process there is also little detectable heat damage. This concentrate may be canned and frozen. For a clear juice these suspended particles have to be removed (McLellan, 1996). Soluble pectin remains in the juice, making it too viscous to filter quickly and enzymatic depectinization is required. Depectinization degrades the viscous soluble pectins and promotes the aggregation of cloud particles. As soluble pectin forms a protective coat, negatively charged, around proteins in suspension, causing particles to repel one another. The effect of pectinolytic enzymes is to degrade pectin and expose part of the positively charged protein beneath. The electrostatic repulsion between cloud particles is thereby reduced so that they clump together. These larger particles will eventually settle out. However, to improve the process flocculating or fining agents can be added.

Fining agents (Table 7.1) work either by sticking to particles making them heavy enough to sink or by using charged ions to cause particles to stick to each other making them settle to the bottom.

Although this conventional clarification was a widely used practice in the clarified juice industry, this technology has been practically replaced by mechanical processes such as ultrafiltration and centrifugal decanters. Conventional clarification left a transparent but by no means clear juice, and further centrifugation and/or filtration is required to obtain the clear juice that many consumers prefer. Yeasts and other microorganisms may also be precipitated also by fining.

Another potential contributor to the haziness of juice, when unripe fruits are processed, is starch. In the case of apples, starch may account as much as 15% in weight. While centrifugation can remove most of the starch, about 5% usually remains. Common practice is to hydrolyze starch with amylases (amyloglucosidases) active at the pH of apple juice, added at the same time as the pectinases.

Most apple juice is concentrated before storage by evaporating up to 75% of the water making both depectinization and destarching essential to avoid gelling and turbidity haze formation during concentration. Increased haze formation occurs when fining with gelatin and bentonite is omitted. Optimization of fining and ultrafiltration steps can help retarding or preventing post-bottling haze development.

Table 7.1 Fruit Juice Clarification Agents

Name	Description
Sparkolloid	A natural albuminous protein extracted from kelp and sold as a very fine powder.
Gelatin	Mixture of gelatins and silicon dioxide, with the active ingredient being animal collagen.
Kieselsol	A liquid in which small silica particles have been suspended. It is usually used in tandem with gelatins. The dosage is 1 milliliter per gram of gelatins. This fining agent aids in pulling proteins out of suspension.
Bentonite	Sold as a powder and as coarse granules. It is refined clay. A better way is to add the same amount to a liter of hot water, stir well, and let stand for 36 to 48 hours. In this time the clay will swell and become almost gelatins like.
Isinglass	Produced from sturgeon swim bladders, isinglass is sold either as a fine white powder or as dry hard fragments. It is protein, extracted from the bladders of these fishes. This product is also available as a prepared liquid called "super-clear."
Filters or Polishers	With a fine porosity pad, filters are very effective at removing particles (yeast cells, proteins, etc.).
Pectic Enzymes	Almost all fruits contain pectin, some more than others. When pectin enzymes are added, it eliminates pectin haze. There is no other way to prevent this condition, and if it is in a juice, the haze will never clear on its own

Source: Lozano, J.E. 2006. Fruit manufacturing: Scientific basis, engineering properties and deteriorative reactions of technological importance. In *Food Engineering Series*; Springer, USA, pp. 1–72. With permission.

7.3 Enzymes in the non-citrus fruit industry

Commercial pectic enzymes (pectinases) and other enzymes are an important part of fruit juice technology practically from the beginning of the industrial processing of this product. Technical enzyme products have been used in the process of making fruit juices since the 1930s (Grampp, 1976). They are used to assist the extraction and clarification of juices from many fruits, including berries, stone fruits, grapes, apples, and pears among others. When clarification is not required, as in the case of cloudy juices, enzymes are still applied to enhance extraction or perform other modifications. Commercial pectinase preparations used in fruit processing generally contains a mixture of pectinesterase (PE), polygalacturonase (PG), and pectinlyase (PL) enzymes (Dietrich et al., 1991). The methods employed are basically the same for many fruits (Rombouts and Pilnik, 1978). Table 7.2 lists the main application of pectolytic enzymes.

Table 7.2 Application of Pectolytic Enzymes to Fruit Processing

Enzymatic process	Examples of application
Clarification of fruit juices	Apple juice; depectinized juices can also be concentrated without gelling and developing turbidity
Enzyme treatment of pulp	Soft fruit, red grapes, citrus, and apples, for better release of juice (and colored material); enzyme treatment of pulp of olives, palm fruit and coconut flesh to increase oil yield
Maceration of fruits and disintegration by cell separation	Used to obtain nectar bases and baby foods
Liquefaction of fruit	Used to obtain products with increased soluble solids content (pectinases and cellulases combined)

7.3.1 Mash enzymatic treatment

Although pectinases were originally used for clarifying and depectiniz-ing juices, at the start of the 1970s, they were used for mash enzymatic treatment of apples. The pectolytic enzymes that are normally used fea-ture polygalacturonases as their main activity. In addition, they can also include amylase, cellulase, and protease activities. The combined use of macerating enzymes (pectinases and cellulases) significantly increases the sugar content in the juice and the yield due to the complete hydrolysis of polysaccharide macromolecules. For this reason, the process of mash liquefaction is legally permitted only in certain countries. Pectolytic mash enzymes primarily hydrolyze the somewhat less esterified pectins of the membranes. Mash enzymatic treatment is used for accelerating the juicing process increasing yields. In particular, ripe apples, which are usually very soft, are very difficult to press economically without the use of enzymes.

Pectinases are often added to fruit pulp, after crushing in a mill. Better results are achieved, however, if the pulp is first stirred in a holding tank for 15–20 minutes so that enzyme inhibitors (polyphenols) are oxidized by the native polyphenol oxidase present in fruits. The pulp is then heated to an appropriate temperature before enzymatic treatment. For apples 30°C is the optimal temperature, whereas stone fruits and berries generally require higher temperatures (approx. 50°C).

Equipment for mash enzymatic treatment includes a unit for control-ling the temperature of the mash, an enzyme dispenser unit, and a tank for the enzymatic reactions. This enzymatic maceration must be done from 15 minutes to 2 hours previous to juice extraction by pressing, depending upon the exact nature of the enzyme and how much is used, the reaction temperature and the variety of fruit. Some varieties such as **Golden Delicious** apples are notoriously difficult to break down. Pectinases also degrade soluble pectin in the pulp, reducing juice viscosity, which facilitates juice extraction. Enzymatic treatment is particularly effective with mature apples and those from cold storage. In the apple juice industry, the juice yield is only 75% without mashing enzymes, rising to the range of 91–96% with first mash enzymes. Moreover, a second enzymatic mashing can increase yields almost to a 99%.

7.3.2 Other enzymes in non-citrus juice production

During the last decades, cellulases, arabanase, and glucose oxidase became commercially available. The addition of cellulases during extraction at 50°C improves the release of color compounds from the skins of fruit. This cellulase effect results particularly useful during the processing of blackcurrants and red grapes. Increasingly cellulases are being used at the time of the initial pectinase addition to totally liquefy fruit tissues.

The polysaccharide araban, a polymer of the pentose arabinose, was found as a component of post-concentration haze in fruit juice. Although commercial pectinase preparations often contain arabanase, fruits with arabans usually require additional arabanase to avoid haze problems.

Glucose oxidase catalyses the breakdown of glucose to produce gluconic acid and hydrogen peroxide. Glucose oxidase, usually coupled with catalase to remove the hydrogen peroxide, is therefore used to remove the oxygen from the headspace above bottled fruit juice drinks, reducing deteriorative oxidation reactions.

7.4 Commercial enzymes activity determination

In general, fruit juice processors are lacking of reliable methods for checking the activity of the different commercial enzymes in use. Complete pectin breakdown during clarification, can only be ensured if all the three types of pectinolytic enzymes (PG, PE, PL) are present in the correct proportion. Moreover, the successful application of a pectinase product also depends on the substrate where they act, and the standardization of a fruit substrate complicates the evaluation of pectinolytic activities. For example, different varieties of apples will give substrates with different acidity, pH, and content of inhibitors or promoters of the enzymatic activity.

7.4.1 Temperature dependence on the pectic enzyme activities

Figure 7.2 shows the residual polygalacturonase (PG) activity of two commercial enzymes after 30 min of heating at different temperatures. It was clearly demonstrated that enzyme started to become inactivated at temperatures higher than 50°C, which is a very well-defined breaking point were the

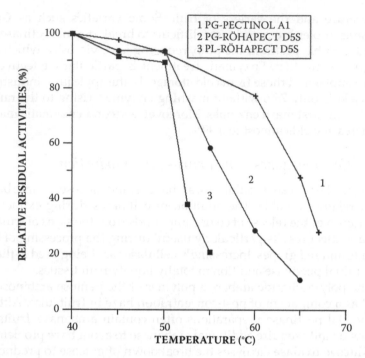

Figure 7.2 Enzymatic residual activities after thermal treatment 30 min at different temperatures of enzyme solutions in 0.1 M citrate/0.2 M phosphate buffers at optimum pH). (Ceci, L. and J.E. Lozano. 1998. Determination of enzymatic activities of commercial pectinases. *Food Chemistry* 31(1/2), 237–241. With permission.)

enzyme rapidly lose its activity. Moreover, the thermal inactivation kinetics for PG has shown a first period characterized as a thermo-labile fraction, and a second period defined as the thermo-resistant fraction of the enzyme (Ceci and Lozano, 1998). Sakai et al. (1993) and Liu and Luh (1978) reported that the optimal temperature for PG activity was in the range 30°–50°C.

Both authors also indicated that inactivation was notable for temperature greater than 50°C after a short period of heating. Figure 7.2 shows the inactivation curve of lyase activity (PL) which also shown a breaking point at approximately 50°C. Inactivation kinetics was mono-phase for PL unlike PG. Liu and Luh (1978) found that commercial enzymes were more heat tolerant than purified fractions, and attributed this phenomenon to the heat protective action of impurities.

7.4.2 pH dependence on the pectic enzymes activities

Ceci and Lozano (1998) studied the effect of pH on commercial pectinase enzymes. Figure 7.3 shows the behavior of PG and PE activities

RÖHAPECT D5S

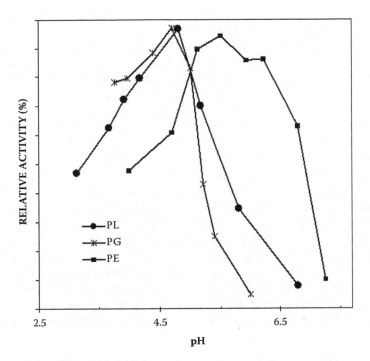

Figure 7.3 Effects of pH on the enzymatic activities of Röhapect D5S. (Ceci, L. and J.E. Lozano. 1998. Determination of enzymatic activities of commercial pectinases. *Food Chemistry* 31(1/2), 237–241. With permission.)

of Röhapect D5S enzyme vs. pH. While in the case of PG activity the optimum pH resulted in approximately 4.6; it was difficult to identify a single optimal value for lyase activity. Instead an optimal range of pH 5–6 was defined.

Pectinolytic enzymes (PG and PE) show a rapid decrease in activity at about pH = 5 and become practically inactivated near neutrality. This problem becomes irrelevant because pH values of fruit juices are in general lower than 4. Therefore, as much as 40% of PG and PE inactivation can be expected during the enzymatic clarification of fruit juices. It was found, however that a broadening of the optimal activities range (Ates and Pekyardimci, 1995) can be obtained after enzyme immobilization on appropriate supports. When the fruit juice clarification is performed at relatively high temperatures (45°–50°C), careful must be taken to avoid excessive inactivation when lyase activity is considered important, because PL is more sensible to thermal inactivation.

7.4.3   *Enzymatic hydrolysis of starch in fruit juices*

Polymeric carbohydrates like starch and arabans may cause difficult filtration and post-process cloudiness. In the case of a positive starch test, the following problems may occur: slow filtration, membrane fouling, gelling after concentration and post concentration haze. Apple juice is one of the juices that can contain considerable amounts of starch, particularly at the beginning of the season. Starch content of apple juice may be high in years when there were relatively low temperatures during the growing season. As the apple mature on the tree, the starch hydrolyzes into sugars, but unripe apples may contain as much as 15% starch (Reed, 1975).

Starch must be degraded by adding starch splitting enzymes, together with the pectinase during depectinization of the juice. Starch is generally insoluble in water at room temperature. Because of this, starch in nature is stored in cells as small granules. Once in the juice, starch granules must be necessarily gelatinized to allow enzymatic hydrolysis. When an aqueous suspension of starch is heated the hydrogen bonds weaken, water is absorbed, and the starch granules swell, and form a gel (Zobel, 1984).

Besides the generalized application of commercial amylase enzymes in the juice industry, there is a lack of information on characteristic and extent of gelatinization of apple starch during juice pasteurization. Starch granules (Figure 7.4) are quite resistant to penetration by both water and

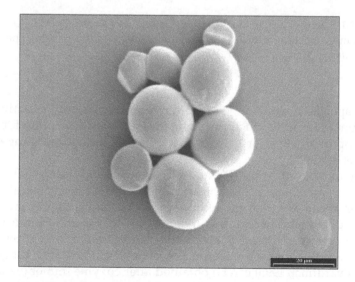

Figure 7.4 Scanning electron photomicrograph of an isolated apple starch granule (5 kV × 4,400). (Carrín, M. E., L. Ceci, and J.E. Lozano. 2004. Characterization of starch in apples and its degradation with amylases. *Food Chemistry* 87, 173–178. With permission.)

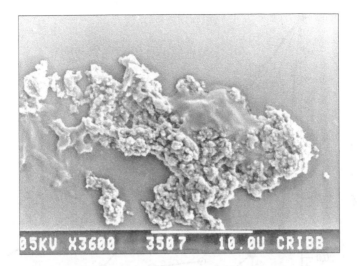

Figure 7.5 Scanning electron photomicrograph of cloudiness precipitated from a pasteurized (5 min at 90° ± 1°C) apple juice (5 kV × 3,600). (Carrín, M. E., L. Ceci, and J.E. Lozano. 2004. Characterization of starch in apples and its degradation with amylases. *Food Chemistry* 87, 173–178. With permission.)

hydrolytic enzymes due to the formation of hydrogen bonds within the same molecule and with other neighboring molecules. When starch granule has not been broken down completely, short-chained dextrin is left. This can lead to a condition known as retrograding. When starch retrogrades, the short-chained dextrins re-crystallize into a form that is no longer susceptible to enzyme attack, regardless of heating.

Figure 7.5 shows a SEM photomicrograph of haze sediment obtained from a nonpasteurized apple juice sample. This scanning electron micrograph (Figure 7.4) shows how apple starch granules collapsed after heat treatment and only gel-like starch fragments dispersed among the other components of turbidity (pectin, cellular wall, etc.) may be observed. Similar behavior was found when wheat starch was gelatinized by heat in excess water (Lineback and Wongsrikasen, 1980).

Carrín et al. (2004) assayed two commercial amylases (solid Rohalase HT, AB Enzymes GmbH, Darmstadt, Germany) and liquid Tyazyme L300 (Solvay Enzimas, Argentina) to degrade starch in pasteurized apple juice. Figure 7.6 shows the kinetics of starch extinction at 50°C. Iodine test resulted negative after no more than 30 min of treatment, and depended on enzyme concentration.

Authors concluded that juices from unripe apples had a high content of soluble and insoluble starch and during industrial pasteurization a practically complete breaking down of the starch granule resulted. In this way, enzymatic starch degradation during apple juice clarification may be

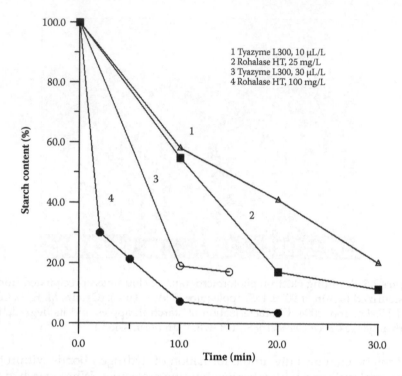

Figure 7.6 Kinetic for starch extinction in pasteurized apple juice treated with different amylase doses at 50°C. (Carrín, M. E., L. Ceci, and J.E. Lozano. 2004. Characterization of starch in apples and its degradation with amylases. *Food Chemistry* 87, 173–178. With permission.)

easily achieved. They also demonstrated that commercial amylases used were highly effective, even in doses lower than those industrially recommended, when they act on gelatinized starch.

7.4.4   *pH and temperature dependences on amylases activities*

The α-amylase activity in Tyazyme L300, measured by an iodometric method, was maximum in pH = 3.4 (Ceci and Lozano, 2002). A noticeable inactivation was observed at temperatures higher than 60°C during 30 min. Two enzymatic fractions (thermo-labile and thermoresistant) were also observed. The enzyme was more susceptible to heat and pH changes when Ca^{2+} sequestering agents, such as citrate ions usually in fruit juices, were present in the reaction medium. Concentrations of Ca^{2+} ions similar to that observed in fortified apple juice, in a reaction media containing citrate ions induced a significant enzyme reactivation. Special attention may be paid to efficiently inactivate the amylases in clarified juices, if they are planned to be fortified with calcium.

7.5 Miscellaneous applications of enzymes in the non-citrus fruit juice industry

7.5.1 Immobilized enzymes

It is well known that immobilization of enzymes offers several advantages that include reuse of the enzyme and its easier separation from the product. Since in food industry it is preferably to avoid the presence of extraneous compounds in the final products the possibility to remove the enzymes is significant advantage. In the literature there is data about immobilization of pectinolytic enzymes on different supports by various methods. Until now, pectinase has been immobilized on various supports including nylon (Lozano et al. 1987), ion exchange resin (Kminkova and Kucera 1983), chitin (Iwasaki et al. 1998) and sodium alginate (Li et al., 2007) among others. Diano et al. (2008) studied the catalytic behavior of a mixture of pectic enzymes, covalently immobilized on different supports (glass microspheres, nylon 6/6 pellets, and PAN beads), in a pectin aqueous solution that simulates apple juice. Commercial pectinase has been immobilized on magnetic duolite-polystyrene particles and assayed for pectin degradation in both batch and magnetic fluidized bed reactors (Demirel and Mutlu, 2005). The application of pectinase immobilized on exchange resin particles has been evaluated for enzymatic mash treatment in carrots to increase the yield of juice (Demirel et al., 2001). Other supports can be used to immobilize pectinases such as acrylic supports (Maxim et al., 1992) and glass functionalized with epoxy and amino-propyl groups (Stratilová, 1995). Endo-pectinlyase isolated from a commercial mixture has been immobilized on a polymer which reversibly soluble-insoluble depending of the pH of the medium (Dinnella et al., 1995). In this way clarified juices free of methanol could be obtained, because the isolated enzyme does not produce methanol, unlike pectinmethylesterase. In spite of the development of an increasing number of works on immobilized enzymes, their application at industrial scale for fruit juices processing not yet has spread.

7.5.2 Application of immobilized enzymes in fruit juice ultrafiltration

While coarse filtration devices are effective in separating particles down to about 20 microns, membrane technology involves the separation of particles below this range extending down to dissolved solutes that are as small as several Angstroms.

Membranes are manufactured with a wide variety of materials including sintered metals, ceramics, and polymers (Zeman and Zydney, 1996). Polymeric UF membranes remove organic compounds in the range of 0.002–0.2 µm and are the most popular cross-flow filtration method in the fruit industry. In cross-flow UF two fluid streams are generated: the ultra-filtered solids free juice (permeate), and the retentate with variable content of insoluble solids which, in the case of apple juice, are mainly remains of cellular walls and pectin. A positive side effect of this method is that rejected particles are continuously carried away from the membrane surface, thereby minimizing contaminant buildup, leaving it free to reject incoming material and to allow free flow of purified liquid. Although membrane cleaning is periodically required, the self-cleaning nature of cross-flow filtration length-ens membrane life enough to make it economically attractive. Membranes are assembled as modules that are easily integrated into systems contain-ing hydraulic components. The module allows to accommodate large filtration areas in a small volume and to resist the pressures required in fil-tration. Tubular, hollow fiber, spiral, and flat plate are the common modules (Cheryan, 1986). Over time, the physical backwash will not remove some membrane fouling. Most membrane systems allow the feed pressure to gradually increase over time to around 30 psi and then perform a cleaning-in-place (CIP) procedure. CIP frequency might vary from around 10 days to several months. Another approach is to use a chemically enhanced back-wash, where on a frequent basis (typically every 1–14 days), chemicals are injected with the backwash water to clean the membrane and maintain sys-tem performance at low pressure without going off-line for a CIP.

The application of UF as an alternative to conventional processes for clarification of apple juice was clearly demonstrated (Heatherbell et al., 1977; Wu et al., 1990). However, the acceptance of UF in the fruit process-ing industry is not yet complete, because there are problems with the operation and fouling of membranes.

Permeate flux (J) results from the difference between a convective flux from the bulk of the juice to the membrane and a counter diffusive flux or outflow by which solute is transferred back into the bulk of the fluid. The value of J is strongly dependent on hydrodynamic conditions, membrane properties and the operating parameters. The main driving force of UF is the transmembrane pressure (ΔPTM). In practice, the J values obtained with apple juice are much less than those obtained with water only. This phe-nomenon is attributable to various causes, including resistance of gel layer, concentration polarization boundary layer (defined as a localized increase in concentration of rejected solutes at the membrane surface due to convec-tive transport of solutes (Constenla and Lozano, 1996) and plugging of pores due to fouling. Some of these phenomena are reversible and disappear after cleaning of the UF membranes and others are definitively irreversible.

As previously indicated, pectin and other large solutes like starch, normally found when unripe apples are processed, tend to form a fairly viscous and gelatinous-type layer on the "skin" of the asymmetric fiber. Flux (J) decline, due to this phenomenon, can be reduced by increasing flow velocity on the membrane.

Carrín et al. (2000) studied the physical immobilization of commercial pectinases on hollow fiber ultrafiltration (HFUF) membranes in view of its possible application in fruit juice clarification. Figure 7.7 shows a TEM micrography of a clean fiber (a), after enzyme immobilization (b), and after pectin filtration (c). As Figure 7.7b shows, a primary adsorption layer of enzyme on the membrane skin was produced after immobilization, which originates an additional resistance (R_e) to permeate flux.

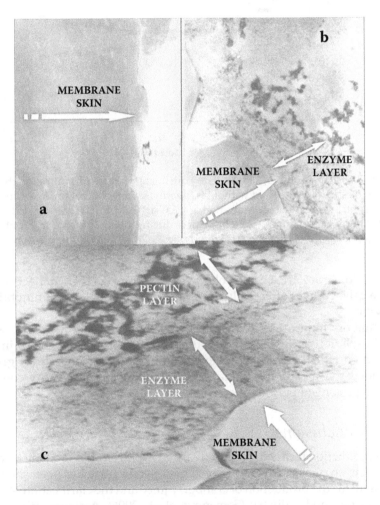

Figure 7.7 TEM micrography of (a) a clean HFUF membrane; (b) after physical immobilization of a commercial pectinase; and (c) after the ultrafiltration of a pectin solution. (Carrín, M. E., L. Ceci, and J.E. Lozano. 2000. Effects of pectinase immobilization during hollow fiber ultrafiltration of apple juice. *Journal of Food Process Engineering* 23: 281–298. With permission.

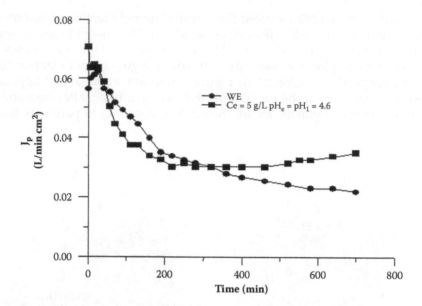

Figure 7.8 Permeate flux (J_p; Lmin⁻1cm⁻²) during apple juice ultrafiltration through membranes with (WE-□) and without (●) pectinase immobilization, in a pilot plant equipment. C_e is enzyme concentration (g/L). (Carrín, M. E., L. Ceci, and J.E. Lozano. 2000. Effects of pectinase immobilization during hollow fiber ultrafiltration of apple juice. *Journal of Food Process Engineering* 23: 281–298. With permission.)

Finally, in Figure 7.7c it can be observed that during HFUF of pectin solution a gel layer was effectively developed on the previously immobilized enzyme.

Authors found that permeate flux was not initially increased when pectin solution or apple juice was ultrafiltered through pectinase immobilized on HFUF membranes. However, enzyme immobilization greatly extended the membrane operation by keeping permeate flux constant during prolonged periods at a reasonable yield (Figure 7.8).

7.6   Conclusions

Juice production technology has advanced rapidly as has enzyme technology. New enzymes preparation of fungal pectin lyase has been shown to be useful for the production of cranberry juice and the clarification of apple juice in the food industry (Semenova et al., 2006). A comparative study showed that the preparation of pectin lyase is competitive with commercial pectinase products, increasing juice yield with same dose. Authors also emphasized that the advantage of pectin lyase is the safety of food, because the activity of commercial clarifying enzymes is related to the presence of polygalacturonase and pectin esterase, the amount of the last being relatively high (8–10% of total protein content). It is well known that catalysis of the substrate with pectin esterase is associated with methanol release. The clarification and haze-diminishing effects of alternative clarification strategies have improved stability of black currant juice. The study of the individual and synergic effects of phenols and anthocyanin on juice turbidity was the base of new clarification treatments involving the use of blends of enzymes, like acid proteases and pectinases. A few other miscellaneous uses of enzymes in fruit industry exist: Fruit juices may be depectinized with PME and used for making low sugar jellies for diabetics.

The technology of immobilized enzymes is going through a phase of evolution and maturation. Probably this is the beginning of the immobilized enzyme technology era, and immobilized enzymes will clearly be more widely used in the future. At the present time, however, direct industrial application of the immobilized enzyme methodology for mash treatment or juice clarification is practically nonexistent.

Abbreviations

CIP:	cleaning-in-place
GRAS:	generally recognized as safe
HFUF:	hollow fiber ultrafiltration
J:	permeate flux
PE:	pectinesterase
PG:	polygalacturonase
PL:	pectinlyase
Re:	resistance
UF:	ultrafiltration
ΔPTM:	transmembrane pressure

References

Ashie, I.N.A. 2003. Bioprocess engineering of enzymes. *Food Technology* **57**(1): 44–51.

Ates, S. and S. Pekyardimci. 1995. Properties of immobilized pectinesterase on nylon. *Macromolecular Reports*, A32, 337–345.

Bump, V.L., 1989. Apple pressing and juice extraction. pp: 53–82. In: *Processed Apple Products*; D.L. Downing, Ed., AVI, Van Nostrand Reinhold, New York.

Carrín, M. E., L. Ceci, and J.E. Lozano. 2000. Effects of pectinase immobilization during hollow fiber ultrafiltration of apple juice. *Journal of Food Process Engineering* 23, 281–298.

Carrín, M. E., L. Ceci, and J.E. Lozano. 2004. Characterization of starch in apples and its degradation with amylases. *Food Chemistry* 87, 173–178.

Ceci, L. and J.E. Lozano. 1998. Determination of enzymatic activities of commercial pectinases. *Food Chemistry* 31(1/2), 237–241.

Ceci, L.N. and J.E. Lozano. 2002. Amylase for apple juice processing: Effects of pH, heat, and Ca^{2+} ions. *Food Technology and Biotechnology* 40(1), 33–38.

Cheryan, M. 1986. *Ultrafiltration Handbook.* Technomic, Lancaster.

Constenla, D. T. and J.E. Lozano 1996. Predicting stationary permeate flux in the ultrafiltration of apple juice. *Lebensmittel Wissenschaft und Technologie* (27), 7–14.

Demir, N., J. Acar, K. Sarioğlu, and M. Mutlu 2001. The use of commercial pectinase in fruit juice industry. Part 3: Immobilized pectinase for mash treatment. *Journal of Food Engineering* 47, 275–280.

Demirel, D. and M. Mutlu. 2005. Performance of immobilized Pectinex Ultra SP-L on magnetic duolite-polystyrene composite particles. Part II: A magnetic fluidized bed reactor study. *Journal of Food Engineering* 70, 1–6.

Diano, N., T. Grimaldi, M. Bianco, S. Rossi, K. Gabrovska, G. Yordanova, T. Godjevargova, V. Grano, C. Nicolucci, L. Mita, U. Bencivenga, P. Canciglia, and D.G. Mita. 2008. Apple juice clarification by immobilized pectolytic enzymes in packed or fluidized bed reactors. *Journal of Agricultural and Food Chemistry* 56 (23), 11471–11477.

Dietrich, H., C. Patz, F. Schoepplein, and F. Will. 1991. Problems in evaluation and standardization of enzyme preparations. *Fruit Processing* 1, 131–134.

Dinnella, C., G. Lanzarini, and P. Ercolessi. 1995. Preparation and properties of an immobilized soluble-insoluble pectinlyase. *Process Biochemistry* 30, 151–157.

FAOSTAT data, 2005. FAO Statistical Databases. www.fas.usda.gov/htp/Presentations/2005.

FDA, 2001. U.S. Food and Drug Administration Center for Food Safety and Applied Nutrition Office of Plant and Dairy Foods and Beverages. *The Juice HACCP Regulation Questions & Answers.*

Food, England. 2003. Statutory Instruments. The Fruit Juices and Fruit Nectars (England) Regulations. LEG 06/325 3.2.03 Parliamentary Under-Secretary of State, Department of Health. pp.1–26.

Grampp, E.A. 1976. New process for hot clarification of apple juice for apple juice concentrate. *Flussiges Obst* 43, 382–388.

Heatherbell, D. A., J.L. Short, and P. Stauebi. 1977. Apple juice clarification by ultrafiltration. *Confructa* 22, 157–169.

Kminkova, M. and J. Kucera. 1983. Comparison of pectolytic enzymes using different methods of binding. *Enzyme and Microbial Technology* 5:204–208.

Lineback, D.R. and E. Wongsrikasem. 1980. Gelatinization of starch in baked products. *Journal of Food Science* 45, 71–74.

Liu, Y.K. and B.S. Luh. 1978. Purification and characterization of endo-polygalacturonase from *Rhizopus arrhizus. Journal of Food Science* 43, 721–726.

Lozano, J.E. 2003. Separation and clarification. In *Encyclopedia of Food Science and Nutrition.* pp: 5187–5196. B. Caballero, L. Trugo, and P. Finglas, Eds. AP Editorial, Elsevier, London, UK.

Lozano, P., A. Manjón, F. Romojaro, M. Canovas, and J.L. Iborra. 1987. A crossflow reactor with immobilized pectolytic enzymes for juice clarification. *Biotechnology Letters* 9, 875–880.

Lozano, J.E., D.T. Constenla, and M.E. Carrín, 2000. Ultrafiltration of apple juice. In *Trends in Food Engineering*, J.E. Lozano, C. Añón, E. Parada/Arias and G. Barbosa-Cánovas, Eds. Food Preservation Technology Series. Technomics, Lancaster, Basel, pp. 117–134.

Lozano, J.E. 2006. Fruit Manufacturing: Scientific basis, engineering properties and deteriorative reactions of technological importance. In *Food Engineering Series*; Springer, USA, pp. 1–72.

Maxim, S., A. Flondor, R. Pasa, and M. Popa. 1992. Immobilized pectolytic enzymes on acrylic supports. *Acta Biotecnológica* 12: 497–507.

McLellan, M.R. 1996. Juice processing. In: *Biology, Principles and Applications*, L.P. Somogyi, H.S. Ramaswmy, and Y.H. Hui, Eds., Technomics, Lancaster, PA, pp. 67–72.

Nagy, S., C.S., Chen, P.E. Shaw, 1993. *Fruit Juice Processing Technology*, AGSCIENCE, Auburndale, FL, pp. 494–495.

Ramaswamy, H.S. and C. Abbatemarco. 1996. Thermal processing of fruits. In *Processing Fruits: Science and Technology*, L.P. Somogyi, H.S. Ramaswamy, and Y.H. Hui, Eds., Technomics, Lancaster, PA, Vol. I, pp. 25–65.

Reed, G. 1975. *Enzyme in Food Processing*, 2nd ed. Academic Press, New York.

Rombouts, F.M., W. Pilnik. 1978. Enzymes in fruit and vegetable juice technology. *Process Biochemistry* 8, 9–13

Sakai, T., T. Sakamoto, J. Hallaert, and E.J. Vandamme. 1993. Pectin, pectinase, and protopectinase: Production, properties, and applications. *Advances in Applied Microbiology* 39, 213–294.

Semenova, M.V., O.A. Sinitsyna, V.V. Morozova, E.A. Fedorova, A.V. Gusakov, and O.N. Okunev. 2006. Use of a preparation from fungal pectin lyase in the food industry, *Applied Biochemistry and Microbiology* 42, 598–602.

Stratilová, E., M. Čapka, M. Czakóová, L. Haladej, and L. Rexová-Benková. 1995. A capillary glass reactor for polygalacturonase immobilization. *Collection of Czechoslovak Chemical Communications* 60, 211–215.

Tuoping, Li T., N. Wang, S. Li, Q. Zhao, M. Guo and C. Zhang. 2007. Optimization of covalent immobilization of pectinase on sodium alginate support. *Biotechnology Letters* 29, 1413–1416.

USDA. 2008. http://www.fas.usda.gov/htp/2008_AppleJuice_WorldMarketTrade.pdf.

Woodroof, J.G. and B.S. Luh. 1986. *Commercial Fruit Processing*, 2nd ed., AVI, Westport, CT.

Wu, M.L., R.R. Zall, and W.C. Tzeng. 1990. Microfiltration and ultrafiltration comparison for apple juice clarification. *Journal of Food Science* 55 (4), 1162–1163.

Zeman, L.J. and A.L. Zydney. 1996. *Microfiltration and Ultrafiltration: Principles and Applications*, Marcel Dekker, New York.

Zobel, H.F. 1984. Starch gelatinization and mechanical properties. In: R.L. Whistler, J.N. BeMiller, E.F. Paschall, Eds. *Starch: Chemistry and Technology*. 2nd ed. Orlando, FL, Academic Press, pp. 285–309.

chapter eight

Enzymes in citrus juice processing

Domenico Cautela, Domenico Castaldo,
Luigi Servillo, and Alfonso Giovane

Contents

8.1   Introduction

In 2007 the world production of citrus fruits exceeded 115 million tons. In terms of economic interchange, the citrus industry is a globally integrated system. Citrus fruits are the most exchanged fruit variety worldwide (UNCTAD, 2003). The main citrus species for juice production are orange, lemon, lime, and grapefruit. The 55% of world production of citrus fruits consists of oranges, of which 80% is processed into orange juice.

There are significant differences in the processing of citrus fruits as compared to pomaceous fruits, stone fruits, and small fruits. The citrus juice extraction is aimed to obtain the best possible separation of juice, peel, and peel oil. Citrus oils are of great importance in the flavor industries. The peel oils are the principal factors affecting the juice taste.

In fact the presence in the juice of essential oils in amounts higher than 0.03% has negative effects on the juice taste and shelf-life due to off-flavors formation during the storage. Therefore, differently from fleshy fruits, citrus fruits cannot be subjected to direct pressure when preparing juice.

Since citrus juice is mostly consumed and appreciated as a cloudy drink, it is of great importance to preserve juice cloudiness over time. Therefore a deeper description will be given of the procedures aimed at the inactivation of endogenous enzymes whose activities are detrimental to cloud stability.

As a final point, processes involving the use of enzymes for the production of some citrus products and by-products will be described.

8.2   Overview of citrus processing technologies

The technology used in citrus juice processing is more or less identical throughout the world. Differences among processing systems and equipment employed in juice production arise both from local traditions and from morphological differences among citrus species, and also according to the kind of final product to be obtained.

A schematic representation of the main process for citrus juice manufacturing is presented in Figure 8.1. Steps that are common to all manufacturing processes before juice extraction include receiving, washing, sizing, and selection of fruits.

The juice extraction occurs through mechanical breakage of juice sacs present in the fruit endocarp. Differences among the extracting machines on the market regard the techniques used for the juice extraction and essential oil recovery.

In the in-line process the extraction of essential oil and juice takes place simultaneously, while in other processes, the juice extraction and oil recovery from the peel are carried out in different steps by employing different equipment.

The *In-Line System* (FMC Corporation, Fairway Avenue Lakeland, FL-USA) is the most used citrus juice extraction equipment worldwide. Before the extraction process, it is necessary to select fruits of uniform size. The "In-Line" extractor simultaneously processes three to eight fruits per cycle. The system consists of three to eight upper and lower cups: the upper cups are movable and descend toward the lower fixed cups. At the base of the bottom cups, there are circular knives mounted on top of the strainer tubes that act as juice pre-finisher elements. Both cups are formed of "finger-shaped" knives that intersect when the upper cups move down to the bottom cups. While the two cups still fit one against the other, the peel is separated from the fruit and cut in strips. The fruit, without peel, is pushed through a strainer tube, where the juice is separated from seeds. Pulp and remaining fruit components are flowed into a collector for the finishing operations. The essential oil is extracted by compression of the peel strips during the descendent movement of the upper cups and is recovered by sprayed water. The water–oil emulsion, bits of peel, and other solid parts of the fruit are transferred to a finisher for the removal of insoluble solids and then to centrifuges for the separation of essential oil.

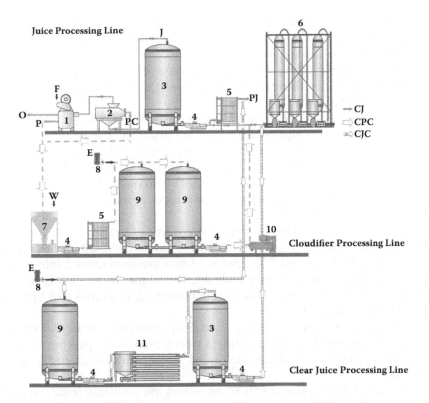

Figure 8.1 Schematic representation of citrus juice manufacturing operations: juice extractor (1); finisher (2); tank (3); pump (4); heat exchanger (5); evaporator (6); peel tank-hammermill-mixing tank (7); enzyme tank and dosing pump (8); reaction tank (9); decanter (10); ultrafiltration unit (11); concentrate juice (CJ); clear concentrate juice (CJC); cloudy peel concentrate (CPC); enzymatic preparation (E); fruit uploading (F); single strength juice (J); peel (P); pulps/cells (PC); pasteurized juice (PJ); cold pressed essential oil (O); water (W).

In equipment such as the "Brown Extractor" (Brown International Corporation, Winter Haven, FL-USA) and the *Birillatrice* (Fratelli Indelicato S.r.l., Catania, Italy), the juice extraction is carried out on half-fruit and precedes that of essential oil that is successively recovered from the juice-exhausted half-fruit peels by various systems.

These extractors require fruit calibration to work properly and operate on the fruits cut in two halves, which are placed in special cups. The juice extraction is accomplished using special bulbs ("reamers"). The reamer can be fixed or can more or less quickly rotate during the squeezing. It is possible to set the distance between the reamer and the fruit wall depending on the type of extraction required. Generally for a soft extraction this distance is about 1.1 cm but it can be 0.2 cm in the case of hard extractions.

The extraction pressure affects not only the yield in juice but also its composition. High pressures, in fact, increase the content of pectin, pulp, and pectin methylesterase enzyme (PME).

The extracted juice goes through a sieve or a vibrating rotary filter, incorporated in the extractor for the removal of seeds, cells, and coarse components of albedo.

The half-peels without juice fall into the oil extracting section. In the *Birillatrice-Sfumatrice* equipment, a rotating drum catches the peels which rolls and are pressed against a stationary profile. This causes repeated deformations of the half-peels and breakage of the oil sacs. A water rain provided by a set of sprayers collects oil. The mixture of water and essential oil falls into a cistern and from there it is conveyed to the next steps for separation and finishing.

In other processes the oil extraction precedes the juice extraction. In this case the essential oil is extracted from the whole fruit by rasping by machines called "peelers." The rasped fruits are then moved into a juice extractor. Machines operating with this technique are manufactured by such Italian companies as Fratelli Indelicato S.r.l. (Catania) and Speciale F. & C. S.r.l. (Catania).

The juice extractor cuts the peeled fruits into two halves that successively are squeezed against a stainless steel screen. After passing through the screen, the juice is collected in closed tanks, while seeds and pulp are expelled.

Detailed descriptions of these extractors are available from the manufacturers (FMC Corporation, Brown International Corporation, Fratelli Indelicato S.r.l., and Speciale F. & C. S.r.l.).

The extracted juice (single strength juice) contains significant amounts of pulp, seeds, rags, and other fibrous materials that are removed during the finishing operations. Orange juice is approximately 11°Brix (% w/w soluble solids), lemon juice has a content of soluble solids of 8°Brix, and grapefruit and tangerine juices have a soluble solid content of about 10°Brix. The pulp content is reduced to approximately 4–5% v/v in downstream finishers. Juices with pulp content of about 1% can be obtained by the use of centrifuges. The discarded pulp can be further processed to produce pulp wash or extracts of water-soluble solids.

The juice obtained must be de-aerated to avoid the ascorbic acid oxidation during pasteurization and storage. The single strength juice is usually pasteurized employing tubular or plate heat exchangers. Generally, the citrus juice is pasteurized to stabilize the product against the microbial growth and to inactivate endogenous enzymes before concentration or aseptic packaging. The pasteurization process is mainly designed to inactivate the heat stable forms of the enzyme pectin methylesterase (PME) as they require higher temperatures than those needed for thermal destruction of microorganisms.

The activity of endogenous pectin methylesterase (EC 3.1.1.11) triggers a sequence of events that lead to the destabilization of the juice cloud resulting in syneresis and/or gelation of concentrate juices. The citrus industry employs different time-temperature combinations for the thermal inactivation of PME depending on the type of juice to be processed. Temperatures and holding times required for the pasteurization of cloudy and pulpy juices are generally higher than that required for the production of clear juice or for juice reconstituted from concentrate before aseptic packaging.

Combinations of time and temperature used to inactivate the PME will be extensively treated in a separate section of this chapter.

Usually the juice is concentrated to overcome the seasonality of citrus production and to reduce the cost of packaging, storage, and shipping. The majority of orange juice produced in Florida and Brazil is converted into concentrated juice.

The evaporators most frequently used in the process of concentration are the "falling film evaporators" in single or multiple steps. Among them, T.A.S.T.E. (Thermally Accelerated Short Time Evaporator) is widely used for concentrating clear and cloudy juice. The operating principle of the T.A.S.T.E. entails use of high temperatures (92–112°C) which, besides fast evaporation, also accomplishes juice pasteurization. Flavors lost during evaporation can be largely restored by the addition to concentrated juice of aqueous aromas recovered from the evaporator condensates and oil-based essences extracted from citrus peel.

Orange juice is marketed mostly as 60 or 65°Brix concentrate, while cloudy lemon juice is sold on the basis of its citric acid content. Generally, lemon juice is concentrated to 400 or to 530 g/L (grams of citric acid per liter). Grapefruit concentrates can be prepared by evaporation to high Brix levels (60–65°Brix), as for the orange and tangerine juices, although 55°Brix grapefruit concentrate is also commonly produced.

Due to the high acidic nature of lemon and lime juices, greater care must be taken in their heat treatment. If the temperature to inactivate pectinases is too high, acid hydrolysis of pectin and sugars can occur. As in the case of pectinase activity, this results in cloud loss and gelation. Furthermore, the excessive acid content accelerates the juice browning due to the Maillard reaction.

8.3   Pectin methylesterase (PME) in citrus juice

Citrus juice is usually consumed as cloudy drinks. The citrus cloud is a fine suspension of particles that gives citrus juice the characteristics of turbidity, color, flavor, aroma, and texture (Mizrahi et al., 1970; Baker and Cameron, 1999). The cloud is composed of high molecular weight polymeric materials including proteins, pectin, hemicellulose, and cellulose (Sinclair 1984).

Cloud particles are between 0.4 and 5 μm in diameter (Klavons et al., 1994), stable cloud is made of particles of about 2 μm diameter (Mizrahi et al., 1970). The cloud stability of citrus juice is related to the molecular weight of pectin (Hotchkiss et al., 2002), its degree of methylesterification (Hills et al., 1949), and the intra-molecular distribution of methylester groups in the pectin molecules (Baker, 1979; Joye and Luzio, 2000; Willats et al., 2001; Wicker et al., 2003).

The pectin methylesterase catalyses the hydrolysis of methylester groups of pectin, causing the formation of negative carboxylate groups and releasing H^+ ions and methanol.

The action of PME modifies high methoxyl pectin into calcium sensitive low methoxyl pectin. Once a critical degree of de-esterification is reached, divalent cations, such as calcium, can cross-link the free acid units on adjacent pectin molecules, forming insoluble calcium pectates.

Cross-linking increases the pectin apparent molecular weight and reduces its solubility, thus leading to flocculation. Cloud is considered definitively broken or lost in orange juice when light transmittance reaches 36% (Redd et al., 1986).

The activity unit of PME (PMEu) is defined as the amount of enzyme that release 1 μmole of carboxylic acid groups in 1 ml of solution for one minute, and is generally measured using the method proposed by Rouse and Atkins (1954), which is based on the titration of carboxyl groups generated by the PME activity during the hydrolysis of 1% pectin solution containing 0.1 M NaCl.

Rouse and Atkins (1954) reported a value of 1.1 PMEu/mL in lemon juice.

For orange juice cv. Navel containing 5% pulp, Ingallinera et al. (2005) reported a pectin methylesterase activity of 1.3 PMEu/mL and found values 2–3 times higher for the Sicilian red orange juices cv. Moro, Tarocco, and Sanguinello with the same pulp content. The juice from cv. Tarocco showed the highest activity (2.85 PMEu/mL).

The technologies of citrus juice extraction affect the composition and the "cloud stability" of the product. An increase of enzyme activity is correlated to a higher content of peel incorporated into the juice during the extraction process (Amstalden and Montgomery, 1994).

Cameron et al. (1999) showed that the PME extracted from Valencia orange peel destabilizes the cloud more rapidly than those extracted from rag or hand-squeezed juice. In addition, PME activity in juice extracted with hard procedures destabilizes the cloud faster than in a juice extracted with a softer process.

The activity of PME in juice varies according to the fruit cultivar and the stage of fruit ripeness. Amstalden and Montgomery (1994) found a higher PME activity in orange juices from cv. Valencia than in those produced from other Brazilian orange cultivars (cv. Natal, Pera, and Pera Rio Coroa).

In citrus fruits, and in general in higher plants, PME is found in multiple isoforms. The isoforms can show marked differences in kinetic properties, activity at low pH, affinity for the pectic substrate, and influence in the process of juice clarification.

Three different PME isoforms were isolated and characterized from Navel oranges (Versteeg et al., 1980), while six PME isoforms were identified in tissue culture cells from Valencia oranges (Cameron et al., 1994).

Evans and McHale (1978) have identified two PME isoforms in Washington Navel oranges: one was localized almost exclusively in the peel, while the other was located in segments covering the juice sacs. Seven putative isoforms with PME activity have been isolated by commercial pectinesterase extracted from Valencia Orange peel (Hang et al., 2000). Cameron and Grohmann (1995) purified and characterized three PME isoforms in red grapefruit finisher pulp, while only two PME isoforms were identified in Marsh White grapefruit pulp (Seymour et al., 1991). McDonald et al. (1993) identified seven fractions with PME activity in lemon and purified two major pectinesterases: one located solely in the peel and the other in the endocarp. Regarding other citrus species, three PME isoforms were identified in bergamot fruits (Laratta et al., 2008), while two forms of pectinesterase were found in West Indian limes (Evans and McHale, 1978).

Data reported in the literature show that some isoforms of PME exhibit considerable resistance to heat treatments. These heat-stable PME isoforms (TS-PME) have particular importance in citrus technology operations for the stabilization of citrus juice and derivatives. However, different standards have been applied to discriminate between thermolabile and thermostable PME isoforms. Cameron and Grohmann (1996) specified a treatment at 80°C for 2 min to inactivate the thermolabile PME isoforms, while Snir et al. (1996) established a heat treatment of 70°C for 5 min to discriminate between thermolabile and thermostable forms. Hang et al. (2000) reported a heat treatment at 90°C for 1 min to discriminate the two isoforms.

The amount of enzyme activity related to thermostable isoforms of PME (TS-PME) with respect to total PME activity represents about 10% in orange juice cv. Valencia and about 5% in orange juice cv. Navel (Carbonell et al., 2006; Cameron and Grohmann, 1996). In red and white grapefruit, the fraction of TS-PME activity ranges between 5.7 and 12.4% depending on the fruit ripening period (Snir et al., 1996).

For Israelis orange juice, Rothschild et al. (1975) indicated a complete inactivation of PME after 45 s of heat treatment at 90°C, while Sadler et al. (1992) measured a residual pectinesterase activity of 0.01% after treatment at 90°C for 1 min. A treatment at 90°C for 1 min was adequate for inactivation of PME from orange juice cv. Pera-Rio (Do Amaral, 2005).

Tests conducted on the heat stable forms of PME extracted from Valencia oranges showed that after 60 s incubation at 90°C the enzyme retained 55% of its activity and a small residual activity still remained after 90 s. A treatment for 2 minutes at that temperature was required for the complete inactivation (Cameron and Grohmann, 1996).

Three of the seven PME isoforms identified by Hang et al. (2000) in Valencia orange peel were heat stable and, among these, the most thermally stable was inactivated by 93% after heat treatment at 90°C for 1 minute.

For Sicilian red orange juice (cv. Tarocco) 3 min treatment at 85°C reduced the PME activity to less than 10% of initial activity (Ingallinera et al., 2005).

For tangerine juice, the treatment at 91°C for 20 s reduced the activity to 0.05% of initial activity (Carbonell et al., 2006), while Rillo et al. (1992) indicated a complete inactivation of the purified PME after thermal treatment at 90°C for 1 min.

Thermostable PME extracted from grapefruit finisher pulp exhibited high thermal stability retaining 66.7% relative activity after 2 min incubation in an 80°C water bath, and 45.2% of its relative activity after 60 set incubation in a 95°C water bath (Cameron and Grohmann, 1995).

McDonald et al. (1993) reported for a PME isoform extracted from lemon endocarp an activity optimum at 70°C and an activity optimum at 60°C for a PME isoform extracted from peel. Both isoforms showed no activity above 88°C.

8.4   Kinetic parameters of PME thermal inactivation in citrus juices

Endogenous PME has detrimental effects on citrus juice stability, therefore, part of this chapter will be dedicated to discuss design and modeling studies on PME inactivation processes by heat treatment.

The kinetic models mathematically describe the evolution of a process over time and are aimed to find out adequate parameters to assess the dependence of reaction rate from temperature. These models provide engineering tools for the assessment, design, and optimization of thermal processes used for biochemical and microbial stabilization of products.

Generally, it is assumed that the thermal inactivation of PME in a citrus juice can be described by a single component first order kinetic model (log-linear model).

In literature many data are reported showing the presence in citrus juices of several isoforms of PME with different thermal stability.

As reported in Chapter 1, in the case of two isoforms, the model used to describe this kind of inactivation kinetics is defined first order kinetic model of two-component (biphasic) systems.

Both models are employed in thermoresistance studies of PME in citrus juices to calculate the Decimal Reduction Time (D_T) and its dependence on temperature (z), as the knowledge of D and z values completely defines the thermoresistance of the enzyme to be inactivated. (For more details see Chapter 1.)

Table 8.1 shows D_T and z parameters for PME thermal inactivation in citrus juices. Some thermostability studies were conducted by measuring the residual activity of the juice subjected to heat treatment, while other studies were performed by conducting thermal stability tests on purified PME isoforms.

Versteeg et al. (1980) reported a D_{90} value of about 1 min for the inactivation of PME in cv. navel orange juice. This value is in agreement with the conditions set in the process recommended by Eagerman and Rouse (1976). These authors reported a value for the decimal reduction time at 90°C (D_{90}) of 60 s for orange juice from cv. Valencia, Hamlin, and Pineapple and z values between 4.9 and 6.8°C. For grapefruit juice $D_{85.6}$ was 1 min and z 5.5°C (Eagerman and Rouse, 1976).

The thermal stability of PME is affected by different factors, including the content of soluble solids. At the same temperature of treatment, higher decimal reduction times were observed when the soluble solid content of the juice increased (Marchall et al., 1985).

To obtain a reliable estimate of the z parameter, thermostability studies have to be performed by exploring a wide range of temperature. In fact, for the red orange juice cv. Sanguinello, De Sio et al. (2001) reported the values of 12.5 s and 9.2°C for $D_{87.8}$ and z, respectively, in the temperature range 75–85°C. But, due to the presence of heat-resistant PME isoforms, the authors found for z the value of 16.4°C in the interval 85–95°C.

Obviously, in an industrial process on such a type of juice, the higher z value has to be utilized to inactivate PME, without concerning if single or multiple enzyme isoforms with different thermoresistance are present in the juice.

Pflug and Odlaug (1978) estimated as a "safety factor" a 30% increase of D and z parameters to compensate for the uncertainty passing from experimental data to practical applications. Kim et al. (1999), in pilot scale experiments, measured in Valencia orange juice the effects on PME inactivation of different thermal treatment conditions. The holding time to obtain a 90% reduction of PME activity was found to range from 33.3 s at 80°C to 17.9 s at 90°C. Tribess and Tadini (2006), applying the first order kinetic model for a two-component system, described the kinetic of PME inactivation in orange juice cv. Pera as a function of pH and at various time-temperature combinations. The data showed a higher PME inactivation rate at temperatures of 85–87.5°C and at pH 3.6–3.7.

Rothschild et al. (1975) and Holland et al. (1976) reported that an orange juice can be considered biochemically stable when the residual PME activity after the thermal treatment becomes lower than 10^{-4} PMEu/mL.

Table 8.1 Kinetic Parameters of Pectin Methylesterase Heat Inactivation

Source	D_T (s)	z (°C)	Notes	References
Orange juice	$D_{55} = 295$	17.5		Tajchakavit and
	$D_{60} = 153$			Ramaswamy (1997 a, b)
	$D_{65} = 79$			
	$D_{70} = 37$			
Orange juice cv. Valencia	$D_{80} = 20$	—		Kim et al. (1999)
	$D_{90} = 11.3$	—		
Orange juice cv. Valencia	$D_{88.3} = 108$	6.8		Eagerman and Rouse
	$D_{90} = 60$			(1976)
Orange juice cv. Hamlin	$D_{87.8} = 60$	4.9		
Orange juice cv. Pineapple	$D_{87.8} = 60$	5.1		
Grapefruit juice cv. Duncan	$D_{85.68} = 60$	5.5		
Thermolabile PME fraction from orange juice cv. Valencia	$D_{75} = 204.8$ $D_{80} = 70.4$ $D_{85} = 22.1$ $D_{90} = 6.3$	9.5		Lee et al. (2003)
Thermostable PME fraction from orange juice cv. Valencia	$D_{75} = 14438$ $D_{80} = 1809$ $D_{85} = 270$ $D_{90} = 33$	5.7		
Thermolabile PME fraction	$D_{90} = 6.5$	17.6		Chen and Wu (1998)
Thermostable PME fraction	$D_{90} = 329$	11		
Thermolabile PME fraction from orange pulp cv. Valencia	$D_{60} = 137.88$ $D_{65} = 41.52$ $D_{70} = 15.97$ $D_{80} = 1.89$ $D_{85} = 0.65$ $D_{90} = 0.23$	10.8		Wicker and Temelli (1988)
Thermostable PME fraction from orange pulp cv. Valencia	$D_{60} > 174$ $D_{65} > 174$ $D_{70} > 174$ $D_{80} > 174$ $D_{85} = 173.25$ $D_{90} = 32.36$	6.5		

Table 8.1 Kinetic Parameters of Pectin Methylesterase
Heat Inactivation (Continued)

Source	D_T (s)	z (°C)	Notes	References
Orange juice cv. Navel	$D_{90} = 0.09$	6.5	PME-I isozyme (pH 4.0)	Versteeg (1979)
	$D_{90} = 0.9$	11	PME-II isozyme (pH 4.0)	Versteeg et al. (1980)
	$D_{90} = 23$	6.5	PME-III isozyme (HMW-PME) (pH 4.0)	
Thermostable PME fraction from orange juice cv. Valencia	$D_{75} = 22.7$ $D_{80} = 3.5$ $D_{85} = 0.46$	5.9		Hou et al. (1997)
Concentrate orange juice cv. Valencia + PME added	$D_{90} = 15$ $D_{90} = 13$ $D_{90} = 14$ $D_{90} = 19$	—	10°Brix 20°Brix 30°Brix 35°Brix	Marshall, Marcy, and Braddock (1985)
Blood orange juice cv. Sanguinello	$D_{87.8} = 12.6$	16.4		De Sio et al. (2001)

On the basis of the experimental results reported above, the choice of temperatures and times normally employed for the stabilization (i.e., to inactivate PME) in the citrus juice industry is in the range 90–98°C for about 60 s for orange and mandarin juices, in the range 85–90°C for 30–40 s for grapefruit juice, and in the range 75–85°C for about 30 s for lemon juice.

8.5 Clear citrus juice processing

The production of clear citrus juices is generally limited to the lemon and lime juices, which are used as acidifiers for nectars and syrups. In this case, the enzymatic treatment is designed for juice depectinization before the filtration and concentration operations.

The traditional method consists of the natural clarification of the product by the action of endogenous pectolytic enzymes (PME, polygalacturonase, pectate lyase, etc.).

The low pH of these acidic juices dramatically reduces the activity of pectolytic enzymes with an increase of holding time in the sedimentation tanks. The prolonged sedimentation step requires additives to prevent microorganism growth and juice oxidation. However, the addition of enzyme preparations coupled with traditional pre-clarifying centrifugation and/or high-performance centrifugation can reduce the holding time to less than 8 hours. Briefly, an initial step provides reduction of juice pulp through finisher and pre-clarifier to a final content of about 2%. Then the juice is pasteurized and a pectinase preparation is added. The commercial pectinase preparations commonly employed are usually extracted from fungi of Aspergillus genus and contain a mixture of the enzymes PME, polygalacturonase (PG), and pectin lyase (PL) (Dietrich et al., 1991).

As lemon and lime juices have high acid content, generally ranging between 44 and 62 g/L (expressed as citric acid), with pH lower than 2, the use of pectolytic enzymes stable at low pH is of primary importance. Unfortunately, many pectolytic enzymes have pH optimum between 3.8 and 4.5, and their activity is considerably reduced when pH drops below 2. Thus only a limited number of pectolytic enzymes retain some residual activity at that pH. The pH of the juice also influences the optimum temperature of pectolytic enzymes; therefore, the temperature must be monitored during the process. Technical specifications of the enzyme preparations (Novozymes Citrozym Ultra L; Rohapect 10) indicate temperature intervals of 25–30°C for juice with pH between 1.8 and 2.2; 30–35°C for juice whose pH is between 2.2 and 2.6; and 35–40°C for juice having pH between 2.6 and 3.0.

The juice is mixed with enzymes under mild agitation in a reaction tank downstream of the heat exchanger. The proper dosage depends on the desired reaction time. Doses of about 60–100 ppm of pectinase allow reaction time lower than 3 hours, while lower doses (10–20 ppm) require longer time of reaction, up to 18 hours. If the process does not exceed three hours, the addition of sulfur dioxide is not required. However, for longer incubation times, the addition of up to 500 mg/kg of SO_2 is needed to prevent oxidation processes.

De Carvalho et al. (2006) studied the effect of enzymatic hydrolysis on the reduction of particle size in lemon juice measuring the distribution of particle size in natural and hydrolyzed lemon juice by laser diffraction technique. Enzymatic treatment with 0.3% of "Cytrozym cloudy" for 40 min of incubation gave the best result as particle size reduction with the lowest enzyme concentration.

At the end of the pectinase treatment, the juice is added with Kieselsol (30% colloidal suspension of silica sol with high surface area). For juice with suspended solid content below 5%, the addition of Kieselsol is between 3000 and 5000 ppm. With lower pulp content of the juice, the Kieselsol addition is lowered accordingly. After mixing with the silica sol, the juice is allowed to stand at least half an hour to allow the precipitation of the suspended particles; then it is separated from the precipitate by decanting/centrifuging with high-performance clarifiers. The clear juice is finally concentrated.

Sometimes, during the storage of a clear juice, opalescence may occur due to precipitation of flavonoids, if they are present in a large amount.

This problem has been overcome by cooling the juice to accelerate the crystallization of hesperidin, the most abundant flavonoid in lemon juice (Fisher-Ayloff-Cook and Hofsommer, 1991).

Specific enzyme preparations are used in the juice treatments that employ membrane technologies (i.e., microfiltration and ultrafiltration). In fact, in such processes the use of enzymes is aimed to reduce the juice viscosity and to improve the permeate flow. For citrus juices the membrane processes can be coupled with the juice treatment with absorbent and/or ionic exchange resins. These processes are called CT (Combined Technologies). The use of the resins is effective in the removal of some components that give the juice a bitter taste, such as limonoids (mainly limonin) and flavonoids (mainly naringin). These technologies are commonly employed in the production of clear non-bitter concentrated grapefruit juice and to remove limonin from orange juice (cv. Navel) which give the juice a bitter taste.

In citrus juice production CT is employed to separate the pulp (cloud) and the clear filtrate (called serum) using cross-flow ultrafiltration. This is followed by the removal of unwanted juice components by passage of the serum through a food-grade adsorbent and/or ion exchange resins. The stream containing pulp can then be re-blended with the upgraded serum or used separately.

The membranes generally used are made of polyvinylidene fluoride (PVDF) or polysulfone (PS), and the most common types of modules are tubular, flat-plate, and hollow fiber modules.

The enzyme treatment before the membrane filtration step increases the permeate flow, reduces the occurrence of fouling and the concentration polarization.

The effectiveness of two pectinase preparations ("Ultrazym 100 G" and "Rohapect C") was tested in an ultrafiltration process of lemon juice using tubular or hollow fiber microfiltration modules. When comparing the permeate flow of treated and untreated juice in clarification tests, it was observed that addition of "Ultrazym 100 G" (200 ppm) or "Rohapect C" (600 ppm) increases the permeate flow from 20 to 40 l/hm^2 with tubular ultrafiltration module and from 50 to 80 l/hm^2 with hollow fiber microfiltration module (Saura et al., 1991). Chamchong and Noomhorm (1991) studied the clarification process of a tangerine juice, pretreated with polygalacturonase, by cross-flow ultrafiltration and microfiltration using polysulfone flat sheet membranes with pore sizes of different molecular weight cut off. The authors reported a depectinization treatment of tangerine juice by the addition of pectinase from Aspergillus aculeatus (10 g/ kg of pulp for 4 h at room temperature) before the ultrafiltration process. Ultrafiltration of the Clementine mandarin juice by hollow fiber membranes was studied by Cassano et al. (2009) who evaluated, as permeate flux, the performance of modified poly(ether ether ketone) (PEEKWC) and polysulfone (PS) hollow fiber membranes.

8.6   Natural cloudifiers from citrus peel processing

Cloudifiers are emulsions designed to add cloudiness at the desired degree to the finished beverages. Cloudy emulsions give natural appearance, opacity, consistency, and rheological properties to citrus beverages. The cloudifiers from citrus peel can be prepared in many ways, but the most used process consists of the enzymatic treatment of peels, which allows the soluble solid extraction with reduction of viscosity and maintenance of cloud stability. Pectolytic enzymes can be used to reduce the molecular weight of pectin and dramatically decreases their ability to gel. However, since the juice has to remain cloudy, this enzymatic treatment must be designed to reduce pectin molecular weight without causing juice clarification.

The peels from the extractor are milled and mixed with water, then the enzymatic preparation is added and the mixture is placed under agitation in a reaction tank. Cloudy peel extract is obtained in a batch process carried out at temperatures of 45–55°C for 60–120 minutes under continuous stirring. In this process pectinase and cellulase are used in combination with specific emicellulase. Some enzyme preparations for these specific applications include cellulase, xylanase, and β-glucanase extracted from Trichoderma reesei (Rohament® CL).

Specific hemicellulases also have a beneficial effect on viscosity reduction, without risk of inducing cloud instability. An adequate combination of those specific pectolytic enzymes and hemicellulases allows fast and efficient viscosity reduction and avoids cloud instability, provided that contact time and temperature are under control.

Several parameters affecting the process are the degree of uniformity and the size of peel particles, the mixing ratio peels/water, the type and the amount of employed enzymes, the temperature optimum of the enzyme preparations, and the speed of mixing. The extracts, prepared as described above, are separated from pulp by decanter and have a soluble solid content of 4–5°Brix. After pasteurization, the extracts are concentrated to about 50°Brix.

8.7   Conclusions

Nowadays enzyme preparations represent an integrating part of the modern industry of fruit juices. The enzyme producers propose a wide range of products to be employed in the preparation of various types of juices and concentrates. Currently, in the citrus industry the enzyme preparations are commonly employed:

- To optimize the production processes of some types of juices, such as clear and semi-cloudy citrus concentrates

- To optimize the debittering processes of orange and grapefruit juices combined with treatments that employ membrane technologies (Combined Technologies)
- To prepare new products starting from citrus by-products (peel cloudifier)
- To reduce the viscosity of concentrated citrus juices without affecting the turbidity

Through the use of specific enzyme preparations, the citrus industries can increase their own flexibility by producing various types of products and by-products and optimizing the operating processes with continuous improvements of the qualitative features of citrus juices and concentrates.

The future perspectives include growing integration of the use of enzyme preparations in the production processes of citrus juices and concentrates aimed to improve the product quality. Moreover, the wide choice of commercially available enzyme preparations with high specificity opens the way for new processes and new types of citrus juices and preparations.

Abbreviations

CT:	Combined Technologies
D_T:	Decimal Reduction Time
PEEKWC:	poly(ether ether ketone)
PG:	polygalacturonase
PL:	pectin lyase
PME:	pectin methylesterase
PMEu:	pectin methylesterase activity unit
PS:	polysulfone
PVDF:	polyvinylidene fluoride
T.A.S.T.E.:	Thermally Accelerated Short Time Evaporator
TS-PME:	heat-stable pectin methylesterase isoforms

References

Amstalden, L. C. and M. W. Montgomery. 1994. Pectinesterase in orange juice: Characterization. *Ciencia e Tecnologia de Alimentos* 14(1): 37–45.

Baker, R. A. 1979. Clarifying properties of pectin fractions separated by ester content. *Journal of Agricultural and Food Chemistry* 27: 1387–1389.

Baker, R. A., and R. G. Cameron. 1999. Clouds of citrus juices and juice drinks. *Food Technology* 53: 64–69.

Brown International Corporation, LLC. 333 Avenue M NW Winter Haven, FL 33881. http://www.brown-intl.com/processing/.

Cameron, R. G., R. P. Niedz, and K. Grohmann. 1994. Variable heat stability for multiple forms of pectin methylesterase from citrus tissue culture cells. *Journal of Agricultural and Food Chemistry* 42(4): 903–908.

Cameron, R. G. and K. Grohmann. 1995. Partial purification and thermal characterization of pectinmethylesterase from red grapefruit finisher pulp. *Journal of Food Science* 60(4): 821–825.

Cameron, R. G., A. R. Baker, A. B. Buslig, and K. Grohmann. 1999. Effect of juice extractor settings on juice cloud stability *Journal of Agricultural and Food Chemistry* **47**(7): 2865–2868.

Cameron, R. G. and K. Grohmann. 1996. Purification and characterization of a thermally tolerant pectin methylesterase from a commercial Valencia fresh frozen orange juice. *Journal of Agricultural and Food Chemistry* 44(2): 458–462.

Carbonell, J. V., P. Contreras, L. Carbonell, and J. L. Navarro. 2006. Pectin methylesterase activity in juices from mandarins, oranges and hybrids. *European Food Research and Technology* 222(1–2): 83–87.

Carvalho de, L. M. J., R. Borchetta, E. M. M. da Silva, C. W. P. Carvalho, R. M. Miranda, and C. A. B. da Silva. 2006. Effect of enzymatic hydrolysis on particle size reduction in lemon juice (*Citrus limon*, L.), cv. Tahiti. *Brazilian Journal of Food Technology* 9(4): 277–282.

Cassano, A., F. Tasselli, C. Conidi, and E. Drioli. 2009. Ultrafiltration of Clementine mandarin juice by hollow fibre membranes. *Desalination* 241: 302–308.

Chamchong, M. and A. Noomhorm. 1991. Effect of pH and enzymatic treatment on microfiltration and ultrafiltration of tangerine juice. *Journal of Food Process Engineering* 14(1): 21–34.

Chen, C. S. and M. C. Wu. 1998. Kinetic models for the thermal inactivation of multiple pectinesterase in citrus juices. *Journal of Food Science* 63(5): 747–750.

De Sio, F., A. Palmieri, L. Servillo, A. Giovane, and D. Castaldo. 2001. Thermoresistance of pectin methylesterase in Sanguinello orange juice. *Journal of Food Biochemistry* 25(2): 105–115.

Dietrich, H., C. Patz, F. Schöpplain, and F. Will. 1991. Problems in evaluation and standardization of enzyme preparations. *Fruit Processing* 1: 131–134.

Do Amaral, S. H., S. A. De Assis, and O. M. M. D. F. Oliveira. 2005. Partial purification and characterization of pectin methylesterase from orange (*Citrus sinensis*) CV. Pera-Rio. *Journal of Food Biochemistry* 29: 367–380.

Eagerman, B. A. and A. H. Rouse. 1976. Heating inactivation temperature-time relationships for pectinesterase inactivation in citrus juice. *Journal of Food Science* 41: 1396–1397.

Evans, R. and D. McHale. 1978. Multiple forms of pectinesterase in limes and oranges. *Phytochemistry* 17(7): 1073–1075.

Fisher-Ayloff-Cook, K. P. and H. J. Hofsommer. 1991. New technological aspects. IV. Processing technology for the production of special citrus products. *Fluessiges Obst* 58(11): 596–600.

FMC Corporation Citrus Systems Division, Box 1708, 400 Fairway Avenue Lakeland, FL 33802. http://www.fmctechnologies.com/upload/extractor-english.pdf.

Hang, Y., S. S. Nielsen, and P. E. Nelson. 2000. Thermostable and theriviolabile isoforms in commercial orange peel pectinesterase. *Journal of Food Biochemistry* 24: 41–54.

Hills, C. H., H. H. Mottern, G. C. Nutting, and R. Speiser. 1949. Enzymedemethylated pectinates and their gelation. *Food Technology* 3: 90–94.

Holland, R. R., S. K. Reder, and D. E. Pritchett. 1976. An accelerated test for residual cloud reducing enzyme activity in citrus juices. *Journal of Food Science* 41:
812–814.

Hotchkiss, A. T., B. J. Savary, R. G. Cameron, H. K. Chau, J. Brouillette, G. A. Luzio,
and M. L. Fishman. 2002. Enzymatic modification of pectin to increase its calcium sensitivity while preserving its molecular weight. *Journal of Agricultural
and Food Chemistry* 50: 2931–2937.

Hou, W. N., Y. Jeong, B. L. Walker, C. I. Wie, and M. R. Marshall. 1997. Isolation
and characterization of pectinesterase from Valencia orange. *Journal of Food
Biochemistry* 21(3): 309–333.

Fratelli Indelicato S.r.l., Giarre (Catania), Italy. http://www.indelicato.it.

Ingallinera, B., R. N. Barbagallo, G. Spagna, R. Palmeri, and A. Todaro. 2005. Effects
of thermal treatments on pectinesterase activity determined in blood oranges
juices. *Enzyme and Microbial Technology* 36(2–3): 258–263.

Joye, D. D. and G. A. Luzio. 2000. Process for selective extraction of pectins from
plant material by differential pH. *Carbohydrate Polymers* 43: 337–342.

Kim, H. B., C. C. Tadini, and R. K. Singh. 1999. Effect of different pasteurization
conditions of enzyme inactivation on orange juice in pilot scale experiments.
Journal of Food Process Engineering 22: 395–403.

Klavons, J. A., R. D. Bennett, and S. H. Vannier. 1994. Physical/chemical nature of
pectin associated with commercial orange juice cloud. *Journal of Food Science*
59: 399–401.

Laratta, B., L. De Masi, P. Minasi, and A. Giovane. 2008. Pectin methylesterase in
Citrus bergamia R.: Purification, biochemical characterisation and sequence
of the exon related to the enzyme active site. *Food Chemistry* 110(4):
829–837.

Lee, J. Y., Y. S. Lin, H. M. Chang, W. Chen, and M. C. Wu. 2003. Temperature-
time relationships for thermal inactivation of pectinesterases in orange juice.
Journal of the Science of Food and Agriculture 83(7): 681–684.

Macdonald, H. M., R. Evans, and W. J. Spencer. 1993. Purification and properties of
the major pectinesterases in lemon fruits (*Citrus limon*). *Journal of the Science of
Food and Agriculture* 62: 163–168.

Marshall, M. R., J. E. Marcy, and R. J. Braddock. 1985. Effect of total solids level on
heat inactivation of pectinesterase on orange juice. *Journal of Food Science* 50:
220–222.

Mizrahi, S., and Z. Berk. 1970. Physico-chemical characteristics of orange juice
cloud. *Journal of the Science of Food and Agriculture* 21: 250–253.

Pflug, I. J., and T. E. Odlaug. 1978. A review of z and F values used to ensure the
safety of low-acid canned food. *Food Technology* 32: 63–70.

Redd, B. J., C. M. Hendrix Jr., and D. L. Hendrix. 1986. In *Quality Control Manual for
Citrus Processing Plants*, Vol. 1. Safety Harbor, FL: Intercit.

Rillo, L., D. Castaldo, A. Giovane, L. Servillo, G. Balestrieri, and L. Quagliuolo.
1992. Purification and properties of pectin methylesterase from mandarin
orange fruit. *Journal of Agriculture and Food Chemistry* 40: 591–593.

Rothschild, G., C. V. Vlist, and A. Karsenty. 1975. Pasteurization condition for juice
and comminuted products of Israel citrus fruits. *Journal of Food Technology*
10: 29–38.

Rouse, A. H. and C. D. Atkins. 1954. Lemon and lime pectinesterase and pectin.
Proceedings of Florida State Horticultural Society 67: 203–206.

Sadler, G. D., M. E. Parish, and L. Wickerm. 1992. Microbial, enzymatic and chemical changes during storage of fresh and processed orange juice. *Journal of Food Science* 57(5): 1187–1197.

Saura, D., J. M. Ros, J. A. Canovas, E. Gomez, and J. Laencina. 1991. Tangential flow filtration with enzymatic treatment in the clarification of lemon juice. *Mededelingen van de Faculteit Landbouwwetenschappen Rijksuniversiteit Gent.* 56: 1705–1707.

Seymour, T. A., J. F. Preston, L. Wicker, J. A. Lindsay, and M. R. Marshall. 1991. Purification and properties of pectinesterases of Marsh white grapefruit pulp. *Journal of Agriculture and Food Chemistry* 39(6): 1080–1085.

Sinclair, W. B. 1984. *The Biochemistry and Physiology of the Lemon.* Oakland, CA: University of California, Division of Agriculture & Natural Resources.

Snir, R., P. E. Koehler, K. A. Sims, and L. Wicker. 1996. Total and thermostable pectinesterases in citrus juices. *Journal of Food Science* 61(2): 349–382.

Speciale, F. & C. S.r.l. Via Torrisi 18 - 95014 Giarre (Catania) Italy. http://www.speciale.it.

Tajchakavit, S. and H. S. Ramaswamy. 1997a. Continuous flow microwave inactivation kinetics of pectin methyl esterase in orange juice. *Journal of Food Processing and Preservation* 21(5): 365–378.

Tajchakavit, S. and H. S. Ramaswamy. 1997b. Thermal versus microwave inactivation kinetics of pectin methylesterase in orange juice under bath mode heating conditions. *Food Science and Technology (Lebensmittel Wissenschaft und Technologie)* 30: 85–93.

Tribess, T. B. and C. C. Tadini. 2006. Inactivation kinetics of pectin methylesterase in orange juice as a function of pH and temperature/time process conditions. *Journal of the Science of Food and Agriculture* 86: 1328–1335.

UNCTAD-Intergovernmental Group on Citrus Fruits. 2003. http://www.fao.org/es/ESC/en/2 0953/20990/index.html.

Versteeg, C., F. M. Rombouts, and W. Pilnik. 1978. Purification and some characteristics of two pectinesterase isozymes from orange. *Food Science and Technology (Lebensmittel Wissenschaft und Technologie)* 11: 267–274.

Versteeg, C. 1979. Pectinesterases from the orange fruit: Their purification, general characteristics and juice cloud destabilizing properties. Ph.D. thesis, *Agricultural Research Reports* 892: 1–109.

Versteeg, C., F. M. Rombouts, C. H. Spaansen, and W. Pilnik. 1980. Thermostability and orange juice cloud destabilizing properties of multiple pectinesterases from orange. *Journal of Food Science* 45: 969–971.

Wicker, L. and F. Temelli. 1988. Heat inactivation of pectinesterase in orange juice pulp. *Journal of Food Science* 53(1): 162–164.

Wicker, L., J. L. Ackerly, and J. L. Hunter. 2003. Modification of pectin by pectinmethylesterase and the role in stability of juice beverages. *Food Hydrocolloids* 17: 809–814.

Willats, W. G. T., C. Orfila, G. Limberg, H. C. Buchholt, G. J. W. M. Van Alebeek, A. G. J. Voragen, et al. 2001. Modulation of the degree and pattern of methylesterification of pectic homogalacturonan in plant cell walls: Implications for pectin methyl esterase action, matrix properties and cell adhesion. *Journal of Biological Chemistry* 276: 19404–19413.

chapter nine

Use of enzymes for wine production

Encarna Gómez-Plaza, Inmaculada Romero-Cascales,
and Ana Belén Bautista-Ortín

Contents

9.1   Introduction

Enzymes play an important role in winemaking. They occur naturally in grapes but they may also arise from yeasts, fungi, and bacteria. The wine-maker can extend the action of these endogenous enzymes by using exo-genous commercial enzymes. Indeed, the application of industrial enzyme preparations in the wine industry is a common practice. Pectinase enzymes were used as far back as 1947 and since that time they have been used to increase juice yield, filtration rate, rate of settling, and clarity of wines (Ough and Crowell, 1979; Brown and Ough, 1981; Felix and Villettaz, 1983; Capdeboscq et al., 1994). Improved knowledge of the nature and structure of the macromolecules found in must and wine has opened new possibilities for using enzymes in winemaking. Nowadays, the changes obtained by enzymatic treatments affect not only clarification and filtration operations but also the extraction and stabilization in both white and red wines such as the increased production of aroma compounds and control of bacteria (Canal-Llaubères, 1993). The most widely used enzymes available for commercial use are pectinases, glucanases, and glycosidases (Lourens and Pellerin, 2000) although other enzymes can also be found, such as lysozymes and ureases.

Enological enzymatic preparations are usually produced by fermenting pure cultures of selected microorganisms. The fungi are grown on a medium, generally a sugar, as carbon source, and are then stimulated to produce the desired enzymes through the correct choice of the substrate. The most commonly used method is culture in immersed medium. The product of the fermentation is an enzymatic preparation containing several enzymatic activities, which can be purified by filtration, ultrafiltration, or other specific treatments. The enzymatic preparations are presented in liquid or solid forms.

The application of commercial enzyme preparations is legally controlled in Europe, mainly by The International Office of the Vine and Wine (OIV, www.oiv.int), that established that only species accepted as GRAS (generally recognized as safe) can be used for enological enzyme production. Both the new International Enological Codex and the OIV Code of Enological Practice authorize the use of enzymes for a whole series of applications (Resolutions 11 to 18/2004). European legislation is more restrictive and is based on the principle of an approved list. Hence, the EC Regulation 1493/1999 only authorizes pectinases from *Aspergillus niger*, β-glucanase produced by *Trichoderma harzianum*, urease from *Lactobacillus fermentum*, and lysozyme usually from egg whites.

9.2   Use of pectinases in winemaking

Grape skin and pulp cell walls are highly complex and dynamic, being composed of polysaccharides, phenolic compounds, and proteins, stabilized by ionic and covalent linkages. Hemicellulose, pectins, and structural proteins are inter-knotted with the network of cellulose microfibrils, the skeleton of cell walls (Huang and Huang, 2001). In grape cell walls, cellulose and pectins account for 30–40% of the polysaccharide components of the cell wall (Nunan et al., 1997). Figure 9.1 shows the percentage of sugar composition of Monastrell skin cell walls.

The enzymes involved in the hydrolysis of fruit cell walls are mainly pectinases, cellulases, and hemicelullases, all of which act in concert. Pectin degradation requires the combined action of several enzymes that can be classified into two main groups: methylesterases, which remove methoxyl groups from pectin, and depolymerases (hydrolases and lyases), which cleave the bonds between galacturonate units. Pectinmethylesterase

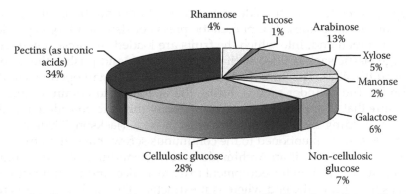

Figure 9.1 Sugar composition of grape skin cell walls (data from Ortega-Regules et al., 2008a).

(EC 3.1.1.11) catalyzes the demethylesterification of galacturonic acid units of pectin, generating free carboxyl groups and releasing protons. Demethylesterified pectin may undergo depolymerisation by glycosidases. Polygalacturonase (EC 3.2.1.15) catalyses the hydrolytic cleavage of β-1,4 linkages of pectic acid. It may be endo (causes random cleavage of β-1,4 of pectic acid) or exo (causes sequential cleavage of β-1,4 linkages of pectic acid from the non-reducing end of the pectic acid chain). Lyases cleave glycosidic bonds by β-elimination, giving rise to unsaturated products. Among these enzymes, pectin lyases show specificity for methyl esterified substrates (pectin lyase (EC 4.2.2.10)), while pectate lyases (EC 4.2.2.2) are specific for unesterified polygalacturonate (pectate). Exogenous pectinases are widely used in winemaking to improve several operations such as grape maceration and the clarification and filtration of musts and wines.

9.2.1   The increase of yield

One of the first uses of enzymes was to increase the yield during the pressing operations. A wine press is a device used to extract must or wine from crushed grapes during winemaking. There are a number of different styles of presses and they can produce juice or wine fractions of different physicochemical properties and qualities. Each style of press exerts controlled pressure to free the juice or wine from the grapes, and they work at room temperature. The vertical presses consist of a large basket that is filled with the crushed grapes. Pressure is applied through a plate that is forced down onto the fruit. The juice flows through openings in the basket. The basket style press was the first type of mechanized press to be developed, and its basic design has not changed in nearly 1000 years. The horizontal press works using the same principle as the vertical press. Instead of a plate being brought down to apply pressure on the grapes, plates from either side of a closed cylinder are brought together to squeeze the grapes. The pneumatic press consists of a large cylinder, closed at each end, into which the fruits are loaded. To press the grapes, a large plastic sack is filled with compressed gas and pushes the grapes against the sides. The juice then flows out through small openings in the cylinder. The cylinder rotates during the operation to obtain a uniform pressure that is applied on the grapes. Today they are considered as the presses yielding the higher quality must or wine (Jackson, 2000). Finally, it should also be mentioned to the continuous screw that differs from the other presses. There is an Archimedes screw to continuously force grapes up against the wall of the equipment to extract juice, and the pomace continues through to the end where is it extracted. This style of press is not often used to produce table wines, and some countries forbid its use in higher-quality wine production.

The degradation of the cell walls of the grape cells by pectinase allows a higher diffusion of the components located inside the vacuoles, facilitating a better extraction of the must during pressing (Uhlig and Linsmaier-Bednar, 1998).

Some studies (Ough and Berg, 1974; Ough et al., 1975) showed that the use of pectinase enzymes to enhance the must yield was more effective in white and rose vinifications than in the elaboration of red wines. In red winemaking it is necessary that a skin-must have contact time to extract polyphenols and other molecules from skins. When the maceration advances, the presence of ethanol also participates in the degradation of the grape cell walls, making the effect of the enzymes on yield at pressing less evident.

9.2.2   Clarification and filterability of musts and wines

During the elaboration of white and rosé wines, and after pressing, grape must is rich in solid particles. Negatively charged pectin molecules form a protective layer around positively charged solid particles, keeping them in suspension. An excessive turbidity in musts induces a herbaceous aroma to the wine, the apparition of sulfur-like off aroma, and a high isoamyl alcohol content (Armada and Falqué, 2007). Therefore, must clarification is a very important operation for improving the quality of white and rosé wines. However, a certain amount of suspended grape particulate is required for fermentation and ester production to proceed, because if clarification is excessive, fermentation becomes difficult and wine flavor is poor (Ferrando et al., 1998). Clarification involves only physical means of removing suspended particulate matter (Jackson, 2000). To accelerate the process, some agents are used to eliminate this matter forming aggregates generally large enough to precipitate. Activated carbon, bentonite, gelatin, casein, polyvinylpyrrolidone, and others are the agents used in this process.

These agents include pectinase enzymes, which break the pectin molecules into smaller components, thereby exposing some of the positively charged grape solid particles underneath this protective layer. This leads to an electrostatic aggregation of oppositely charged particles (positive proteins and negative tannins and pectin) and the flocculation of the cloud and subsequent clarification of the must. The enzymes used for clarification are the most basic commercial enzymes. They have three main activities: pectin lyase (PL), pectin methylesterase (PME) and polygalacturonase (PG). Clarification enzymes work mainly on the soluble pectins (mainly homogalacturonans) of the pulp of grapes. The first stage consists of destabilization of the cloud by PL, which results in a strong decrease in the viscosity. PG becomes active only after the action of PME. Because of its high molecular weight, PG cannot hydrolyze pectin with a high degree of methylation as a result of steric hindrance. PME cleaves the methyl units of the polygalacturonic acid chain and pectin becomes pectate. When the methyl groups are removed, PL is unable to recognize its substrate. At this stage pectin hydrolysis is mainly due to the action of PG. The second stage is the cloud flocculation. The cloud is composed of proteins that are positively charged at the pH of the juice. These proteins are bound to hemicelluloses that are surrounded by pectin as a negatively charged protective colloid layer. The hydrolysis of pectin results in the aggregation of positive-charged proteins and negative-charged tannins and pectin, flocculation of the cloud, and subsequent clarification of the must.

The pectolytic enzymes can also be used after fermentation. Filtration is employed in the wine industry to produce clear wines and to improve the visual quality, a very important characteristic in white and rose wines, and in many cases, to obtain microbe-free wines at the moment of bottling. Three different types of filtration are used in the wineries: conventional filtration that removes particles down to diameter of 1 μm, microfiltration (1.0–0.1 μm), and ultrafiltration (0.2–0.05 μm). An excess of colloids is capable of hindering or impeding filtration. When added before filtration, the level of enzyme supplemented must be adjusted to allow for the inhibitory effect of alcohol on pectinases (Kashyap et al., 2001). The effects of the enzymes are evident during the wine filtration process, less filter material is needed, less wine is lost, and less filtration time is employed (Brown and Ough, 1981). Microfiltration experiments with model solutions and wines have confirmed the fouling properties of wine polysaccharides (Vernhet et al., 1999; Vernhet and Moutounet, 2002).

The use of pectinase preparations can also be necessary for red wines made with thermovinification, since the heating of the harvest has as a consequence a complete inactivity of the natural enzymes of the grape and an increase in the content of pectins in must (Mourgues y Bérnard, 1980; Doco et al., 2007). In this case, very concentrated preparation of pectolytic enzymes needs to be employed for the clarification of press wines and must from heated grapes.

9.2.3   Maceration of grapes for red wine vinification

During winemaking, the grape skin cell walls form a barrier that hinders the diffusion of components that is important for the aroma and color of wine. The color of red wines is mainly due to anthocyanins and tannins. Anthocyanins are located in the skin cell vacuoles and transferred from grape skins to must/wine during the maceration stage. Tannins can be found in seeds and grape skins. In the skins, they are found in free form inside the vacuole or bound to the cell wall.

Romero-Cascales et al. (2005) showed that the anthocyanin content of a given cultivar is not always correlated with the anthocyanin concentration in the wine produced from it, since some varieties show greater difficulty in releasing their anthocyanins and tend to retain them in the skins, even after the maceration step. A certain correlation between the characteristics of the cell wall and the easiness of anthocyanin extraction has been found (Ortega-Regules et al., 2006).

The extraction of anthocyanins during maceration requires that the pectin-rich middle lamella has to be degraded to release the cells, and the cell walls to be broken to allow the cell vacuole contents to be extracted or to diffuse into the wine (Barnavon et al., 2000). Cell walls loosening requires the breakdown of chemical bonds among the structural components (Huang and Huang, 2001), and so, changes in the cell-wall polysaccharide structure could affect the solubility and the tissue disassembling mechanisms (Shiga et al., 2004). Figure 9.2 shows how the use of commercial pectinase preparations (Lafase HE Gran Cru, Laffort Enologie, Bordeaux, France) affects the composition of polysaccharides of the skin cell walls during the maceration process, especially the concentration of uronic acids. Although the level of those acids and total sugars decreases as the maceration time increases, the decrease was always higher when the commercial enzyme was used, illustrating the action of the enzyme on cell walls. Figure 9.3 shows how the enzymes act on the skin cell walls.

Besides pectinases, the commercial preparations used for maceration usually contain other activities that can help disgregate the cell wall, such as cellulase (EC 3.2.1.4) and hemicellulases (xylanases and galactanases) (EC 3.1.1.73), which also are effective in increasing wine color (Gump and Halght, 1995) and for liberating tannins bound to the cell walls (Amrani Joutei et al., 2003).

The use of exogenous pectolytic enzyme preparations, comprising a mix of the previously mentioned activities, may help in the disgregation of the cell wall structural polysaccharides, as shown in Figure 9.3, facilitating

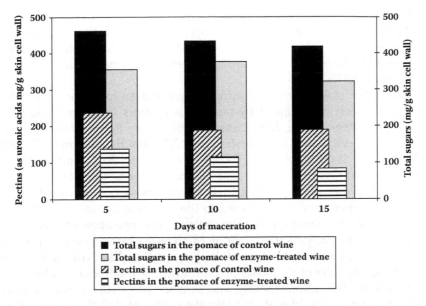

Figure 9.2 Effect of the use of a commercial pectinase preparation (Lafase HE Gran Cru, Laffort Enologie, Bordeaux, France) on the concentration of sugars and pectins (expressed as uronic acids) in the skin cell walls after 5, 10, or 15 days of maceration.

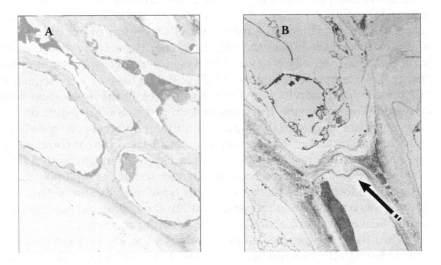

Figure 9.3 Dissembling effect of a commercial enzyme (Enozym Vintage, Agrovin, Spain) on the skin cell wall structure A: control grapes, B: using a commercial enzyme preparation.

polyphenol extraction (Parley, 1997; Gil and Vallés, 2001; Clare et al., 2002) and it has become a common practice in enology. Macerating enzymes not only affects wine color but may also modify the stability, taste, and structure of red wines and increase mouthfeel sensations (Canal-Llaubères and Pouns, 2002).

Since the claim that macerating enzymes may improve wine color, they have been investigated by numerous authors (Watson et al., 1999; Pardo et al., 1999; Delteil, 2000; Canal-Llaubères and Pouns, 2002; Zimman et al., 2002; Revilla and Gonzalez-San José, 2003; Bautista-Ortín et al., 2004; Alvarez et al., 2005; Sacchi et al., 2005; Bautista-Ortín, 2005; Bautista-Ortín, 2007; Romero-Cascales et al., 2008; Romero-Cascales, 2008). However, contradictory results regarding their effectiveness can be found in the literature, which may be attributed to the different nature and different activities of the commercial preparations, the presence of some side activities in the enzyme preparations (such as β-glucosidase or cinnamyl esterase), or the different nature of the grape cell walls of different varieties (Ortega-Regules et al., 2008a). Some studies have pointed to substantial increases in wine color, stability, and sensory properties using the maceration enzymes (Sacchi et al., 2005; Bautista-Ortín et al., 2005; Kelebek et al., 2007; Romero-Cascales et al., 2008) while others reported very limited effects (Zimman et al., 2002; Alvarez et al., 2005).

Several different enzymatic preparations can be obtained commercially, all of them a mix of pectinases, cellulases, and/or hemicellulases. They are usually added to the must just after crushing and the doses depend on the supplier. Usually 2–5 g/HL are recommended, although the correct doses will depend on the concentration of the commercial preparation, the must pH, and the temperature. Optimum pH of pectinases is around 4.5, so the higher the pH, the higher the activity. For low pH (<3.2), enzymatic activity is reduced, so it is important to increase enzyme dosage. Low temperatures (under 15°C) will also have as a consequence that the enzymes' activity is reduced and therefore the doses should be increased. The enzymatic spectrum of each preparation depends on the microorganism strain and the culture conditions, and these are specific to each enzyme producer. The activity profiles of six different commercial pectinase preparations (E1–6) are shown in Figure 9.4. It can be seen how enzyme preparation E1 presented the highest activity of total polygalacturonase, endo-polygalacturonase, pectin lyase, and cellulase. This last enzyme was not present in preparations E2 and E4, while preparations E5 and E6 showed very low activities for all enzymes concerned.

When these six commercial preparations were used for the vinification of Monastrell wines (Romero-Cascales et al., 2008), the results showed that the enzyme-treated wines had a higher phenolic and tannin content at the end of maceration and alcoholic fermentation, and a higher color

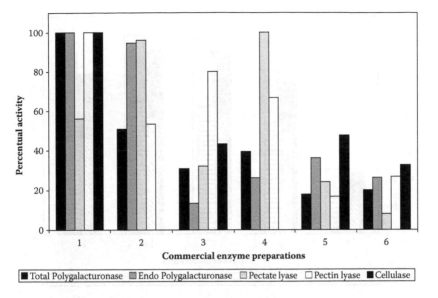

Commercial enzyme preparations

| ■ Total Polygalacturonase | ■ Endo Polygalacturonase | □ Pectate lyase | □ Pectin lyase | ■ Cellulase |

Figure 9.4 Comparison of the activity of each of the most important enzyme activities present in six different commercial preparations (E1–6). The highest activity of a given enzyme among the six commercial preparations is attributed a value of 100% of activity and the activity of this same enzyme in the other preparations is expressed as a percentage of the one showing the maximum value. (Adapted from Romero-Cascales et al., 2008.)

intensity after 12 months of bottle storage, compared with the control wine. Although differences in the chromatic characteristics of the control and enzyme-treated wines were quite pronounced, only small differences were found among the enzyme-treated wines themselves, despite the differences found in their respective enzymatic characteristics. Similar results were found by Revilla and González-San José (2003). The use of pectolytic enzymes gave wines better chromatic characteristics and these were more stable over time than the control wines, whether the so-called clarifying enzymes (pectinase preparations specifically prepared for clarification) or maceration enzymes were used.

Since the color of red wines is such an important characteristic for wine quality, attention has been paid to the effect of enzymes depending on grape characteristics (especially maturity at the moment of harvest) or vinification conditions (mainly maceration time and temperature). For example, Romero-Cascales (2008) reported the effect of a commercial preparation on wine color, depending on maceration time (5, 10, and 15 days). The results suggested that the use of enzymes seems to be a good tool to shorten the time spent by the mass of wine in the tanks during the

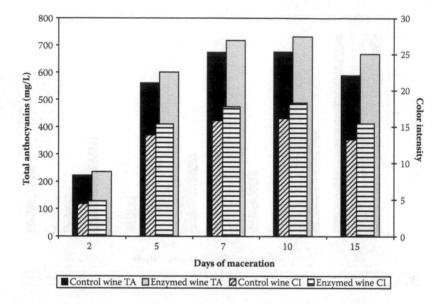

Figure 9.5 Evolution of total anthocyanins (TA) and color intensity (CI) in control and enzyme-treated wine (Lafase He Gran Cru, Laffort Enologie, Bordeaux, France) during 15 days of maceration.

maceration step, achieving a breakthrough in the extraction of phenolic compounds of about 3 days, compared with wines produced without the addition of the enzyme (Figure 9.5). In addition, the chromatic characteristics were better maintained for a longer time.

The effect of the degree of grape maturation on the effectiveness of the enzyme has also been studied. Ortega-Regules et al. (2008b) stated that the quantity of the cell wall material was higher in grapes at the beginning of the ripening process than when they reached maturity. Bearing this in mind, the use of macerating enzymes could be very interesting in the vinification of less than fully ripe grapes, since they would help skin disgregation and favor better extraction of the phenolic compounds into the must. The results showed that the enzyme was less effective in the most mature grapes. These grapes were over-ripe, and their skin showed a high degree of natural degradation, and therefore the effect of the enzyme was less evident. Better effect was found in the early and mid-term harvested grapes (Romero-Cascales, 2008).

Pectinase preparations may contain some enzymatic side-activities that could be beneficial or detrimental for wine quality, depending on the style of the wine. Among them, special attention has been paid to β-glucosidases and cinnamyl esterases, whose effects and mechanism of action will be discussed in the next sections.

9.3   β-*Glucosidases*

Wine aroma, a very important sensory parameter, is composed of a wide variety of compounds with different aromatic properties. More than 800 volatile compounds have been identified in wine. Among them, there are the flavor compounds originating from fruits, the varietal aroma compounds. These compounds are mainly located in the internal cell layers of the skin and in lower concentration in the pulp and juice (Gómez et al., 1994). In grapes, apart from free flavor components, a significant part of the important flavor compounds is accumulated as non-volatile and flavorless glycoconjugates, known as glycosidic aroma precursors and identified for the first time by Bayonove et al. (1984) in Muscat of Alexandria. In several grape varieties, volatiles originating from glycosides have been detected in concentrations several-fold greater than their free counterparts (Pogorzelski and Wilkowska, 2007). They may also appear in non-aromatic varieties (Gómez et al., 1994) and their release can also modify wine aroma (Figure 9.6).

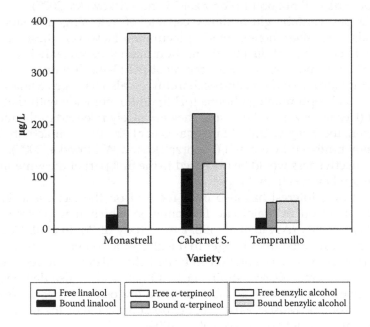

Figure 9.6 Free and bound volatile compounds in non-aromatic varieties. (Adapted from Gómez et al., 1994.)

The aglycone part of glycosides is often formed by monoterpenes, C13-norisoprenoids, benzene derivatives, and long-chain aliphatic alcohols. The sugar moiety includes glucose or disaccharides (Mateo and Di Stefano, 1997). In *Vitis vinifera* there are mainly diglycosides, which means that the aroma compounds are bound to glucose and other carbohydrate residue such as arabinose, rhamnose, or apiose (Mateo and Di Stefano, 1997). The sugar breakdown must be sequential and the other sugars must be removed first before glucose can be released, therefore glucosidase alone is not effective in releasing the aromatic components from the di-glycosylated precursors. The action of exoglycosidases is necessary, according to the sugar moieties of the substrates (Pogorzelski and Wilkowska, 2007). After cleavage of the intersugar linkage, which releases the corresponding sugars and β-glucosides, β-glucosidase liberates the aglycone and glucose.

It is now well established that the glycosidically bound fraction forms a reserve of aroma that may be worth exploiting. Upon acid hydrolysis, these odorless non-volatile glycosides can give rise to odorous volatiles during the winemaking process or wine storage, although the studies have shown that acid hydrolysis liberation of glycosides occurs quite slowly in the winemaking conditions and the liberation of this reserve of aroma compounds will not occur (Pogorzelski and Wilkowska, 2007).

Grapes contain glucosidases capable of releasing aromatic compounds from their non-aromatic precursors. However, these enzymes are not very efficient during vinification mainly because their optimum pH (5.0) does not coincide with the must pH (3–4). Certain wine yeasts, both Saccharomyces and non-Saccharomyces, also have glycosidase activity but their optimum conditions (pH 5) do not coincide with that of the must (Hernandez et al., 2003) and they are rarely released in the medium (Palmeri and Spagna, 2007; Manzanares et al., 2000). Their activity is also strongly inhibited by ethanol (Pogorzelski and Wilkowska, 2007), and so their effectiveness would be restricted to the first part of the wine-making process, when no ethanol is present.

Recently, interest has also been focused on the lactic acid bacteria strains involved in malolactic fermentation which may present a glucosidase activity (Aryan et al., 1987). In a study of Boido et al. (2002), the authors maintained that changes in the glycoside content of Tannat wines during malolactic fermentation indicated the existence of such activity in the commercial *O. oeni* strains used. Other studies also demonstrated that some *O. oeni* strains are able to act on glycosides extracted from the highly aromatic Muscat variety (Ugliano et al., 2003) or the nonaromatic Chardonnay variety (D'Incecco et al., 2004).

Due to the limited effect of glucosidases from grape and *Saccharomyces cerevisiae* in winemaking, a large part of glycosides is still present in young wines. In terms of industrial production, the commonest application is represented by commercial preparations of *Aspergillus niger.* The exogly-cosidases and β-glucosidase from *Aspergillus niger* have shown good stability at the acidic wine pH, contrary to the enzymes from *Saccharomyces cerevisiae.* Fungal β-glucosidase activity is partly inhibited by glucose. For this reason, enzymes from fungi should be used in the final stages of winemaking. Another point of interest is that fungal glycosidases also contain all four enzyme activities—namely glucosidase, arabinosidase, rhamnosidase, and apiosidase—to complete the release of the aglycon (Pogorzelski and Wilkowska, 2007). The enzymatic action must be stopped after one to four months depending on the desired effect that is required. The enzymes can be removed with bentonite.

Nowadays, one of the most common options for applying β-glucosidase is to use commercial pectinase preparations from Aspergillus with β-glucosidase as side activity. This could be interesting for releasing wine aroma compounds but it could be a problem with color stability in red wines since this activity could affect the molecule of anthocyanins, leading to the formation of unstable anthocyanidin; in fact, this side activity of most of pectinase preparations is considered undesirable for red wine production. Le Traon-Masson et al. (1998) studied the specificity of two β-glucosidases purified from an *Aspergillus niger* and found that they had totally different activity spectra. One of them was highly active on aroma glucosides but very slowly degraded malvidin-3-glucoside, and therefore its effect on color was negligible and could be used to reveal aroma compounds even in red wines. The other enzyme can be defined as a specific anthocyanase since its activity is directly related to the release of glucose from anthocyanidin-3-glucosides. Therefore, the release of aroma compounds in red wines should be based on enzyme preparations whose β-glucosidase activity would be ineffective towards anthocyanins.

9.4   The presence of cinnamyl esterase activity in enzyme preparations

Cinnamyl esterase (EC 3.1.1.73), an enzyme activity present in certain pectinase enzyme preparations (as a side activity), together with the cinnamyl decarboxylase produced by certain yeast strains, can be responsible for the production of volatile vinyl-phenols (Gerbeaux et al., 2002), mainly 4-vinylphenol and 4-vinylguaiacol. These volatile phenols can cause off-flavors in wine, which are described as phenolic or medicinal when present at high concentrations (Suarez et al., 2007). The first step in the formation of these volatile phenols is the action of cinnamyl esterase on the hydroxy-cinnamic acid tartaric esters present in the must. These esters are a major group of polyphenols in white grape musts and are also abundant in reds. The second reaction is the transformation of the hydroxycinnamic acids into vinylphenols by the action of a yeast cinnamate decarboxylase (Figure 9.7).

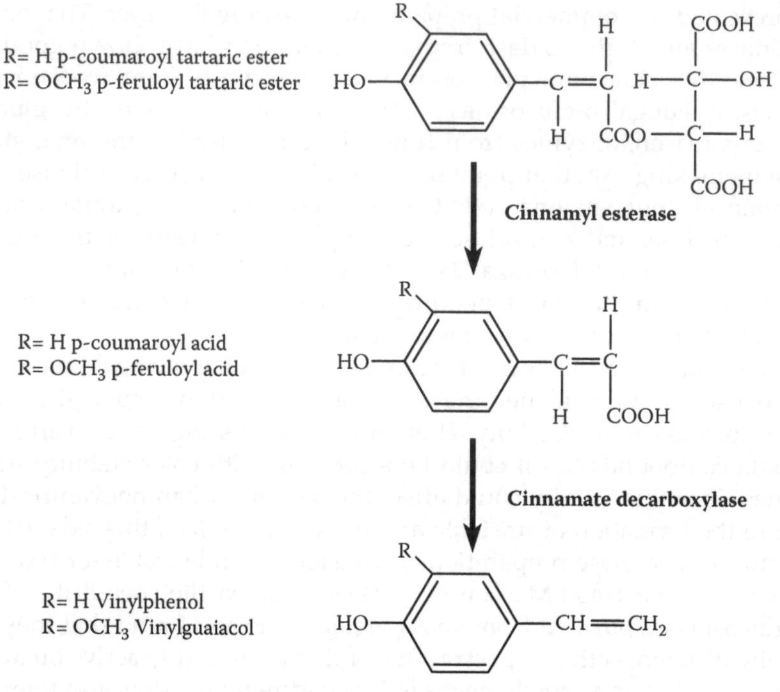

Figure 9.7 Mechanism of vinylphenols formation from hydroxycinnamic esters.

This last enzyme is produced by some *Saccharomyces cerevisiae* strains known as POF(+) (Phenyl Off Flavors (+)) as well as by contaminating yeasts. This problem can largely be solved, especially in white wines, by using purified enzyme preparations free of cinnamyl esterase activity.

The problem with red wines is different. Cinnamate-decarboxylase from *S. cerevisiae* is inhibited by phenolic compounds (Gerbeaux et al., 2002), and this may explain the low concentrations of vinylphenols in red wines. But red wines can be contaminated with other yeasts such as the Brettanomyces. This yeast also presents cinnamate decarboxylase enzyme activity and can therefore also decarboxylase hydroxycinnamic acids to vinyl-phenols. However, in addition to the decarboxylase, *Brettanomyces* also contain the enzyme vinyl-phenol reductase which converts vinyl-phenols into ethyl-phenols (Figure 9.8). This is a unique characteristic of Brettanomyces, and causes a deeper problem due to the much stronger off-flavors of the ethylphenols (Suarez et al., 2007). Through microbiological studies performed in Burgundy on Pinot Noir wine, it has been shown that these yeasts are present in about 50% of wines undergoing maturation and in about 25% of those in bottles (Gerbaux et al., 2000).

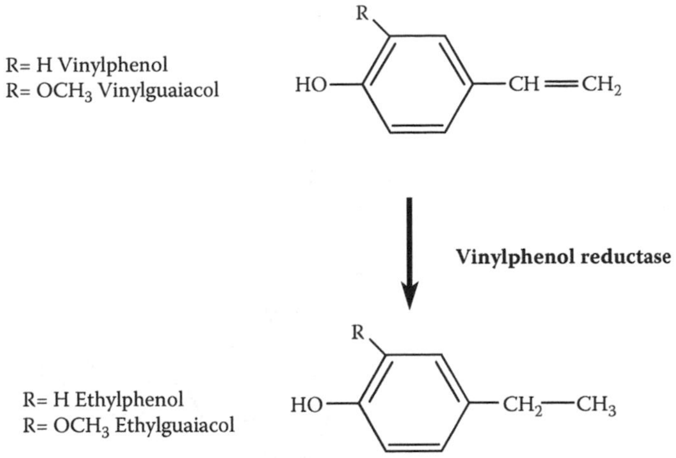

R= H Vinylphenol
R= OCH₃ Vinylguaiacol

$$HO-\bigcirc-CH=CH_2$$

Vinylphenol reductase

R= H Ethylphenol
R= OCH₃ Ethylguaiacol

$$HO-\bigcirc-CH_2-CH_3$$

Figure 9.8 Mechanism of ethylphenols formation mediated by a vinylphenol reductase.

This problem can be especially severe in oak-matured wines since oak wood can be easily contaminated with this yeast and they are very difficult to clean (Suarez et al., 2007; Pérez-Prieto et al., 2002). Due to their lower pH and higher SO_2 content, *Brettanomyces* are rarely found in white wines (Gerbaux et al., 2002).

Anyway, other considerations must be taken into account in red wine technology. The reactions between malvidin-3-glucoside, the main anthocyanin in red wines and vinylphenols, can lead to the formation of pyranoanthocyanins-vinylphenols, very stable compounds that may ensure the color stability of mature red wines (Figure 9.9). The formation of vinylphenol anthocyanidin adducts is favored by the presence of the substrates (hydroxycinnamic acids) and fermenting the must with yeasts that show hydroxycinnamate decarboxylase activity. This enzyme turns hydroxycinnamic acids in the grape into vinylphenols, which readily condense with malvidin-3-O-glucoside or other anthocyanins to form pyranoanthocyanins-vinylphenols (Morata et al., 2007).

Thus, according to Morata et al. (2007), the presence of cinnamyl esterase activity (which could provide the substrate for these reactions), together with the use of yeasts selected for high hydroxycinnamate decarboxylase activity, can raise the concentration of stable pigments in wine, favoring color intensity and reducing hue values. This could be particularly important in hot winemaking areas where grape anthocyanins synthesis is reduced, or when wines are destined for long ageing in barrels, as long as the wines are protected from the presence of Brettanomyces.

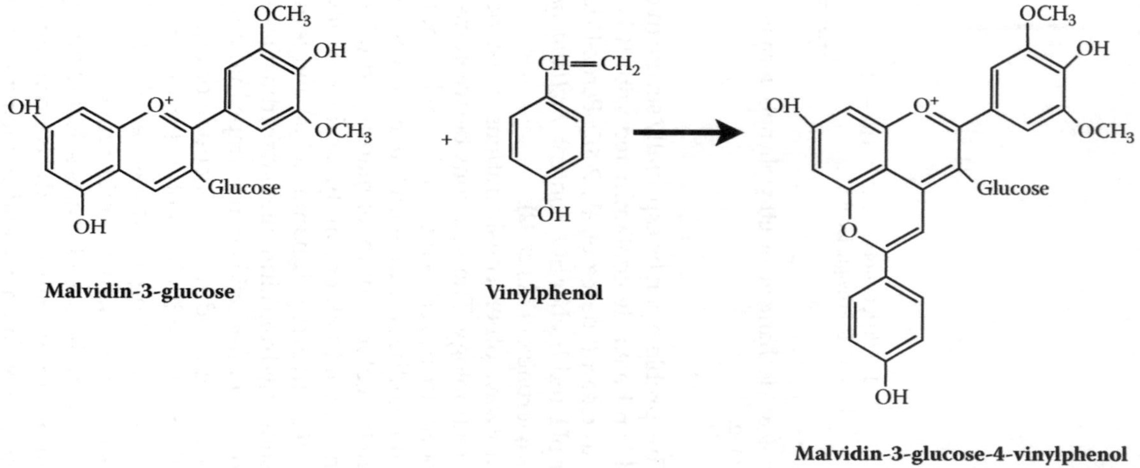

Figure 9.9 Mechanism of formation of malvidin-3-glucose-4-vinylphenol.

9.5   Glucanases

Glucanases (EC 3.2.1) are mainly used in winemaking to avoid the problems caused by glucans, which appear in wines from *Botrytis cinerea* infected grapes, during the clarification and filtration processes, especially in sweet white wines. This was the first use to which glucanases were put, although they were later used to favor maturation of wines on lees.

The glucans are polysaccharides, more specifically polymers of glucose that exist in the cell walls of some yeasts, fungi, bacteria, fruit, and algae (Volman et al., 2008; Xu et al., 2008; Shih et al., 2008; Rhee et al., 2008). The glucans present in musts and wine are usually of microbial origin and generally show some structural variability. Glucans from *Saccharomyces cerevisiae* are mainly β-D-1,3-linked glucose units with β-D-1,6-linked lateral glucose chains. Some branched β-1,6-glucans with some β-1,3-links are also present (Manners et al., 1973a, 1973b). *Pediococcus dammosus*, a spoilage bacteria causing ropy wines, secretes a trisaccharide repeating unit: →3)-β-D-Glc*p*-(1→3)-[β-D-Glc*p*-(1→2)]-β-D-Glc*p*-(1→ (Llabères et al., 1980). *Botrytis cinerea* releases a high-molecular-weight glucan (approximately 800,000 Daltons) that consists of a β-D-1,3-linked backbone with very short β-D-1,6-linked side chains (Montant et al., 1977; Dubourdieu et al., 1981; Villettaz et al., 1993).

In a similar manner to pectin, the glucans present in musts and wine from botrytised grapes (gray rot/noble rot) can prevent the natural sedimentation of the cloud particles. This can cause premature filter fouling during the filtration processes. Fining agents such as bentonite or centrifugation, which are used to accelerate settling, are generally unable to overcome this effect since glucans are not removed, and therefore, filtration problems remain. The filtration problems occur only in the presence of alcohol, so the filtration problems related to the presence of Botrytis β-glucan are less severe for grape juice (Villettaz et al., 1984). The high molecular weight glucan produced by *Botrytis cinerea* can only be hydrolyzed by β-glucanases releasing glucose and gentiobiose (Dubourdieu et al., 1981).

Glucanases, classified as endo- and exo-glucanases, act synergistically in hydrolyse the β-*O*-glycosidic linkages of β-glucan chains leading to the release of glucose and oligosacccharides (Dubourdieu et al., 1981). β-glucan-degrading enzymes are classified according to the type of β-glucosidic linkages: 1,4-β-glucanases (including cellulases); 1,3-β-glucanases; and 1,6- β-glucanases (Pitson et al., 1993).

Shortening of the glucan chain avoids problems during filtration (Villettaz et al., 1984; Humbert-Goffard et al., 2004). The alcohol seems to act as an aggregation factor, inducing some kind of polymerization of the glucan molecules. It has been reported that the filtration problems exponentially increase as soon as the alcohol level rises in the wine (Dubourdieu et al., 1981); therefore, the most severe problems occur at the end of the alcoholic fermentation. Glucanase enzymes can be applied to musts before fermentation or on wine after the alcoholic fermentation.

The commercial β-glucanase preparations authorized for use in wine-making are produced by *Trichoderma* (e.g. *T. harzianum and reesei*) species. These enzymatic preparations are a mixture of endo- and exo-β-1,3 glu-canases and can also contain hemicellulases and cellulases as secondary activities. These enzymes show good activity at the temperature, pH, SO_2, and alcohol levels of wine (Molina- Úbeda, 2000).

The other use of β-glucanases is for the maturation of wines. Aging on lees has been used in the manufacture of white wines fermented in barrels (Bourgogne wines), natural sparkling wines (Champagne, Cava), and aged biological wines produced with flor yeast (Sherry) (Ribéreau-Gayon et al., 1998a). During aging on lees, the wines become enriched in volatile aro-matic compounds, their phenolic compounds are protected from oxidation by the reducing nature of the lees, and their density is increased through the release of high molecular weight compounds from the cell walls of the dead yeast (Fornairon-Bonneford et al., 2002; Fornairon-Bonnefond and Salmon, 2003; Charpentier et al., 2004; Pérez-Serradilla and Luque de Castro, 2008). In recent years, this winemaking technique has been used by many wineries for red wine vinification since it leads to good qual-ity products with better structure, aromatic profile, and color stability. Manoproteins liberated during the autolysis of yeast are responsible for increasing wine quality since they participate in processes such as the pre-vention of potassium bitartrate precipitation (Lubbers et al., 1993; Ribéreau-Gayon et al., 1998b); the prevention of protein precipitation (Ledoux et al., 1992; Waters et al., 1993; Dupin et al., 2002); aroma stabilization (Feuillat et al., 1987; Lubbers et al., 1994); and interactions with phenolic compounds (Escot et al., 2001; Riou et al., 2002), leading to a decrease in astringency, an increased roundness and volume in mouth, and greater color stability.

Aging on lees begins with the autolysis of the yeasts. This involves the hydrolysis of their cell walls by the enzyme β-1,3 glucanase (Ledoux et al., 1992) and the release of manoproteins into the medium. The yeast pro-duces β-1,3 glucanase, which is active in the biomass and can be secreted into the surrounding media. By selecting strains that undergo rapid autol-ysis or accelerating the effect with the addition of exogenous β-glucanase enzymes, shorter ageing times can be achieved, with a better wine struc-ture and stability (Charpentier and Feuillat, 2008; Palomero et al., 2007; Palomero et al., 2009).

9.6   Ureases

Ethyl carbamate, a naturally occurring component in all fermented foods and beverages, is being spontaneously produced by the reaction between urea and ethanol. Because ethyl carbamate has shown potential carcinogenic activity when at high doses, there is great interest in reducing its levels in food products. Urea, once a widely used nutrient supplement in wine fermentation to avoid stuck fermentation, is known to form ethyl carbamate in the presence of alcohol, and therefore urea is no longer recommended for use in winemaking (Tegmo-Larsson and Henick-Kling, 1990).

The hydrolysis of urea to ammonia and carbon dioxide by a highly specific enzyme, such as acid urease (EC 3.5.1.5), seems to be a suitable way to avoid the formation of ethyl carbamate from such a precursor (Fidaleo et al., 2006). Most ureases present an optimum pH of 6–7, so they are not suitable for acting in wine. A urease with an optimum pH at 2.4, closer to that of the wine, was first found in *Lactobacillus* sp. (Suzuki et al., 1979). Urease from *Lactobacillus fermentum* was partially purified, characterized, and named *acid urease* by Takebe and Kobashi (1988). The enzyme is applicable to the elimination of urea in fermented beverages. Its use has been accepted by the OIV (Resolution OENO 5/2005). According to this resolution, urease must be obtained from the fermentation of *L. fermentum* in a synthetic environment. When it is complete, the culture is filtered and washed in water, and the cells are killed in 50% (v/v) alcohol. The suspension is freeze-dried or dried by pulverization. Therefore, the preparation consists of a powder made up of whole dead cells containing enzymes.

Urease must be used in wine if it contains more than 3 mg/L of urea. When the decrease in urea is noticeable all enzymatic activity must be eliminated by filtering the wine. The elimination of urease is necessary to avoid health problems such as kidney stones.

9.7 Utilization of lysozyme

Lysozyme is an enzyme found in tears, nasal secretions, and the white of avian eggs. It has been used for years as a biopreservative in the processing and storage of hard cheese (Carini et al., 1985). The use of lysozyme in winemaking was approved by OIV in 1994 (resolution OENO 10/97) at a maximum rate of 500 mg/L. Lysozyme is a peptidoglycan N-acetylmuramylhydrolase (EC 3.2.1.17) (Marchal et al., 2002), which cleaves the peptidoglycan in the cell wall, leading to fractures in the cell membrane by lysis, after which a fine sediment resembling the texture of fine lees will occur. This enzyme attacks the cell walls of Gram-positive bacteria, such as lactic bacteria, and causes the cell to rupture and die but has no effect on yeasts or Gram-negative bacteria.

The main use of lysozyme in winemaking is to prevent or delay malolactic fermentation and to diminish the quantities of SO_2 added to the must or the wine. Some wines may have a brighter, fresher flavor profile if the wine only undergoes partial malolactic fermentation. In other cases, the wine-maker wants to prevent or stop the fermentation entirely. There are cases in red vinification where the malolactic fermentation occurs very early; then, lactic spoilage problems may occur, and the skin maceration has to be stopped, even if it is too early for the quality of the wine, this problem being particularly common in the case of carbonic maceration with whole grapes.

The inhibition of malolactic fermentation by addition of lysozyme to the must or to the wine has been investigated by several authors. Gerbeaux et al. (1997) added lysozyme at different doses to Chardonnay and Sauvignon Blanc musts and observed that malolactic fermentation was inhibited with the addition of 250–500 mg/L of lysozyme. Similarly, Nygaard et al. (2002) reported that in the Merlot wines treated with lysozyme, the populations of indigenous lactic acid bacteria decreased from 10^4 to 10^3 cfu/mL during alcoholic fermentation, in contrast to untreated wines where populations were $>10^4$ cfu/mL. However, malolactic fermentation was successfully induced in wines after alcoholic fermentation, indicating minimal residual enzymatic activity. Green et al. (1995) stated that the use of lysozyme and lysozyme with SO_2 could delay malolactic fermentation in Pinot Noir and Chardonnay wines. The inhibition of malolactic fermentation is more delicate than the delay and cannot be effective in all cases. Lysozyme cannot always prevent the start of malolactic fermentations; it is only a tool to diminish the bacteria population, and has to be used in combination with SO_2 and good hygienic conditions. The addition of lysozyme may also prevent the increase of volatile acidity during stuck/sluggish alcoholic fermentation. Finally, even after fermentation, eliminating residual bacteria may help prevent the formation of undesirable volatile acids, histamines, and other off aromas or flavors. It is important to note that lysozyme has no effect on acetic bacteria or on yeasts, so, in this respect, wines must be surveyed with regard to the concentration of these two types of microorganisms; lysozyme treatment has to be used in combination with SO_2 if a contamination is suspected (especially with *Brettanomyces*).

Also, the formation of complexes between lysozyme and polysaccharides results in a substantial enhancement in foamability. Marchal et al. (2002) observed the effect on the foaming characteristics of Champagne base wine of adding lysozyme to must and wines before and after bentonite or charcoal treatment. They noted that lysozyme had a protective effect on foaming properties when added before bentonite treatment, restoring the correct foamability, even if the deproteinization treatment was severe. However, the effect was smaller after charcoal treatment, probably because of low protein adsorption.

Gerbaux et al. (1997) carried out an experiment to evaluate the ability of lysozyme to suppress the formation of acetic acid and biogenic amines by wine spoilage microorganisms. They observed that the addition of

250 mg/L to red wines, after malolactic fermentation, promoted microbiological stabilization. The control lots (without lysozyme) had higher bacterial populations. In the wines to which lysozyme was added, there was no increase in the content of acetic acid and biogenic amines over a period of six months of storage at 18°C. After fermentation and processing, control lots had 20% higher volatile acidity levels and a four times higher cumulative value for histamine, tyramine, and putrescine.

Gao et al. (2002) investigated the efficacy of lysozyme to control histamine production by spoilage bacteria during winemaking. Results indicated that when used preventively at the beginning of alcoholic fermentation, lysozyme was able to inhibit the growth of the wine-spoilage lactic acid bacteria and subsequently to prevent the formation of histamine by the spoilage bacteria.

The action of lysozyme may be stopped by adding bentonite or metatartaric acid. If the lysozyme is not eliminated, it can induce a precipitate during aging by enhancing protein instability.

9.8 Conclusions

Commercial enzyme preparations are widely used in winemaking (see Table 9.1). The future of the production of enzymes for its use in enology is focused in the production of purified enzymes with no side activities. Pectinase preparations degrade grape pectin but also introduce compositional changes that are detrimental to the quality of the wine from both health and sensory perspectives. Side activities, such as excessive pectin methyl esterase activity, β-glucosidase and cinnamyl esterase activities, of some commercial enzyme preparations may lead to the formation of methanol (which can be oxidized in the body to the toxic compounds formaldehyde and formic acid), anthocyanin degradation, the occurrence of 4-vinylphenol and 4-vinylguaiacol (which can give rise to a medicinal off-flavor), and of oxidation artifacts during the hydrolysis of authentic norisoprenoid glycosides (Pretorius and Hoj, 2005). Obtaining genetically modified microorganisms that do not produce those activities is attracting a lot of attention. Genetically modified microorganisms produce only the desired enzymes, as they are genetically manipulated to continually produce a specific enzyme. The enzyme structure is not modified and does not differ from the one produced by the un-manipulated organism, because the genetic modification techniques apply only to productive microorganism strains and not to enzymes themselves.

Other studies focus on the screening and production of yeasts that contain sufficient pectinase and glycosidase activities to act in the wine during fermentation (Blanco et al., 1999; Takayanagi et al., 2001; Rodríguez et al., 2004).

Table 9.1 Summary of the Main Enzyme Preparations Used in Winemaking

Objective	Enzyme	Moment of application	Contact time (Approximately)
Clarification	Pectinases	White and rosé must: clarification before alcoholic fermentation	1 day
		Clarification of white and rosé wine: before fining and bottling	1–8 days
		Clarification of red press wines: before fining and bottling	5–20 days
Maceration	Pectinases	When filling the maceration tank after crushing	3–8 days
Maturation	Pectinases/ Glucanases	At the end of alcoholic fermentation	6 weeks
Filtration	Pectinases/ Glucanases	Before stabilization and fining	3–20 days
Increased aroma	β-glucosidase	At the end of alcoholic fermentation	3 weeks
Elimination of urea	Urease	At the end of alcoholic fermentation	2 weeks
Elimination of bacteria	Lysozyme	Before the onset of malolactic fermentation	Its effects can last several months in low tannin wines. It can be eliminated with bentonite
		After malolactic fermentation for wine stabilization	

Abbreviations

E1–6: six different commercial pectinase preparations
OIV: the International Office of the Vine and Wine
PG: polygalacturonase (PG)
PL: pectin lyase
PME: pectin methylesterase

References

Aryan, A. P., B. Wilson, C. R. Strauss, and P. J. Williams. 1987. The properties of glycosidases of *Vitis vinifera* and comparison of their β-glucosidase activity with that of exogenous enzymes. An assessment of possible applications in enology. *American Journal of Enology and Viticulture* 38: 182–188.

Álvarez, I., J. L. Aleixandre, M. J. García, and V. Lizama. 2005. Impact of prefermentative maceration on the phenolic and volatile compounds in Monastrell red wines. *Analytica Chimica Acta* 563: 109–115.

Amrani Joutei, K., F. Ouazzani Chahdi, D. Bouya, C. Saucier, and Y. Glories. 2003. Electronic microscopy examination of the influence of purified enzymatic activities on grape skin cell wall. *Journal International des Sciences de la Vigne et du Vin* 37: 23–30.

Armada, L., and E. Falqué. 2007. Repercussion of the clarification treatment agents before the alcoholic fermentation on volatile composition of white wines. *European Food Research and Technology* 225: 553–558.

Barnavon, L., T. Doco, N. Terrier, A. Ageorges, C. Romieu, and P. Pellerin. 2000. Analysis of cell wall neutral sugar composition, β-galactosidase activity and a related cDNA clone throughout the development of *Vitis vinifera* grape berries. *Plant Physiology and Biochemistry* 38: 289–300.

Bautista-Ortín, A. B., J. I. Fernández-Fernández, J. M. López-Roca, and E. Gómez-Plaza. 2004. Wine-making of high colored wines: Extended pomace contact and run-off of juice prior to fermentation. *Food Science and Technology International* 10: 287–295.

Bautista-Ortín, A. B., J. M. López-Roca, A. Martínez-Cutillas, and E. Gómez-Plaza. 2005. Improving color extraction and stability in red wines: The use of maceration enzymes and enological tannins. *International Journal of Food Science and Technology* 40: 867–878.

Bautista-Ortín, A. B., J. I. Fernández-Fernández, J. M. López-Roca, and E. Gómez-Plaza. 2007. Enological practices to improve wine color and their dependence on grape characteristics. *Journal of Food Composition and Analysis* 20: 546–552.

Bayonove, C., Y. Gunata, and R. Cordonnier. 1984. Mise en evidence de l'intervention des enzymes dans le developpement de l'arome du jus de Muscat avant fermentation: La production des terpenols. (Demonstration of the role of enzymes in the development of the aroma of Muscat must prior fermentation: Terpenols production.) *Bulletin de l'O.I.V.* 57: 741–758.

Blanco, P., C. Sieiro, and T. G. Villa. 1999. Production of pectic enzymes in yeasts. *FEMS Microbiology Letters* 175: 1–9.

Boido, E., A. Lloret, K. Medina, F. Carrau, and E. Dellacassa. 2002. Effect of beta-glucosidase activity of *Oenococcus oeni* on the glycosylated flavor precursors of Tannat wine during malolactic fermentation. *Journal of Agricultural and Food Chemistry* 50: 2344–2349.

Brown, M. R. and C. S. Ough. 1981. A comparison of activity and effects of two commercial pectic enzyme preparations on white grape musts and wines. *American Journal of Enology and Viticulture* 32: 272–276.

Canal-Llaubères, R. M. 1993. Enzymes in winemaking. In *Wine microbiology and biotechnology*, ed. G. H. Fleet, 477–506. Philadelphia: Hardwood Academic.

Canal-Llaubères, R. M. and J. P. Pouns. 2002. Les enzymes de maceration en vinification en rouge. Influence d'une nouvelle preparation sur la composition des vins. (The maceration enzymes during red winemaking. Influence of a new preparation of wine composition.) *Revue des Oenologues* 104: 29–31.

Capdeboscq, V., P. Leske, and N. Bruer. 1994. An evaluation of some winemaking characteristics of commercial pectic enzyme preparations. *Australian Grapegrower and Winemaker*, 366 A: 146–150.

Carini, S., G. Mucchetti, and E. Neviani. 1985. Lysozyme, activity against clostridia and use in cheese production: A review. *Microbiologie, Aliments, Nutrition* 3: 299–320.

Charpentier, C., A. M. Dos Santos, and M. Feuillat. 2004. Release of macromolecules by *Saccharomyces cerevisiae* during ageing of French flor sherry wine "Vin jaune." *International Journal of Food Microbiology* 96: 253–262.

Charpentier, C. and M. Feuillat. 2008. Élevage des vins rouges sur lies. Incidence de l'addition d'une β-glucanase sur la composition en polysaccharides et leurs interactions avec les polyphénols. (Red wine aging on lees. Effect of the addition of a β-glucanase on the polysaccharide composition and its interaction with polyphenols.) *Revues des Œnologues* 129: 31–35.

Clare, S., G. Skurray, and L. Theaud. 2002. Effect of a pectolytic enzyme on the colour of red wine. *The Australian & New Zealand Grapegrower & Winemaker* 456: 29–35.

D'Incecco, N., E. J. Bartowsky, S. Kassara, A. Lante, P. Spettoli, and P. A. Henschke. 2004. Release of glycosidically bound flavour compounds from Chardonnay by *Oenococcus oeni* during malolactic fermentation. *Food Microbiology* 21: 257–265.

Delteil, D. 2000. Effet d'une préparation enzymatique sur l'évolution du profil polyphénolique et sensoriel d'un vin rouge de Mourvédre. (Effect of an enzymatic preparation on the evolution of the polyphenol and sensory profile of a Monastrell red wine.) *25ème Congrès Mondial de la Vigne et du Vin OIV*, Paris, France.

Doco, T., P. Williams, and V. Cheynier. 2007. Effect of flash release and pectinolytic enzyme treatments on wine polysaccharide composition. *Journal of Agricultural and Food Chemistry* 55: 6643–6649.

Dubourdieu, D., C. Desplanques, J.-C. Villettaz, and P. Ribèrau-Gayon. 1981. Investigations of an industrial β-D-glucanase from *Trichoderma harzianum*. *Carbohydrate Research* 144: 277–287.

Dupin, I. V. S., V. J. Stockdale, P. J. Williams, G. P. Jones, A. J. Markides, and E. J. Waters. 2002. *Saccharomyces cerevisiae* mannoproteins that protect wine from protein haze: Evaluation of extraction methods and immunolocalization. *Journal of Agriculture and Food Chemistry* 48: 1086–1095.

Escot, S., M. Feuillat, L. Dulau, and C. Charpentier. 2001. Release of polysaccharides by yeasts and the influence of released polysaccharides on colour stability and wine astringency. *Australian Journal of Grape and Wine Research* 7: 153–159.

Felix, R., and J.-C. Villettaz. 1983. Wine. In *Industrial Enzymology: The applications of enzymes in industry*, eds. T. Godfrey and J. Reichelt, 410–421. New York: The Nature Press.

Ferrando, M., C. Güell, and F. López. 1998. Industrial wine making: Comparison of must clarification treatments. *Journal of Agricultural and Food Chemistry* 46: 1523–1528.

Feuillat, M., D. Peyron, and J. L. Berger. 1987. Influence de la microfiltration tangentielle des vins sur leur composition physicochimique et leurs caractères sensoriales. (Influence of wine cross-flow microfiltration on its physico-chemical composition and sensory characteristics.) *Bulletin de l'O.I.V* 60: 227–244.

Fidaleo, M., M. Esti, and M. Moresi. 2006. Assessment of urea degradation rate in model wine solutions by acid urease from *Lactobacillus fermentum. Journal of Agricultural and Food Chemistry* 54: 6226–6235.

Fornairon-Bonnefond, C., C. Camarasa, M. Moutounet, and J. M. Salmon. 2002. New trends on yeast autolysis and wine ageing on lees: A bibliographic review. *Journal International des Sciences de la Vigne et du Vin* 36: 49–69.

Fornairon-Bonnefond, C., and J. M. Salmon. 2003. Impact of oxygen consumption by yeast lees on the autolysis phenomenon during simulation of win aging on lees. *Journal of Agriculture and Food Chemistry* 51: 2584–2590.

Gao, Y., S. Krentz, G. Zhang, S. Darius, J. Power, and G. Lagarde. 2002. Inhibition of spoilage lactic acid bacteria by lysozyme during wine alcoholic fermentation. *Australian Journal of Grape and Wine Research* 8: 76–83.

Gerbaux, V., A. Villa, C. Monamy, and A. Bertrand. 1997. Use of lysozyme to inhibit malolactic fermentation and to stabilize wine after malolactic fermentation. *American Journal of Enology and Viticulture* 48: 49–51.

Gerbaux, V., S. Jeudy, and C. Monamy. 2000. Etude des phénols volatils dans les vins de Pinot noir en Bourgogne. (Study of the volatile phenols of Pinot noir wines from Bourgogne.) *Bulletin de l' OIV* (835–836): 581–599.

Gerbaux, V., B. Vincent, and A. Bertrand. 2002. Influence of maceration temperature and enzymes on the content of volatile phenols in Pinot Noir wines. *American Journal of Enology and Viticulture* 53: 131–137.

Gil, J. V., and S. Valles. 2001. Effect of macerating enzymes on red wine aroma at laboratory scale: Exogenous addition or expression by transgenic wine. *Journal of Agricultural and Food Chemistry* 49: 5515–5523.

Gómez, E., A. Martínez, and J. Laencina. 1994. Localization of free and bound aromatic compounds among skin, juice and pulp fractions of some grape varieties. *Vitis* 33: 1–4.

Green, J. L., B. T. Watson, and M. A. Daeschel. 1995. Efficacy of lysozyme in preventing malolactic fermentation in Oregon Chardonnay and Pinot Noir wines (1993 and 1994 vintages). ASEV 46th Annual Meeting Abstracts. *American Journal of Enology and Viticulture* 46: 410.

Gump, B. H., and K. G. Halght. 1995. A preliminary study of industrial enzyme preparations for color extraction/stability in red wines. *Cati publication 950901*, Viticulture and Enology Research Centre, California State University, Fresno.

Hernández, L., J. Espinosa, M. Fernández-González, and A. Briones. 2003. β-glucosidase activity in a *Saccharomyces cerevisiae* wine strain. *International Journal of Food Microbiology* 80: 171–176.

Huang, X. M., and H. B. Huang. 2001. Early post-veraison growth in grapes: Evidence for a two-step mode of berry enlargement. *Australian Journal of Grape and Wine Research* 7: 132–136.

Humbert-Goffard, A., C. Saucier, V. Moine-Ledoux, R. M. Canal-Llaubères, D. Dubourdieu, and Y. Glories. 2004. An assay for glucanase activity in wine. *Enzyme and Microbial Technology* 34: 537–543.

Jackson, R. 2000. *Wine Science: Principles, Practice and Perception*. San Diego, CA: Academic Press.

Kashyap, D. R., P. K. Vohra, S. Chopra, and R. Tewari. 2001. Applications of pecti-
    nases in the commercial sector: a review. *Bioresource Technology* 77: 215–227.
Kelebek, H., A. Canbas, T. Cabaroglu, and S. Selli. 2007. Improvement of antho-
    cyanin content in the cv. Öküzgözü wines by using pectolytic enzymes. *Food
    Chemistry* 105: 334–339.
Ledoux, V., L. Dulau, and D. Dubourdieu. 1992. Interprétation de l'amélioration de
    la stabilité protéique des vins au cours de l'élevage sur lies. (Interpretation
    of the improvement of the protein stability of wines aged on lees.) *Journal
    International des Science de Vigne et Vin* 26: 239–251
LeTraon-Masson, M. P., and P. Pellerin. 1998. Purification and characterization of
    two β-d-glucosidases from an *Aspergillus niger* enzyme preparation: affinity
    and specificity toward glucosylated compounds characteristic of the process-
    ing of fruits. *Enzyme and Microbial Technology* 22: 374–382.
Llaubères, R. M., B. Richard, A. Lonvaud, D. Dubourdieu, and B. Fournet. 1990.
    Structure of an exocellular β-D-glucan from *Pediococcus* sp., a wine lactic bac-
    teria. *Carbohydrate Research* 203: 103–107.
Lourens, K. and P. Pellerin. 2000. Enzymes in winemaking. *Wynboer, A Technical
    Guide for Wine Producers* http://www.wynboer.co.za/recentarticles/0411
    enzymes.php3.
Lubbers, S., B. Leger, C. Charpentier, and M. Feuillat. 1993. Essai colloides protect-
    eurs d'extraits de parois de levures sur la stabilité tartrique d'un vin modêde.
    (Assay of colloids from yeast cell walls on the tartaric stability of a model
    wine.) *Journal International des Science de Vigne et Vin* 27: 13–22.
Lubbers, S., C. Charpentier, M. Feuillat, and A. Voilley. 1994. Influence of yeast
    walls on the behavior of aroma compounds in a model wine. *American Journal
    Enology and Viticulture* 45: 29–33.
Manners, D. J., A. J. Masson, and J. C. Patterson. 1973a. The structure of a β-(1→3)-
    D-glucan from yeast cell walls. *Biochemistry* 153: 19–30.
Manners, D. J., A. J. Masson, and J. C. Patterson. 1973b. The structure of a β-(1→6)-
    D-glucan from yeast cell walls. *Biochemistry* 153: 31–36.
Manzanares, P., V. Rojas, S. Genovés, and S. Vallés. 2000. A preliminary search
    of anthocyanin-β-glucosidase activity in non-Saccharomyces wine yeasts.
    International Journal of Food Science and Technology 35: 95–103.
Mateo, J. and R. Di Stefano. 1997. Description of the β-glucosidase activity of wine
    yeasts. *Food Microbiology* 14: 583–591.
Marchal, R., D. Chaboche, R. Douillard, and P. Jeandet. 2002. Influence of
    lysozyme treatments on Champagne base wine foaming properties. *Journal
    of Agricultural and Food Chemistry* 50: 1420–1428.
Molina Úbeda, R. 2000. *Teoría de la clarificación de mostos y vinos y sus aplicaciones
    prácticas.* (Must and wine clarification theory and its practical applications.)
    Madrid: Mundi Prensa.
Montant, C. and L. Thomas. 1977. Structure d'un glucose exocellulaire produit par
    le Botrytis cinerea. (Structure of an exocellular glucose product of *Botritis
    cinerea.*) *Annals Science Nature* 18: 185–192.
Morata, A., C. González, and J. A. Suarez-Lepe. 2007. Formation of vinylphenolic
    pyranoanthocyanins by selected yeasts fermenting red grape musts supple-
    mented with hydroxycinnamic acids. *International Journal of Food Microbiology*
    116: 144–152.

Mourgues, J. and P. Bénard. 1980. Le chauffage de la vendange et ses conséquences. (Thermovinification and its consequences.) *Comptes Rendus de l'Academie d'Agriculture de France* 66: 823–827.

Nunan, K. J., I. M. Sims, A. Bacic, S. P. Robinson, and G. Fincher. 1997. Isolation and characterization of cell walls from the mesocarp of mature grape berries (*Vitis vinifera*). *Planta* 203: 93–100.

Nygaard, M., L. Petersen, E. Pilatte, and G. Lagarde. 2002. Prophylactic use of lysozyme to control indigenous lactic acid bacteria during alcoholic fermentation. The ASEV 53rd Annual Meeting, Portland, Oregon.

Ortega-Regules, A. E., I. Romero-Cascales, J. M. Ros-García, J. M López-Roca, and E. Gómez-Plaza. 2006. A first approach towards the relationship between grape skin cell-wall composition and anthocyanin extractability. *Analytica Chimica Acta* 563: 26–32.

Ortega-Regules, A., J. M. Ros-García, A. B. Bautista-Ortín, J. M. López-Roca, and E. Gómez-Plaza. 2008a. Differences in morphology and composition of skin and pulp cell walls from grapes (*Vitis vinifera* L.). Technological implications. *European Food Research and Technology* 227: 223–231.

Ortega-Regules, A., J. M. Ros-García, A. B. Bautista-Ortín, J. M. López-Roca, and E. Gómez-Plaza. 2008b. Changes in skin cell wall composition during the maturation of four premium wine grape varieties. *Journal of the Science of Food and Agriculture* 88: 420–428.

Ough, C. S. and H. W. Berg. 1974. The effect of two commercial pectic enzymes on grape musts and wines. *American Journal of Enology and Viticulture* 25: 108–211.

Ough, C. S., A. Noble, and D. Temple. 1975. Pectin enzyme effects on red grapes. *American Journal of Enology and Viticulture* 26: 195–200.

Ough, C. S. and E. A. Crowell. 1979. Pectic-enzyme treatment of white grapes: temperature, variety and skin-contact time factors. *American Journal of Enology and Viticulture* 30: 22–27.

Palmeri, R. and G. Spagna. 2007. β-glucosidase in cellular and acellular form for winemaking application. *Enzyme and Microbiological Technology* 40: 382–389.

Palomero, F., A. Morata, S. Benito, M. C. González, and J. A. Suárez-Lepe. 2007. Conventional and enzyme-assisted autolysis during ageing over lees in red wines: Influence on the release of polysaccharides from yeast cell walls and on wine monomeric anthocyanin content. *Food Chemistry* 105: 838–846.

Palomero, F., A. Morata, S. Benito, F. Calderón, and J. A. Suárez-Lepe. 2009. New genera of yeasts for over-lees aging of red wines. *Food Chemistry* 112: 432–441.

Pardo, F., R. Salinas, G. Alonso, G. Navarro, and M. D. Huerta. 1999. Effect of diverse enzyme preparations on the extraction and evolution of phenolic compounds in red wines. *Food Chemistry* 67: 135–142.

Parley, A. 1997. The effect of pre-fermentation enzyme maceration on extraction and color stability in Pinot noir wine. Master's thesis, University of Lincoln, Lincoln, New Zealand.

Pérez-Prieto, L. J., J. M. López-Roca, A. Martínez-Cutillas, F. Pardo Mínguez, and E. Gómez-Plaza. 2002. Maturing wines in oak barrels. Effects of origin, volume, and age of the barrel on the wine volatile composition. *Journal of Agricultural and Food Chemistry* 50: 3272–3276.

Pérez-Serradilla, J. A., and M. D. Luque de Castro. 2008. Role of lees in wine production: A review. *Food Chemistry* 111: 447–456.

Pitson, S. M., R. J. Seviour, and B. M. McDougall. 1993. Non cellulolytic fungal ß-glucanases: their physiology and regulation. *Enzyme and Microbial Technology* 15: 178–192.

Pogorzelski, E. and A. Wilkowska. 2007. Flavour enhancement through the enzymatic hydrolysis in juices and beverages: a review. *Flavour and Fragance Journal* 22: 251–254.

Pretorius, I. S. and P. B. Hoj. 2005. Grape and wine biotechnology: challenges, opportunities and potential benefits. *Australian Journal of Grape and Wine Research* 11: 83–108.

Revilla, I. and M. L. González-San José. 2003. Addition of pectolytic enzymes: an enological practice which improves the chromaticity and stability of red wines. *International Journal of Food Science and Technology* 38: 29–36.

Rhee, S. J., S. Y. Cho, K. M. Kim, D.-S. Cha, and H.-J. Park. 2008. A comparative study of analytical methods for alkali-soluble β-glucan in medicinal mushroom, Chaga (*Inonotus obliquus*). *LWT-Food Science and Technology* 41: 545–549.

Ribéreau-Gayon, P., D. Dubourdieu, B. Donèche, and A. Lonvaud. 1998a. *Traité d'Enologie 1. Microbiologie du vin. Vinifications. (Handbook of Enology 1. Wine microbiology. Vinifications)* Paris: Dunod.

Ribéreau-Gayon, P., Y. Glories, A. Maujean, and D. Dubourdieu. 1998b. *Traité d'Enologie 2. Chimie du vin. Stabilisation et traitements. (Handbook of Enology 2. Wine chemistry. Stabilization and Treatments.)* Paris: Dunod.

Riou, V., A. Vernhet, T. Doco, and M. Moutounet. 2002. Aggregation of grape seed tannins in model wine-effect of wine polyssacharides. *Food Hydrocolloids* 16: 17–23.

Rodríguez, M. E., C. A. Lopes, M. van Broock, S. Valles, D. Ramón, and A. C. Caballero. 2004. Screening and typing of Patagonian wine yeast for glycosidase activities. *Journal of Applied Microbiology* 96: 84–95.

Romero-Cascales, I., A. E. Ortega-Regules, J. M. López-Roca, J. I. Fernández-Fernández, and E. Gómez-Plaza. 2005. Differences in anthocyanins extractability from grapes to wines according to variety. *American Journal of Enology and Viticulture* 56: 212–219.

Romero-Cascales, I., J. I. Fernández-Fernández, J. M. Ros-Garcia, J. M. López-Roca, and E. Gómez-Plaza. 2008. Characterisation of the main enzymatic activities present in six commercial macerating enzymes and their effects on extracting colour during winemaking of Monastrell grapes. *International Journal of Food Science and Technology* 43: 1295–1305.

Romero-Cascales, I. 2008. Extracción de compuestos fenólicos de la uva al vino. Papel de los enzimas de maceración. (Phenolic compound extraction from grapes to wine. Role of maceration enzymes.) Ph.D. thesis. University of Murcia, Spain.

Sacchi, K., L. Bisson, and D. O. Adams. 2005. A review of the effect of winemaking techniques on phenolic extraction in red wines. *American Journal of Enology and Viticulture* 56: 197–206.

Shiga, T. M., F. M. Lajolo, and T. M. Filisetti. 2004. Changes in the cell wall polysaccharides during storage and hardening of beans. *Food Chemistry* 84: 53–64.

Shih, I., B. Chou, C. Chen, J. Wu, and C. H. Shih. 2008. Study of mycelial growth and bioactive polysaccharide production in batch and fed-batch culture of *Grifola frondosa*. *Bioresource Technology* 99: 785–793.

Suarez, R., J. A. Suarez-Lepe, A. Morata, and F. Calderón. 2007. The production of ethylphenols in wine by yeasts of the genera Brettanomyces and Dekkera: A review. *Food Chemistry* 102: 10–21.

Suzuki, K., Y. Benno, S. Mitsuoka, S. Takebe, K. Kobashi, and J. Hase. 1979. Urease-producing species of intestinal anaerobes and their activities. *Applied and Environmental Microbiology* 37: 379–382.

Takayanagi, T., T. Uchibori, and K. Yokotsuka. 2001. Characteristics of yeast poly-galacturonases induced during fermentation on grape skins. *American Journal of Enology and Viticulture* 52: 41–44.

Takebe, S. and K. Kobashi. 1988. Acid urease from *Lactobacillus* of rat intestine. *Chemical and Pharmaceutical Bulletin* 36: 693–699.

Tegmo-Larsson, I. and T. Henick-Kling. 1990. Ethyl carbamate precursors in grape juice and the efficiency of acid urease on their removal. *American Journal of Enology and Viticulture* 41: 189–192.

Ugliano, M., A. Genovese, and L. Moio. 2003. Hydrolysis of wine aroma precursors during malolactic fermentation with four commercial starter cultures of *Oenococcus oeni*. *Journal of Agricultural and Food Chemistry* 51: 5073–5078.

Uhlig, M. and E. Linsmaier-Bednar. 1998. Industrial enzymes and their applications. Los Alamitos, CA: Wiley-IEEE Press.

Vernhet, A., P. Pellerin, and M. P. Belleville. 1999. Relative impact of major wine polysaccharides on the performances of an organic microfiltration membrane. *American Journal of Enology and Viticulture* 50: 51–56.

Vernhet, A. and M. Moutounet. 2002. Fouling of organic microfiltration membranes by wine constituents: importance, relative impact of wine polysaccharides and polyphenols and incidence of membrane properties. *Journal of Membrane Science* 201: 103–122.

Villettaz, J.-C., D. Steiner, and H. Trogus. 1984. The use of a beta-glucanase as an enzyme in wine clarification and filtration. *American Journal Enology and Viticulture* 35: 253–256.

Volman, J. J., J. D. Ramakers, and J. Plat. 2008. Dietary modulation of immune function by β-glucans. *Physiology and Behavior* 94: 276–284.

Waters, E., W. Wallace, M. E. Tate, and P. J. Williams. 1993. Isolation and partial characterization of a haze protective factor from wine. *Journal of Agriculture and Food Chemistry* 41: 724–730.

Watson, B., N. Goldberg, H. P. Chen, and M. McDaniel. 1999. Effects of macerating pectinase enzymes on color, phenolic profile, and sensory character of Pinot Noir wines. In: *2nd Joint Burgundy-California-Oregon Winemaking Symposium*, ed. C. Butzke, 36–44. Davis, CA: American Vineyard Foundation.

Xu, X., J. Xu, Y. Zhang, and L. Zhang. 2008. Rheology of triple helical Lentinan in solution: Steady shear viscosity and dynamic oscillatory behaviour. *Food Hydrocolloid* 22: 735–741.

Zimman, A., W. Joslin, M. Lyon, J. Meier, and A. Waterhouse. 2002. Maceration variables affecting phenolic composition in commercial-scale Cabernet Sauvignon winemaking trials. *American Journal of Enology and Viticulture* 53: 93–98.

chapter ten

Effect of novel food processing on fruit and vegetable enzymes

Indrawati Oey

Contents

10.1 Introduction

Enzymes are of importance in catalyzing chemical reactions in almost all biological processes. Due to its substrate specificity, the set of enzymes can speed up a few reactions among all possibilities resulting in a very specific reaction pathway and mechanism. In fruits and vegetables, enzymes play an important role in determining the quality attributes such as texture, flavor, color, nutritional value and sensorial properties. After harvest, most enzymes are still active and the continuation of enzymatic activity during storage affects the quality of fruit and vegetables. To increase the palatability and to prolong the shelf life, mechanical treatments (e.g., squeezing, crushing, mixing, slicing) and food processing/preservation (e.g., cooking, canning, freezing) are carried out. The aforementioned treatments cause cell disruption facilitating contact between enzymes, substrates, and other (bio)molecules (e.g., cofactors, coenzymes, activators, and inhibitors) and enabling enzyme-catalyzed reactions, which could affect the overall food quality properties.

Active endogenous enzymes and their substrates are complementary natural food ingredients of fruit and vegetables. From a food engineering point of view, it is essential to be able to control both activity and stability of food-quality-related enzymes during food processing to obtain food products with certain targeted quality properties. These endogenous (bio) compounds can be used as natural ingredients for example to improve the quality of fruit- and vegetable-based food products. Therefore, optimization of (conventional) thermal processing conditions such as blanching and development of novel processing technologies are currently being investigated to obtain a balance between food safety and quality.

In the last two decades, possible applications of novel processing technologies (e.g., high hydrostatic pressure/HP, high-intensity pulsed electric field/PEF, ultrasound, pulsed UV light, irradiation) for food processing and preservation have been investigated. Current findings have shown that these technologies could meet the consumer desire for high-quality foods in the sense of preserving more fresh-like properties (i.e., close to the original natural properties), being healthier, and using less additives/ preservatives than foods that have previously been available. With regard to effects of novel processing on fruit and vegetable enzymes, detailed studies on HP and PEF have been carried out and reported in literature more intensively than other novel processing techniques. In the following sections, novel processing being discussed will be limited to HP and PEF. First, general technological aspects and effects of novel processing on biomaterials will be briefly discussed. Second, understanding the mechanism of fruit and vegetable enzyme stability during novel processing will be given based on the knowledge that is currently available on protein stability. Since a conformational alteration of protein molecule (mostly known as enzyme denaturation) or a modification of functional groups with or even without conformational changes entails either a reversible or an irreversible loss of enzyme ability to perform a certain biological function, enzyme inactivation will be referred to the resulting reduction of enzyme activity measured after processing. Moreover, the reversibility of enzyme inactivation will be mainly concerned in evaluating the effects of novel processing on the stability of fruit and vegetable enzymes. Since enzyme catalyzed reaction can be enhanced during processing, several case studies will be additionally given to illustrate the possibilities of controlling enzyme-catalyzed reactions during novel processing, such as during HP processing to improve the quality of fruit-and-vegetable-based food products.

10.2   Novel processing: technology and effects on biomaterials

10.2.1   High hydrostatic pressure (HP) processing

The basic construction of an HP unit consists of a cylindrical pressure vessel, a pressure generator, a temperature control device, and a pressure handling system. HP processing is conducted in a closed system (batch), however, multiple closed system vessels can be used to maintain the production continuity (semicontinuous). A pressure-transmitting medium (water, oil or water/oil mixture) is pumped to the pressure vessel using a pressure intensifier to reach the desired pressure level and afterwards, the pressure is isolated in the vessel. From this moment, the processing time is counted and stopped by releasing pressure. The whole HP processing cycle normally has a duration of minutes.

Based on the Pascal principle, hydrostatic pressure at a given point is the same in all directions. Pressure is transformed immediately and uniformly to all directions independent of product size and geometry. Therefore, this technique is mostly promoted as a "uniform" processing. However, it must be taken into consideration that the classical limitation of heat transfer still exists during HP processing. During pressurization and decompression, an increase and a decrease in pressure result in an increase (adiabatic heating) and a decrease (adiabatic cooling) in temperature of the whole vessel content, respectively. The extent of temperature increase and decrease respectively due to pressurization and decompression is dependent on pressurization rate and the physical properties of pressure transmitting medium and food products/samples (e.g., heat conductivity, density, compressibility). To identify the process intensity of HP processing, pressure, temperature, and treatment time are mainly used as process parameters.

The feasibility of using HP technology for food applications has been studied in a broad temperature range from subzero to elevated temperatures. Pressure has a limited effect on the covalent bonds of low-molecular-mass molecules such as vitamins, pigments, and volatile compounds in contrast to that of high-molecular-mass molecules such as proteins/enzymes in which various covalent and non-covalent interactions stabilize its complex three-dimensional structure architecture. However, different processes can occur simultaneously during HP processing (100–1000 MPa/–20 to 60°C); that is, (1) cell wall and membrane disruption enabling contact between enzymes and their substrates, (2) enhancement and retardation of enzymatic and chemical reactions, (3) microorganism inactivation, and (4) modification of biopolymers including protein denaturation, enzyme inactivation, and gel formation (Oey et al., 2008).

Pressure effects on enzymatic or chemical reactions and physical changes (e.g., protein denaturation, phase transition) depend on the resulting total volume changes during processing. According to the *Le Chatelier* principle, pressure shifts the reaction equilibrium to the state having the smallest volume. As a consequence, the aforementioned reactions and changes can only be enhanced if pressure decreases the total reaction volume (a negative change of partial molar volume between initial and final state at constant temperature). Since pressure favors reactions accompanied with a volume decrease and vice versa, it indirectly implies that mechanism and kinetics of enzymatic and chemical reactions during HP processing could differ from those occurring at atmospheric pressure. Hereto, a better understanding of pressure effect on biomaterials is still a great challenge.

10.2.2   High-intensity pulsed electric field (PEF) processing

The basic construction of a PEF unit consists of a pulsed generator, treatment chamber(s) (e.g., cylindrical or rectangular chamber, electrodes), temperature and pulse monitoring systems, and a fluid handling system for food product in case of continuous mode. PEF processing can be conducted both in batch and continuous systems (Min et al., 2007). Circuitry of pulsed power supply system influences the resulting shape of the pulses such as (monopolar) rectangular/square-wave, bipolar of rectangular, exponential decay, damped oscillating shape etc (de Haan, 2007). Similar to other processing technologies, PEF treatment also has problems with process uniformity; for example of temperature, electrical current. In this case, the design of the treatment inside the chamber plays an important role in the distribution of temperature inside the PEF chamber.

In literature, electric field strength (expressed in kV/cm) and total treatment time/energy are usually referred to as important process parameters to identify the intensity of PEF processing. The duration of PEF treatment normally ranges from micro- to milliseconds shorter than HP processing.

The calculation of total treatment time or energy varies depending on pulse number, pulse frequency, pulse delay, pulse width, pulse shape, and so on. However, it should be taken into account that other process parameters such as pulse polarity, pulse shape, pulse frequency, etc. involved in the processing could also affect the stability of biomaterials. These aforementioned process parameters are dependent on the design of PEF equipment and experiment. PEF equipment and the concomitant process parameters are not yet well standardized. Equipment specifications (such as pulsed power supply covering voltage rating, power rating, pulse duration and pulse repetition rate, geometry of the treatment chamber, circuitry, etc.) and the details of experimental setup are not fully documented. Therefore, appropriate comparison and evaluation between studies found in literature are limited.

The feasibility of using PEF technology for food applications has been studied at temperatures above subzero (mostly from moderate to elevated temperatures). Alteration of the cell membrane electropermeabilization is one of the key elements of PEF treatment. Depending on the process intensity, the occurrence of reversible and irreversible pore formation and cell disintegration could (1) affect the cell vitality such as resulting in inactivation of microorganisms (Min et al. 2007); (2) induce stress response reactions at low intensity; and (3) affect permeabilization of plant and animal tissues enhancing enzymatic reaction and improving mechanical separation, extractability, or mass transport. Furthermore, several studies have shown that PEF affects the stability of enzymes (for a detailed discussion, see Section 10.4) and preheating might be incorporated to optimize the PEF effects on biomaterials. Hereto, PEF applications open up a lot of new possibilities, particularly for handling raw materials and microbial decontaminants (Knorr et al., 2008).

10.3 Effect of HP processing on fruit and vegetable enzymes

10.3.1 Understanding of HP processing effect on stability of enzyme as a protein molecule

Pressure-temperature stability of proteins reveals an elliptical contour (Suzuki, 1960; Brandts et al., 1970; Hawley, 1971; Zipp and Kauzmann, 1973), as schematically illustrated in Figure 10.1. It depicts the possibility of protein denaturation by low temperature (cold inactivation), elevated pressure, high temperature (heat inactivation), or a combination of these factors (Mozhaev et al., 1994; Hayashi et al., 1998).

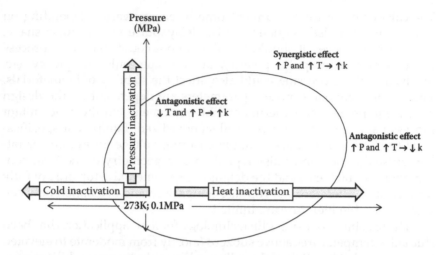

Figure 10.1 Schematic pressure-temperature diagram of protein stability (*P*: pressure, *T*: temperature, *k*: reaction rate).

Regarding the hierarchy of protein structure, four structural levels involving different bonds and interactions can be distinguished. The primary structure of protein is limitedly affected under pressure since pressure has a limited effect on covalent bonds (Cheftel, 1991; Heremans, 1992; Mozhaev et al., 1994). Hydrogen bonds, which are responsible for maintaining the secondary, tertiary, and quaternary structure levels of a protein, are rather stable toward pressure, and very high pressure levels (> 700 MPa) can disrupt these bonds, affecting the secondary structure. The changes in secondary structure inevitably lead to irreversible protein denaturation (Balny and Masson, 1993). In contrast to the unaltered primary and secondary structural levels due to pressure, the tertiary and quaternary structure of proteins is lost due to pressure (Heremans, 1993; Mozhaev et al., 1996) predominantly caused by (1) a disturbance of hydrophobic and electrostatic interactions beyond 150–200 MPa (Cheftel, 1991; Balny and Masson, 1993); (2) a dissociation of oligomeric enzymes into subunits at about 150–200 MPa (Balny and Masson, 1993); (3) imperfect packing of atoms at the subunit interface together with the disruption of hydrophobic and electrostatic interactions in the inter-subunit area leading to large volume changes (Cheftel, 1991; Mozhaev et al., 1994); and (4) unfolding of dissociated subunits at high pressure (Silva and Weber, 1993).

Reversible protein denaturation could occur at low pressure (< 200 MPa) (Cheftel, 1991; Heremans, 1992; Masson, 1992). A slow refolding process and both conformational drift and hysteric behaviors occur after pressure release. Pressure beyond 300 MPa could result in irreversible effects including chemical modifications or unfolding of single-chain proteins and protein denaturations. Changes of protein structure under pressure are also governed by the *Le Chatelier* principle in which pressure favors reactions accompanied by negative volume changes. It is also suggested that elevating pressure level increases the degree of protein molecule ordering (referred to the principle of microscopic ordering) (Cheftel, 1991; Heremans, 1992; Masson, 1992; Mozhaev et al., 1994).

At moderate temperature and atmospheric pressure, most enzymes are stable for example between 20 and 45°C. Elevating pressure or temperature respectively at constant temperature or pressure might enhance enzyme inactivation. In literature, this phenomenon is generally termed *synergistic pressure and temperature effect* (Figure 10.1). In this case, the rate constants (*k* values) of enzyme inactivation increase with elevating pressure at constant temperature or with elevating temperature at constant pressure.

At high temperatures (close to the temperatures resulting in thermal inactivation of enzymes at atmospheric pressure), elevating pressure (mostly <200 MPa) could retard the thermal inactivation of enzymes (lower *k* values when pressure is increased, Figure 10.1). Such antagonistic effect of pressure on thermal inactivation can be explained by the fact that at atmospheric pressure, elevating temperature (related to heat inactivation) affects non-covalent as well as covalent bonds (above 70°C), resulting in aggregated or incorrectly folded enzymes and chemically altered enzymes, respectively. An enzyme will lose its activity if the active site becomes inaccessible or disassemble due to protein unfolding. From a thermodynamic point of view, enzyme denaturation leads to a very large change in entropy (caused by a less ordered conformational structure) exceeding the absolute value of the enthalpy change and making the change in Gibbs free energy negative in which the denaturation is favorable. Since elevating pressure can increase the degree of protein molecule ordering, it could reassemble the appropriate protein structure, especially the active site leading to partial/complete recovery of enzyme activity.

Some enzymes could undergo significant denaturation during freezing and thawing; however, many are unaffected. At subzero and low temperatures, elevating pressure could affect the enzyme stability resulting in enzyme activity loss such as lipoxygenase (Indrawati et al., 1999, 2000a, 2000b, 2001; Van Buggenhout et al., 2006). In cases of cold denaturation/inactivation at atmospheric pressure, the inverse thermodynamic explanation at high temperature (i.e., a negative entropy change corresponding to a higher ordered protein molecule) cannot solely explain the phenomenon because the interactions between protein molecules and adjacent water must be taken into consideration (Meersman et al., 2008). At low temperature, water molecules may form a shell or a layer around adjacent nonpolar molecules resulting in different nonpolar entities and loss of the ability to interact among each other. As a consequence, low temperature promotes an exposure of nonpolar side chains to water and hydrophobic associations become less stable (Privalov, 1990; Da Poian et al., 1995; Silva et al., 1996). The nonpolar interactions are more affected by pressure because they are more compressible (Weber, 1995). It explains the additive effect of high pressure and low temperature to reduce the entropy (Silva et al., 1996). As a consequence, temperature decrease under pressure enhances the enzyme inactivation. Such antagonistic effects of pressure and low temperature on enzyme inactivation (Figure 10.1) have been noticed for lipoxygenase (Indrawati et al., 1999, 2000a, 2000b, 2001) and myrosinase (Van Eylen et al., 2007, 2008a, 2008b).

10.3.2   Effect of HP processing on stability of fruit and vegetable enzymes

It is a challenge to thoroughly elucidate pressure effects on enzyme stability in fruit and vegetables because (1) enzymes in fruit and vegetables are present in a complex system as illustrated in Figure 10.2 and (2) at the same time pressure enhances/retards the chemical and enzymatic reactions depending on the reaction volume. Hence, mechanistic and kinetic studies on enzyme stability at different molecular levels and different complexities of food matrix have been conducted. To decrease the complexity of enzyme system or to eliminate the presence of other endogenous (bio)compounds, fruit and vegetable enzymes are (partially) purified and afterwards dissolved in controlled buffer medium (e.g., certain pH, buffer, ion strength) or in fruit/vegetable juices (Table 10.1). Pressure effects on endogenous enzymes have been studied in fruit and vegetables with different intensities of matrix disruption such as in juices, purees, or in intact food matrices (Table 10.2). The latter allows an overall evaluation

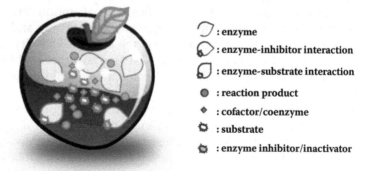

Figure 10.2 Schematic illustration of complex enzyme systems in fruit and vegetables.

Table 10.1 Effect of Combined High Pressure and Temperature Processing on the Stability of (Partially) Purified Fruit and Vegetables Enzymes in Buffer Solution or in Fruit/Vegetable Juices

Enzymes	Enzyme and buffer solution/medium	Processing condition	Enzyme stability[b]	References
Lipoxygenase (LOX)	Partially purified[a] tomato LOX; MOPS/KOH (10 mM; pH 6.8)	10 to 60°C; 100 to 650 MPa; various treatment time up to 60 min.	Irreversible inactivation T≥20°C, synergistic effect of increases in pressure and temperature on inactivation T<20°C, lowering temperature enhanced pressure inactivation First-order inactivation kinetics Antagonistic effect of pressure (<550 MPa) on thermal inactivation (50 and 60°C)	Rodrigo et al. (2006a)
	Commercial purified soybean LOX; Tris HCl buffer (0.4 mg/mL 10 mM; pH 9)	−15 to 68°C; 0.1 up to 650 MPa; various treatment time	Irreversible inactivation T≥30°C, synergistic effect of increases in pressure and temperature on inactivation T<30°C, lowering temperature enhanced pressure inactivation First-order inactivation kinetics Antagonistic effect of low pressure (<200 MPa) on thermal inactivation (65°C)	Indrawati et al. (1999)

(continued)

Table 10.1 Effect of Combined High Pressure and Temperature Processing on the Stability of (Partially) Purified Fruit and Vegetables Enzymes in Buffer Solution or in Fruit/Vegetable Juices (Continued)

Enzymes	Enzyme and buffer solution/medium	Processing condition	Enzyme stability[b]	References
Myrosinase (MYR)	Partially purified broccoli MYR; phosphate buffer (0.1 M; pH 6.55)	20°C; 350 to 500 MPa and 35°C; 150 to 450 MPa; various treatment time up to 80 min.	Irreversible inactivation At 35°C, antagonistic effect of low pressure (<350 MPa) on thermal inactivation Inactivation kinetics described by consecutive step model	Ludikhuyze et al. (1999)
	Partially purified mustard seed MYR; broccoli juice (pH adjusted to 6.5)	40 to 60°C; up to 700 MPa; max. treatment time=2 h	Very pressure stable No inactivation at 55°C and 600 MPa for 2 h.	Van Eylen et al. (2008a)
Pectinmethylesterase (PME)	Commercial purified orange peel PME; clear apple juice	25°C; 200–400 MPa; various treatment time intervals up to 180 min.	400 MPa/25°C/25 min.: highest enzyme inactivation (1 log unit reduction)	Riahi and Ramaswamy (2003)
	Tomato PME purified with cation exchange chromatography; Na acetate buffer (40 mM; pH 4.4)	25°C; up to 8500 MPa; 17 min. (including 2 min. equilibration time)	Pressure stable 850 MPa/25°C/17 min.: 50% inactivation Tomato varieties gave no influence of PME pressure stability	Rodrigo et al. (2006b)
	Tomato PME purified with affinity chromatography; citrate buffer (50 mM; pH 4.4)	25°C and 66°C; 550 to 700 MPa; various treatment time	Pressure stable Antagonistic effect of pressure on thermal inactivation	Fachin et al. (2002a)

Strawberry PME purified with affinity chromatography; Tris-HCl buffer (20 mM; pH 7)	10°C; 850 to 1000 MPa; various treatment time up to 600 min.	Very pressure stable Pressure labile and stable fractions observed 10% pressure stable fraction Only pressure labile fraction inactivated Elevating pressure enhanced the inactivation	Ly-Nguyen et al. (2002a)
White grapefruit PME purified with affinity chromatography; Tris buffer (20 mM; pH 7)	10 to 62°C; 100 to 800 MPa; various treatment time	Pressure labile and stable fractions observed 20% pressure stable fraction Only pressure labile fraction inactivated Synergistic effect of increases in pressure and temperature on inactivation Antagonistic effect of low pressure (up to 200 MPa) on thermal inactivation	Guiavarc'h et al. (2005)
Green pepper PME crude extract and purified with affinity chromatography; citrate buffer (pH 5.6)	25 to 60°C; 400 to 800 MPa; 15 min.	Pressure stable Purified PME in citrate buffer was more pressure stable than PME in crude extract	Castro et al. (2005)

(continued)

Table 10.1 Effect of Combined High Pressure and Temperature Processing on the Stability of (Partially) Purified Fruit and Vegetables Enzymes in Buffer Solution or in Fruit/Vegetable Juices (Continued)

Enzymes	Enzyme and buffer solution/medium	Processing condition	Enzyme stability[b]	References
	Green pepper PME purified with affinity chromatography; citrate buffer (pH 5.6)	10 to 62°C, 100 to 800 MPa, various treatment time	Pressure labile and stable fractions observed Effective to inactivate pressure labile fraction At 10 to 30°C, pressure stable fraction could be inactivated at 800 MPa Synergistic effect of increases in pressure and temperature on inactivation Antagonistic effect of low pressure (up to 350 MPa) on thermal inactivation (>54°C)	Castro et al. (2006a)
	Carrot PME purified with affinity chromatography; Tris buffer (20 mM; pH 7)	10°C; 600 to 700 MPa; various treatment time up to 20 h	Pressure labile and stable fractions observed 5–10% pressure stable fraction Only pressure labile fraction effectively inactivated Synergistic effect of increases in pressure and temperature on inactivation	Ly-Nguyen et al. (2002b)

Carrot PME purified with affinity chromatography; Tris buffer (20 mM; pH 7)	10 to 65°C; 100 to 825 MPa; various treatment time	Pressure labile and stable fractions observed 5–6% pressure stable fraction Only pressure labile fraction effectively inactivated Synergistic effect of increases in pressure and temperature on inactivation Antagonistic effect of low pressure (up to 300 MPa) on thermal inactivation (>50°C)	Ly-Nguyen et al. (2003)
Banana PME purified with affinity chromatography; Tris buffer (20 mM; pH 7)	10°C; 600 to 700 MPa; various treatment time	Pressure labile and stable fractions observed 8% pressure stable fraction Only pressure labile fraction effectively inactivated Synergistic effect of increases in pressure and temperature on inactivation	Ly-Nguyen et al. (2002c)
Plums PME purified with affinity chromatography	25°C; 650 to 800 MPa; various treatment time	First-order kinetic inactivation	Nunes et al. (2006)

(continued)

Table 10.1 Effect of Combined High Pressure and Temperature Processing on the Stability of (Partially) Purified Fruit and Vegetables Enzymes in Buffer Solution or in Fruit/Vegetable Juices (Continued)

Enzymes	Enzyme and buffer solution/medium	Processing condition	Enzyme stability[b]	References
	Tomato PME purified by affinity chromatography and followed by cation exchange chromatography; citrate buffer (0.1 M; pH 6)	20 and 40°C; 100 to 800 MPa; various treatment time up to 30 min.	Pressure labile isozyme found First-order kinetic inactivation 600 MPa/40°C/6 min: one log unit of inactivation 600 MPa/20°C/18 min.: one log unit of inactivation	Plaza et al. (2007)
Peroxidase (POD)	Partially purified kiwi POD	10 to 50°C; 200 to 500 MPa; various treatment time up to 30 min.	Different isozymes had different resistance towards pressure At 30 and 50°C, synergistic effect of increases in pressure and temperature on inactivation At constant pressure and 50°C, prolonging treatment time remarkably enhanced the inactivation 600 MPa/50°C/30 min.: max. 70% inactivation	Fang et al. (2008)
	Commercial purified horseradish POD; Tris buffer (50 mM; pH7) with H_2O_2 (50 mM) and guaiacol (0.23 M)	25 and 40°C; 100 to 500 MPa; various treatment time up to 5 min.	Irreversible inactivation	Garcia et al. (2002)

Polygalacturonase (PG)	Tomato PG purified with cation exchange chromatography; Na acetate buffer (40 mM; pH 4.4)	25°C; up to 500 MPa; 17 min. (including 2 min. equilibration time)	PG1 and PG2 were pressure labile Up to 300 MPa, no inactivation of PG1 and PG2 300 to 500 MPa, PG1 and PG2 inactivation found PG1 was completely inactivated by mild pressure treatment Tomato varieties gave no influence of PG pressure stability	Rodrigo et al. (2006b)
	Partially purified tomato PG; NaCl solution (0.5 M)	5 to 50°C; 300 to 600 MPa, various treatment time	Pressure labile and stable fractions observed Only pressure labile fraction effectively inactivated Synergistic effect of increases in pressure and temperature on inactivation	Fachin et al. (2002b)
	Tomato PG purified with cation exchange chromatography; Na acetate buffer (40 mM; pH 4.4)	25°C; 500 to 800 MPa; 15 min.	PG1 and PG2 were pressure labile PG2 had a higher pressure stability than PG1 β-subunit was pressure stable protein	Peeters et al. (2004)
Polyphenoloxidase (PPO)	Partially purified grape PPO; Mc Illvaine buffer (pH 4 and 5) and grape juice	10 to 60°C; 400 to 800 MPa; 17 min. (including 2 min equilibration time)	Synergistic effect of increases in pressure and temperature on inactivation Different stability of PPO in buffer (pH 4 and 5) and in grape juice	Rapeanu et al. (2006)

(continued)

Table 10.1 Effect of Combined High Pressure and Temperature Processing on the Stability of (Partially) Purified Fruit and Vegetables Enzymes in Buffer Solution or in Fruit/Vegetable Juices (Continued)

Enzymes	Enzyme and buffer solution/medium	Processing condition	Enzyme stability[b]	References
	Partially purified Victoria grape PPO; Mc Illvaine buffer (pH3 to 6)	25°C; 400 to 800 MPa; 15 min.	Increasing pH stabilized the enzyme	Rapeanu et al. (2005a)
	Partially purified Victoria grape PPO; Mc Illvaine buffer (pH3 to 6)	10 to 55°C; 0.1 to 800 MPa; various treatment time	First-order inactivation kinetics Synergistic effect of increases in pressure and temperature on inactivation Antagonistic effect of low pressure (up to 600 MPa) on thermal inactivation (>45°C)	Rapeanu et al. (2005a)
	Partially purified strawberry PPO; phosphate buffer (0.1 M; pH 7)	10 to 50°C; 100 to 750 MPa; various treatment time	Pressure labile and stable fractions observed Pressure labile fraction completely inactivated 19–81% inactivation due to pressure build up and adiabatic heating Synergistic effect of increases in pressure and temperature on inactivation of pressure stable fraction Antagonistic effect of low pressure (up to 200 MPa) on thermal inactivation (>50°C)	Dalmadi et al. (2006)

[a] Partially purification was done mostly using ammonium sulfate precipitation.
[b] Enzyme stability referring to the residual enzyme activity after HP processing (*post factum*) measured at atmospheric pressure.

Table 10.2 Effect of Combined High Pressure and Temperature Processing on the Stability of Endogenous Enzymes in Fruit and Vegetables

Enzymes	Fruit and vegetable matrix	Processing condition	Enzyme stability[a]	References
Hydroperoxide lyase (HPL)	Tomato juice	25 to 90°C;100 to 650 MPa, 15 min. (including 3 min. of equilibration time)	Irreversible inactivation <300 MPa/25°C/15 min.: 20% inactivation 650 MPa/25°C/15 min.: 80% inactivation Synergistic effect of increases in pressure and temperature on inactivation	Rodrigo et al. (2007)
Lipoxygenase (LOX)	Soy milk and crude extract centrifugated from soy milk	5 to 60°C; 0.1 to 650 MPa; various treatment time	Irreversible inactivation T≥10°C, synergistic effect of increases in pressure and temperature on inactivation T<10°C, lowering temperature enhanced pressure inactivation First-order inactivation kinetics	Wang et al. (2008)
	Tomato juice	25 to 90°C;100 to 650 MPa, 15 min. (including 3 min. of equilibration time)	Irreversible inactivation 550 MPa/25°C/15 min.: complete inactivation Synergistic effect of increases in pressure and temperature on inactivation	Rodrigo et al. (2007)

(continued)

Table 10.2 Effect of Combined High Pressure and Temperature Processing on the Stability of Endogenous Enzymes in Fruit and Vegetables (Continued)

Enzymes	Fruit and vegetable matrix	Processing condition	Enzyme stability[a]	References
	Tomato pericarp tissue	−26 to 20°C; 100 to 500 MPa; 13 min.	Irreversible inactivation Under pressure, subzero temperature enhanced the inactivation At 20°C, pressure up to 500 MPa hardly inactivates LOX 400–500 MPa/ −10 and −20°C/13 min.: complete inactivation 100 to 500 MPa/−26°C/13 min.: max. 25% inactivation	Van Buggenhout et al. (2006)
	Green bean juice	−10 to 60°C; 200 up to 700 MPa; various treatment time up to 210 min.	Irreversible inactivation T≥20°C, synergistic effect of increases in pressure and temperature on inactivation T<20°C, lowering temperature enhanced pressure inactivation	Indrawati et al. (2000a)
	Green bean juice	−10 to 70°C; 50 up to 650 MPa; various treatment time up to 210 min.	Irreversible inactivation T≥10°C, synergistic effect of increases in pressure and temperature on inactivation T<10°C, lowering temperature enhanced pressure inactivation First-order inactivation kinetics	Indrawati et al. (2000b)

Green beans	−10 to 70°C; 50 up to 550 MPa; various treatment time up to 150 min.	Irreversible inactivation T≥10°C, synergistic effect of increases in pressure and temperature on inactivation T<10°C, lowering temperature enhanced pressure inactivation First-order inactivation kinetics	Indrawati et al. (2000b)
Green pea juice	−15 to 70°C; 50 up to 625 MPa; various treatment time	Irreversible inactivation T≥10°C, synergistic effect of increases in pressure and temperature on inactivation T<10°C, lowering temperature enhanced pressure inactivation First-order inactivation kinetics Antagonistic effect of pressure on thermal inactivation at <200 MPa and >60°C	Indrawati et al. (2001)
Green peas	−10 to 70°C; 100 up to 500 MPa; various treatment time	Irreversible inactivation T≥10°C, synergistic effect of increases in pressure and temperature on inactivation T<10°C, lowering temperature enhanced pressure inactivation First-order inactivation kinetics Antagonistic effect of pressure on thermal inactivation at <200 MPa and >60°C	Indrawati et al. (2001)

(continued)

Table 10.2 Effect of Combined High Pressure and Temperature Processing on the Stability of Endogenous Enzymes in Fruit and Vegetables (Continued)

Enzymes	Fruit and vegetable matrix	Processing condition	Enzyme stability[a]	References
Myrosinase (MYR)	Broccoli heads	15 to 60°C; 50 to 500 MPa; various treatment time	P≥150 MPa, 15 to 50°C: synergistic effect of increases in pressure and temperature on inactivation At 55 and 60°C and up to 100 MPa: antagonistic effect of pressure on thermal inactivation First-order inactivation kinetics	Van Eylen et al. (2008)
	Broccoli juice	10 to 60°C; 100 to 600 MPa; various treatment time	3.4–86.8% inactivation after pressure buildup Synergistic effect of increases in pressure and temperature on inactivation (up to 40°C) but not at high temperature (50 to 60°C) and low pressure (<200 MPa)	Van Eylen et al. (2007)
Pectinmethylesterase (PME)	Tomato pericarp tissue	−26 to 20°C; 100 to 500 MPa; 13 min.	No/limited inactivation	Van Buggenhout et al. (2006)
	Tomato juice	60 to 75°C; 0.1 to 800 MPa	Antagonistic effect of pressure on thermal inactivation (75°C)	Stoforos et al. (2002)

Tomato juice	50°C; 550,600,700 MPa; various treatment time intervals up to 60 min. 62°C; 500 MPa; various treatment time intervals up to 70 min.	Very pressure resistant Antagonistic effect of pressure on thermal inactivation	Fachin et al. (2002b)
Tomato juice	4, 25, 50°C; 100 to 500 MPa; 10 min.	Pressure resistant 200MPa/25°C/10min.: 27.8% inactivation	Hsu et al. (2008)
Golden delicious and Judaine apples	15 to 65°C; 200 to 600 MPa; various treatment time up to 10.5 min.	Incomplete inactivation (mostly only inactivation of pressure labile isozymes)	Baron et al. (2006)
Greek navel orange juice	30 to 60°C; 100 to 700 MPa; various treatment time intervals up to 30 min.	Irreversible inactivation of pressure labile isozymes Synergistic effect of increases in pressure and temperature on inactivation but not at high temperature and low pressure Antagonistic effect of pressure on thermal inactivation	Polydera et al. (2004)
Valencia and Navel orange juice	20°C; 600 MPa; 60 s	Incomplete inactivation	Bull et al. (2004)

(continued)

Table 10.2 Effect of Combined High Pressure and Temperature Processing on the Stability of Endogenous Enzymes in Fruit and Vegetables (Continued)

Enzymes	Fruit and vegetable matrix	Processing condition	Enzyme stability[a]	References
	Orange juice	300, 350, 400 MPa; 1 to 3 pressure cycles; 20 to 120 min.	Pressure cycling had statistically less significant effects on PME inactivation rather than pressure level, temperature, and pressure holding time	Basak et al. (2001)
	Orange juice	25, 37.5, 50°C; 400, 500, 600 MPa, various treatment time up to 30 min.	Less than 1 log unit irreversible inactivation	Nienaber and Shellhammer (2001)
	Valencia Navel orange juice-milk based beverage	25 to 65°C; 0.1 to 700 MPa; 2 to 75 min.	Two fractions observed: pressure labile and stable fractions 7% pressure stable fraction Up to 550 MPa and up to 55°C: only inactivation of pressure labile fraction At constant temperature, inactivation rate constant of pressure labile fraction was more pressure sensitive than that of pressure stable fraction (for ≥550 MPa) PME in orange juice-milk beverage was more stable than in orange juice alone (without milk)	Sampredo et al. (2008)

Green bell pepper and pepper puree	25, 40 and 60°C; 0.1 to 500 MPa; 15 min.	At 25°C, residual PME activity increased by elevating pressure. At 40°C, residual PME activity increased by elevating pressure up to 300 MPa and decreased by further pressure increase. At 60°c, around 40% inactivation. PME in puree was more stable than intact tissue	Castro et al. (2005)
Green bell pepper	Room temperature (18–20°C); 100 and 200 MPa; 10 and 20 min.	Slight increase in activity probably due to pressure induced enzyme extractability	Castro et al. (2008)
Peroxidase (POD)			
Carrot	−26 to 20°C; 100 to 500 MPa; 13 min.	No/limited inactivation	Van Buggenhout et al. (2006)
Carrot	20 to 50°C; 250 to 450 MPa/various treatment time up to 60 min.	At 20°C, no clear tendency of inactivation was observed by increasing pressure	Akyol et al. (2006)
Carrot	25 to 45°C; 0.1 to 600 MPa; 15 min.	506 MPa/ 25, 35, 40°C/15 min.: reversible inactivation. 600 MPa/40°C/15 min.: max. 91% irreversible inactivation	Soysal et al. (2004)

(continued)

Table 10.2 Effect of Combined High Pressure and Temperature Processing on the Stability of Endogenous Enzymes in Fruit and Vegetables (Continued)

Enzymes	Fruit and vegetable matrix	Processing condition	Enzyme stability[a]	References
	Green bell pepper	Room temperature (18–20°C); 100 and 200 MPa; 10 and 20 min.	70% enzyme inactivation	Castro et al. (2008)
	Green beans	20 to 50°C/250 to 450 MPa/various treatment time up to 60 min.	350 MPa/50°C: enzyme activity increased from 55 to 84% by prolonging treatment time from 30 to 60 min.	Akyol et al. (2006)
	Green peas	20 to 50°C/250 to 450 MPa/various treatment time up to 60 min.	No effective inactivation was observed 30°C/300 and 350 MPa: higher enzyme activity was found	Akyol et al. (2006)
	Lychee	20, 40, 60°C; 200, 400, 600 MPa; 10 and 20 min.	At 200 MPa, increase in enzyme activity after HP treatment. Effect at 40°C was greater than at 20 and 60°C 600 MPa/ 60°C/20 min.: 50% inactivation	Phunchaisri a and Apichartsrangkoon (2005)
	Red bell pepper	Room temperature (18–20°C); 100 and 200 MPa; 10 and 20 min.	40% enzyme inactivation	Castro et al. (2008)

Polygalacturonase	Tomato pericarp tissue	−26 to 20°C; 100 to 500 MPa; 13 min.	At 20°C, pressure >300 MPa enhances enzyme inactivation 500 MPa/20°C/13 min.: 89% inactivation 100–500 MPa/−10 to −26°C/13 min.: No/limited inactivation	Van Buggenhout et al. (2006)
	Tomato juice	5 to 55°C; 200 to 550 MPa; various treatment time intervals up to 210 min.	First-order inactivation model Synergistic effect of increases in pressure and temperature on inactivation	Fachin et al. (2003)
	Tomato juice and pieces	25°C; 01 to 500 MPa; 17 min. (including 2 min of equilibration time)	PG in tomato juice was more stable than in tomato pieces 400 MPa/25°C/17 min.: 70% inactivation Complete inactivation at 600 MPa Dissociation of stable PG fraction (PG1) under pressure The behavior of PG inactivation under pressure different from during thermal treatment at 1 atm	Peeters et al. (2004)
	Tomato juice	4, 25, 50°C; 100 to 500 MPa; 10 min.	Irreversible inactivation Beyond 400 MPa/4 and 25°C/10 min.: up to 90% inactivation 100 to 500 MPa/50°C/10 min.: max. approximately 30% inactivation	Hsu et al. (2008)

(continued)

Table 10.2 Effect of Combined High Pressure and Temperature Processing on the Stability of Endogenous Enzymes in Fruit and Vegetables (Continued)

Enzymes	Fruit and vegetable matrix	Processing condition	Enzyme stability[a]	References
Polyphenoloxidase	Potato	−26 to 20°C; 100 to 500 MPa; 13 min.	No/limited inactivation	Van Buggenhout et al. (2006)
	Victoria grape must	20 to 70°C; 100 to 800 MPa; various treatment time up to 120 min.	Inactivation of labile and stable fractions Kinetics described by a biphasic model Synergistic effect of increases in pressure and temperature on inactivation Antagonistic effect of low pressure on thermal inactivation	Rapeanu et al. (2005b)
	Mango puree	Room temperature; 207, 345, 483 and 552 MPa; 2 s and 0, 1, 3, 5, 10 and 15 min	Limited inactivation	Guerrero-Beltrán et al. (2006)

Lychee	20, 40, 60°C; 200, 400, 600 MPa; 10 and 20 min.	Remarkable pressure inactivation at 60°C 600 MPa/ 60°C/20 min.: 90% inactivation	Phunchaisri a and Apichartsrangkoon (2005)
Red bell pepper	Room temperature (18–20°C); 100 and 200 MPa; 10 and 20 min.	Pressure stable	Castro et al. (2008)
Green bell pepper	Room temperature (18–20°C); 100 and 200 MPa; 10 and 20 min.	50% inactivation	Castro et al. (2008)

[a] Enzyme stability referring to the residual enzyme activity after HP processing (*post factum*) measured at atmospheric pressure.

of pressure effect on enzyme stability in food matrix while the effects of other reactions that occurred at the same time during processing on enzyme stability are also incorporated.

In general, pressure stability of enzymes is largely varied dependent on enzyme type, enzyme source, food matrix, and so on. At constant pressure and temperature, decrease in enzyme activity could be enhanced by prolonging treatment time. Therefore, first-order kinetic model is mostly used as an approach to describe the time dependency of enzyme inactivation (Tables 10.1 and 10.2). To eliminate quality degradations due to further enzymatic reaction after processing and during storage, a complete and irreversible inactivation of undesired food-quality related enzymes is targeted. In most cases described in the literature, pressure does not always result in irreversible (fruit and vegetable) enzyme inactivation for example peroxidase (POD) at moderate temperature (up to 50°C) combined with moderate pressure (up to 500 MPa) in contrast to other enzymes such as hydroperoxide lyase (HPL), lipoxygenase (LOX), myrosinase (MYR), pectin methyl esterase (PME), polygalacturonase (PG) and polyphenoloxidase (PPO) (Tables 10.1 and 10.2). In this temperature and pressure region, PME and PPO can be irreversibly but incompletely inactivated due to the existence of isozymes having different pressure stability. In contrast, PG, MYR, LOX, and HPL can be irreversibly and completely inactivated in this pressure and temperature area. It indicates that pressure stability of MYR, LOX, and HPL is lower than PME and PPO.

At elevated temperatures (closed to the temperature area in which thermal degradation at ambient pressure occurs), elevating pressure (mostly at low pressure level up to 200 MPa) retards the thermal inactivation of enzymes for example MYR, LOX, PME, and PPO inactivation at low pressure (<200 MPa). In this case, the inactivation rate constant (k values) at low-pressure level is lower than at ambient pressure. The antagonistic effect of pressure and temperature on enzyme is also observed at low temperature (<20°C) such as MYR and LOX. The inactivation of these enzymes at constant pressure is enhanced by lowering temperatures and can be completely inactivated at subzero temperatures under pressure. The latter provides a possibility to replace two sequential process steps of blanching and freezing with a single HP processing at subzero temperature, particularly when LOX inactivation is targeted as blanching index such as for blanching and freezing of legume vegetables.

10.3.3 *Understanding of pressure effect on enzymatic reactions in fruit and vegetables*

HP processing could instantaneously affect both enzyme stability and activity. Various research approaches have been conducted to enable a better understanding of pressure effects on fruit and vegetable enzymes. In the case of enzyme stability, pressure effects are mostly evaluated based on the degree of enzyme inactivation after the treatment. It is determined by measuring the residual enzyme activity after the treatment at ambient pressure compared to the initial enzyme activity. However, in the case of enzyme activity, the rate of enzyme-substrate reactions is mostly considered. Enzyme, substrate, and other (bio)molecules are exposed together to HP processing. The amount of substrate consumed or degradation products formed after enzymatic reaction at elevated pressure is followed as a function of reaction time.

In fruit and vegetables, enzymes coexist with their substrates, activators, cofactors, inhibitors, etc., as schematically illustrated in Figure 10.2. Enzymes and those (bio)compounds could be located in different cell compartments or bound to the membranes. To enable the enzyme-substrate interactions, it is necessary to know the pressure-temperature-time window of HP processing in which (1) enzyme and substrate(s) are still stable and (2) alteration of the cell membranes or the membranes of intracellular organelles resulting in a release of enzymes and their substrates occurs.

The conformational alterations in the protein structure of enzymes due to pressure could result in changes of both the stability and the catalytic activity of enzyme for example due to (1) reversible or irreversible enzyme inactivation and denaturation; (2) partial or complete enzyme inactivation; (3) changes in the substrate affinity; and (4) changes in the interactions with activators, inhibitors, cofactors, and so on. The extent of these changes is influenced not only by extrinsic factors including HP process conditions such as pressure level, temperature level, processing time, and rate of pressure ramp, but also by intrinsic factors including the enzyme molecular structure, and the surrounding microenvironment, such as pH, ionic strength, ion types, water activity, and so on. Similarly, pressure affects the stability of other (bio)molecules involving in the enzymatic reactions such as substrates. Unfolding of macromolecular substrates (e.g., protein, starch), changes in ion charges, and substrate degradation due to chemical reactions could influence the susceptibility of substrate, inhibitors, or enzymes towards enzymatic reactions.

At the same time, pressure can enhance and retard enzymatic and chemical reactions depending on the resulting volume change (*Le Chatelier* principle). Due to the instantaneous occurrence of several (complex) reactions under pressure, only the overall catalytic reaction could be observed and measured. In the literature, effects of combined high pressure and temperature processing on the catalytic activity of fruit and vegetable enzymes have been studied in different types of buffer solution and in fruit/vegetable matrices as summarized in Table 10.3. It is shown that enzyme-substrate reactions are influenced by intrinsic (e.g., presence of

Table 10.3 Effect of Combined High Pressure and Temperature Processing on the Catalytic Activity[a] of Fruit and Vegetable Enzymes in Buffer Solution and Fruit/Vegetable Matrices

Enzymes	Enzyme, substrate, and matrices	Processing condition	Remarks	References
Myrosinase (MYR)	Endogenous MYR, commercial sinigrin, broccoli juice	20 to 50°C; 50 to 250 MPa; 2, 10 and 15 min.	Enzyme activity increased with increasing pressure up to 100 MPa and decreased with further pressure increase. Maximal activity found at 45°C and 100–150 MPa for 15 min.	Van Eylen et al. (2008b)
	Endogenous MYR and glucosinolates, intact broccoli florets	20 to 50°C; 50 to 250 MPa; 2, 10, and 15 min.	Maximal activity found at 40°C and 250 MPa for 30 min.	Van Eylen et al. (2008b)
	Endogenous MYR and glucosinolates, intact broccoli florets	20 to 40°C; 100 to 500 MPa; 15 and 30 min.	HP induced glucosinolate hydrolysis and promote the formation of isothiocyanates and indole oligomers	Van Eylen et al. (2009)
	Partially purified mustard seed MYR, commercial sinigrin, broccoli juice (pH adjusted to 6.5)	20 to 60°C; 100 to 400 MPa; 3 and 13 min.	Maximal activity found at 60°C and 200 MPa	Van Eylen et al. (2008a)
	Partially purified mustard seed MYR, commercial sinigrin, Bis-Tris buffer (2 mM; pH 6.5)	20 to 60°C; 100 to 600 MPa; 3 and 13 min.	Maximal activity found at 40°C and 200 MPa. No activity found at 600 MPa at all temperatures studied, however, no process loss in activity was observed	Van Eylen et al. (2008a)

	Partially purified broccoli MYR, commercial sinigrin, phosphate buffer (0.1 M; pH 6.55)	20 to 50°C; 50 to 200 MPa	At all temperatures, activity slightly increased with increasing pressure up to 50–75 MPa. No activity found at 150 and 200 MPa most probably due to enzyme inactivation	Ludikhuyze et al. (2000)
Pectinmethylesterase (PME)	Endogenous PME, endogenous pectin, intact shredded carrots (*in situ*)	Up to 600 MPa; 30–60°C; total treatment time = 18 min. (including 3 min. of equilibration time)	Most pronounced PME catalytic activity at 50°C in the pressure range of 200–400 MPa (optimum at 380 MPa and 50°C). At constant T, increase in released methanol with increasing P, followed by a plateau and a decline	Sila et al. (2007)
	Endogenous PME, endogenous pectin, carrot pieces (*in situ*)	Up to 500 MPa; 30–60°C; total treatment time = 15 min.	Most pronounced PME catalytic activity at 60°C in the pressure range of 100–400 MPa (optimum at 380 MPa and 50°C). At constant T, increase in released methanol with increasing P, followed by a plateau	Sila et al. (2007)

(continued)

Table 10.3 Effect of Combined High Pressure and Temperature Processing on the Catalytic Activity[a] of Fruit and Vegetable Enzymes in Buffer Solution and Fruit/Vegetable Matrices (Continued)

Enzymes	Enzyme, substrate, and matrices	Processing condition	Remarks	References
	Endogenous PME, commercial apple pectin (75% degree of methylation); Golden delicious and Judaine apple juices	15 to 65°C; 200 to 600 MPa; various treatment time up to 10.5 min.	Rates of methanol release varied from 6.2 to 12.4±1.0 mM/day for HP treated juices	Baron et al. (2006)
	Tomato PME purified with affinity chromatography (1.1–1.2 U/mL), commercial apple pectin, Na-acetate buffer (0.1 M; pH 4.4; 0.117 M NaCl)	30 to 70°C; 0.1 to 500 MPa; 20 min.	In presence of tomato PG, PME catalyzed hydrolysis of pectin accelerated with increasing pressure up to 300 MPa and temperature up to 60°C and decreased with increasing both process parameters In absence of tomato PG, increasing pressure (up to 400 MPa) and temperature (up to 70°C) only enhanced PME catalyzed hydrolysis of pectin Up to 300 MPa combined with T up to 60 °C, tomato PME was more active in presence of tomato PG than in absence of PG	Verlent et al. (2007)

Green bell pepper PME purified with affinity chromatography, commercial apple pectin, citrate buffer (20 mM; pH 5.6 containing 0.4 M NaCl)	30 to 60°C, 200 to 600 MPa; various treatment time	Maximal activity at 200 MPa and 55°C.	Castro et al. (2006b)
Tomato PME purified with affinity chromatography, commercial apple pectin, Tris HCl (0.1 M, pH 8, 0.4 M NaCl) and Na-acetate buffer (0.1 M; pH 4.4; 0.4 M NaCl)	30 to 65°C; 100 to 500 MPa; up to 35 min. (for pH 8) 35 to 35°C; 150 to 600 MPa; 35 min.	At both pH values, catalytic PME activity at elevated pressure was higher than at 1 atm. At pH 8, maximal activity at 55°C and 300 MPa. At pH 4.4, maximal activity at 47°C and 450 MPa. Under pressure, the catalytic activity at pH 8 was higher than pH 4.4	Verlent et al. (2004a)

(continued)

Table 10.3 Effect of Combined High Pressure and Temperature Processing on the Catalytic Activity[a] of Fruit and Vegetable Enzymes in Buffer Solution and Fruit/Vegetable Matrices (Continued)

Enzymes	Enzyme, substrate, and matrices	Processing condition	Remarks	References
Polygalacturonase (PG)	Tomato PG purified with cation exchange chromatography, polygalacturonic acid (PGA), Na-acetate buffer (0.1 M; pH 4.4)	25 to 65°C; 100 to 500 MPa; various reaction time intervals	Pressure retarded PG-catalyzed PGA depolymerization Pressure sensitivity of the reaction rate increased with elevating temperature Temperature sensitivity of the reaction rate decreased with elevating pressure Lower enzymatic activity at elevated pressure that at 1 atm	Verlent et al. (2004b)
	Tomato PG purified with cation exchange chromatography, commercial apple pectin with different patterns of methyl esterification, Na-acetate buffer (0.1 M; pH 4.4)	30 to 65°C, 100 to 300 MPa; various reaction time intervals of 5 min (0–35 min.)	Catalytic activity of tomato PG under pressure was higher on pectin deesterified by tomato PME (block wise) than by fungal PME (random) At all T, enzyme activity decreased with increasing pressure	Verlent et al. (2005)

Tomato PG purified with cation exchange chromatography (0.65–0.8 U/mL), commercial apple pectin, Na-acetate buffer (0.1 M; pH 4.4; 0.117 M NaCl)	30 to 70°C; 0.1 to 400 MPa; 20 min.	In presence of tomato PME, PG catalyzed pectin depolymerization accelerated with increasing pressure up to 300 MPa and temperature up to 50°C and decreased with increasing both process parameters	Verlent et al. (2007)

ᵃ The term *enzyme catalytic activity* refers to the catalytic activity of enzyme to convert its substrate by incorporating pressure effects on the whole enzyme system during HP processing (*in situ*). In this case, the enzyme activity is ceased after HP processing and the products of substrate conversion are identified and quantified at atmospheric pressure.

other enzymes (Verlent et al., 2007), type of substrates (Verlent et al., 2005; 2007), pH (Verlent et al., 2004a), ionic strength, degree of matrix disruption (Sila et al. 2007; Van Eylen et al., 2008b, 2009), and extrinsic (e.g., pressure, temperature) factors.

10.3.3.1   Case study on texture improvement of fruit and vegetables under pressure

In fruit and vegetables, PME and PG play an important role in pectin degradation. PME catalyzes the deesterification of pectin and PG catalyzes further the hydrolytic cleavage of the a-1,4-glycosidic bonds of the deesterified pectin. This depolymerization results in a drastic loss in rheology for example decrease in viscosity. In tomato-based products, it is a challenge to control both PME and PG activity and stability at the same time during processing. PME and PG have different stability towards pressure and temperature. At ambient pressure, PME is more temperature labile than PG, while their stability is the reverse at elevated pressure. Therefore, HP processing can interestingly be used to knock out undesired enzymes while enhancing the activity of desired enzyme. For example, under pressure the activity of PME can be enhanced while PG is inactivated.

Different types/patterns of substrate and the existence of other enzymes participating in the same enzymatic reaction pathway could affect the pressure and temperature combination optimal for enzyme activity. Based on the data reported by Verlent et al. (2004b; 2007), isorate contour diagrams of catalytic reactions of PME and PG as a function of pressure and temperature are depicted (Figure 10.3). In case of tomato PME, the optimum PME activity takes place at high pressure and temperature levels, i.e., around 60–65°C and 300–500 MPa in presence of pectin (favorable substrate of PME) (Figure 10.3a) while adding PG next to pectin shifts the optimum condition of PME activity drastically to lower pressure and temperature combination (Figure 10.3b). In case of tomato PG, the optimum activity occurs at low pressure (around 200 MPa) and temperature (45–55°C) in presence of polygalacturonic acid (PGA, favorable substrate of PG, Figure 10.3c). When PME is used to deesterify pectin facilitating substrate for PG, the optimum activity occurs at a pressure level somewhat lower (around 100 MPa) but in a broad temperature range (45–65°C) (Figure 10.3d). It can be concluded that tomato PME activity during HP processing is indirectly influenced by tomato PG and vice versa. To improve or to preserve the rheological properties of tomato-based products, different pressure and temperature combinations can be used to reduce tomato PG activity while maintaining sufficient PME activity.

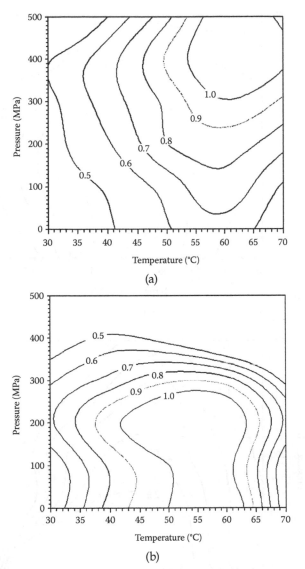

Figure 10.3 Pressure-temperature iso rate contour diagram of tomato PME and PG activity in Na acetate buffer (0.1, pH 4.4). The lines indicate the pressure and temperature combinations resulting in the same rate of catalytic reaction. The numbers indicate the rate of catalytic reaction expressed in µM methanol/min and mM reducing groups/min respectively for PME and PG. (a) Tomato PME catalyzed pectin deesterification in absence of PG (pectin as substrate); (b) tomato PME catalyzed pectin deesterification in presence of PG (pectin as substrate); (c) tomato PG catalyzed PGA depolymerization (PGA as substrate); (d) tomato PG catalyzed PGA depolymerization (in presence of PME and pectin as substrate). These figures are depicted based on the data of Verlent et al. (2004b and 2007).

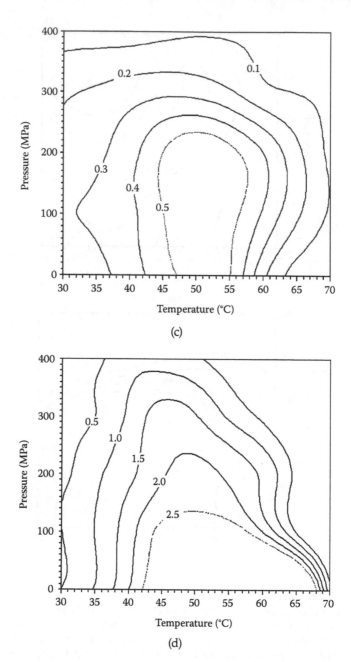

(c)

(d)

Figure 10.3 (Continued)

10.3.3.2 Case study on health benefit enhancement of Brassicaceae *vegetables under pressure*

In *Brassicaceae* vegetables such as broccoli, myrosinase (EC 3.2.1.147), a hydrolytic and temperature labile enzyme, coexists with glucosinolates in different cell compartments. Hence, the enzymatic reaction can be facilitated after cell disruption yielding glucose and an aglucone, which spontaneously decomposes into a wide range of products depending on the reaction conditions. Some of the hydrolysis products are believed to have an anticarcinogenic effect, such as sulforaphane (Zhang et al., 1992) and indole-3-carbinol (Qi et al., 2005). Those hydrolysis products are temperature labile but relatively stable at elevated pressure. Besides health effects, the hydrolysis products affect the odor and taste of broccoli, which are important for consumer acceptance. Hereto, it is necessary to control the activity and stability of MYR.

Pressure/temperature of 100–400 MPa/10–60°C induces cell leakage in broccoli florets (Van Eylen et al., 2008b). Figure 10.4 depicts the integrated

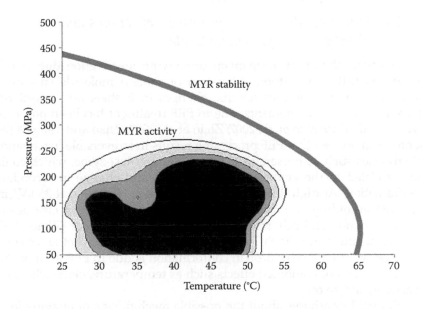

Figure 10.4 Simulated pressure and temperature diagram of endogenous myrosinase stability and activity in intact broccoli florets (based on the data of Van Eylen et al., 2008b). The lines of MYR stability indicate pressure and temperature combinations resulting in 90% enzyme inactivation for total process time of 60 min. The dark color of MYR activity indicates the pressure and temperature combinations resulting in faster enzyme reaction kinetics.

ISO rate contour diagrams of MYR degradation and MYR activity as a function of pressure and temperature. The optimal MYR activity is situated at a pressure level up to 200 MPa in the temperature range of 30–50°C. To have a balance between enhancing the health benefits and eliminating the undesired odor and taste of broccoli, kinetic information on MYR activity and stability can be used as an approach to design and optimize HP processing. In this case, processing with step-wise conditions can be suggested; that is, first process condition at pressure/temperature/time combination optimal for MYR activity subsequently followed by process condition at pressure/temperature/time combinations necessary to inactivate the residual MYR activity. However, maintaining the stability of health-beneficial hydrolysis products formed during the first process condition should be taken into precautions in determining process parameters of the latter process condition.

10.4 Effect of PEF processing on fruit and vegetable enzymes

10.4.1 Understanding of PEF processing effect on stability of enzyme as a protein molecule

At this time, the exact mechanism of enzyme inactivation due to PEF is not yet fully understood. Alteration of protein molecule structure/conformations (i.e., loss of α-helix, changes in β sheet and unordered secondary structural elements) due to PEF treatment has been observed (Yeom et al., 1999; Min et al., 2007; Zhao et al., 2007; Zhao and Yang, 2008). It has been noticed that PEF processing results in irreversible inactivation of enzymes such as horseradish POD, PME, and lysozyme, which could be correlated to the evidence of α helix loss and change in secondary conformation. At high PEF process intensity (e.g., lysozyme at 35 kV/cm; 1200 μs), unfolding of enzyme structure accompanied with a cleavage of disulfide bonds and self-association aggregation can occur (Zhao et al., 2007; Zhao and Yang, 2008). However, it is still not clear whether enzyme inactivation or change of protein conformation is due to PEF itself or due to other effects or combined effects such as temperature, electrochemical reactions, and so on.

Several hypotheses about the possible mechanisms of enzyme inactivation during PEF treatment have been postulated. According to Perez and Pilosof (2004), PEF could polarize the protein molecule that affects the electrostatic forces and tends to attract each other. If non-covalently linked protein sub-units are dissociated, the quaternary structure of protein could be interfered. As a consequence of changes in protein conformation, buried hydrophobic amino acids or sulfydryl groups could be exposed, and in the case of a severe PEF process such as long exposure to electric field, hydrophobic interactions, or covalent bonds (i.e., disulfide bonds) could be disturbed forming aggregates. Pilar and Vercet (2006) have also proposed that changes in electrostatic interactions of protein molecules due to PEF might lead to changes in redox potentials at the active site of the enzymes, ionization of some chemical groups, or breakage of electrostatic interaction inside a polypeptide chain of protein molecules, and so on. However, further studies are still required to prove the aforementioned hypotheses since these suggestions are quite descriptive.

10.4.2   Effect of PEF on stability of fruit and vegetable enzymes

Similar to pressure effects on enzyme inactivation, the stability of fruit and vegetable enzymes is dependent not only on the enzyme itself but also on the intrinsic (e.g., food matrix, medium in which the enzyme is dissolved) and extrinsic (e.g., process parameters, batch/continuous system) factors. So far, effects of PEF processing on fruit and vegetable enzymes have been more focused on the enzyme stability rather than on the enzymatic catalytic reaction during processing. The sensitivity of fruit and vegetable enzymes such as LOX, papain, PME, PG, POD, and PPO towards PEF treatment is summarized in Table 10.4. In general, it has been found that (1) irreversible inactivation can be obtained during PEF treatment; (2) the degree of enzyme inactivation depends on the process intensity; (3) increasing electric field strength and treatment time enhances the enzyme inactivation; (4) pulse polarity could affect the degree of enzyme inactivation such as for PME and POD; and (5) increasing electrical energy reduces the enzyme activity. Furthermore, several kinetic models using first-order reaction kinetic model, polynomial function, Weibull distribution, Fermi model, Hülsheger models, and model based on Bayesian framework have been tried to describe the kinetics of enzyme inactivation during PEF processing.

Since a complete enzyme inactivation is hardly obtained by PEF treatment, shelf-life studies have been conducted to evaluate the quality of PEF-treated food products, such as tomato juice (Min et al., 2003b), orange juice (Elez-Martínez et al., 2006), strawberry juice (Aguiló-Aguayo et al., 2009), and orange juice-milk beverage (Sampedro et al., 2009). Those studies have shown that PEF-treated food products have similar or even better quality properties (e.g., color, odor) than thermally treated products, even though the residual enzyme activity after the PEF and thermal treatments situates in the same order of magnitude.

Table 10.4 Effect of PEF Processing on the Stability of Fruit and Vegetables Enzymes in Buffer Solution or in Fruit/Vegetable Juices

Enzymes	Enzyme and buffer solution/ medium	Processing condition	Enzyme stability[b]	References
Lipoxygenase (LOX)	Green bean juice	Square wave pulse; monopolar mode; electric field strength: 2.5–20 kV/cm; pulse width: 1 µs; pulse numbers: 100 to 400; pulse frequency: 1 Hz	No inactivation	Van Loey et al. (2002)
	Soymilk	Square wave pulse; bipolar mode; electric field strength: 20 to 42 kV/cm; pulse width: 1–5 µs; treatment time for up to 1036 µs[c]	Irreversible inactivation Max. inactivation: 88% treated at 42 kV/cm for 1036 µs with 400 Hz of pulse frequency and 2 µs of pulse width at 25 °C Increasing electric field strength and treatment time enhanced the enzyme inactivation At 30 kV/cm for 345 µs, increasing pulse frequency up to 600 Hz significantly enhances the inactivation maximally up to 55% At 30 kV/cm, 345 µs and 400 Hz, increasing pulse width (0 to 5 µs) enhanced inactivation maximally up to 64 % Several kinetic models (first-order reaction, Weibull distribution, Fermi) were used to describe enzyme inactivation	Li et al. (2008)

Soymilk	Square wave pulse; electric field strength: 20, 30, and 40 kV/cm; treatment time: 25, 50, and 100 µs; pre-treatment heating temperatures: 23, 35, and 50°C	Irreversible inactivation Max. inactivation: 84.5% at 40 kV/cm for 100 µs with preheating at 50°C Increasing electric field strength and treatment time enhanced the enzyme inactivation First-order kinetic model to describe enzyme inactivation	Riener et al. (2008a)
Tomato juice	Square wave pulse; bipolar mode; 4 co-field tubular treatment chambers; electric field strength: 0 –35 kV/cm; treatment time: 20–70 ms; pulse width: 3 µs; PEF treatment temperature: 10–50°C; flow rate: 1 mL/s; pulse delay time: 20 µs	Irreversible inactivation Max. inactivation: 80% treated at 35 kV/cm for 50 or 60 ms at 30 °C Increasing electric field strength, temperature, and treatment time enhanced the enzyme inactivation Applied electric field strength was the primary variable for tomato LOX inactivation at a given energy input Several kinetic models (first-order reaction, Hülsheger, Fermi) were used to describe enzyme inactivation	Min et al. (2003a)

(*continued*)

Table 10.4 Effect of PEF Processing on the Stability of Fruit and Vegetables Enzymes in Buffer Solution or in Fruit/Vegetable Juices (Continued)

Enzymes	Enzyme and buffer solution/ medium	Processing condition	Enzyme stability[b]	References
	Tomato juice	Square wave pulse; bipolar mode; 6 co-field tubular treatment chambers; electric field strength: 40 kV/cm; treatment time: 57 μs; pulse width: 2 μs; temperature before PEF treatment: 45°C (temperature increase of 8°C during treatment)	Irreversible inactivation Max. inactivation: 54%	Min et al. (2003b)
Papain	Commercial purified papaya papain initially activated by reducing agents (cysteine 20 mM and DIT 5 mM) and diluted in 1 mM EDTA	Square wave pulse; electrical field strength: 20, 30, 40, and 50 kV/cm; pulse width: 4 μs; continuous system; flow rate: 0.77 ml/s; pulse numbers: up to 1500	Irreversible inactivation Max inactivation: up to 40% immediately after PEF treatment at 50 kV/cm and 500 pulses Increasing pulse numbers enhanced the enzyme inactivation Storage at 4°C for 24 and 48 h decreased further the activity of PEF treated enzymes regardless of electric field strength Oxidation of papain active site was not the major cause of papain inactivation by PEF treatment Inactivation of PEF-treated papain was related to the loss of α-helix structure	Yeom et al. (1999)

Pectinmethylesterase (PME)	Commercial purified[a] orange PME, details on medium and enzyme concentration not completely described	Exponential decay pulse; coaxial and parallel-plane treatment chamber, electric field strength: 5–25 kV/cm; treatment time: 0.5–15 μs; pulse frequency: 1 Hz, pulse numbers : 207 to 1449	Irreversible inactivation. Max. inactivation: 17% and 16.46% in coaxial cylinder and parallel-plane chamber, respectively, treated at 207 pulses of 25 kV/cm. Increasing electric field strength and treatment time enhanced the enzyme inactivation	Zhang et al. (2006)
	Partially[a] purified banana, carrot, orange and tomato PME dissolved in Tris buffer (20 mM; pH 7.0)	Square wave pulse of 40 μs; electric field strength: 13.2 and 19.1 kV/cm; treatment time: 1.6 ms; pulse frequency: 0.5 or 5 Hz	Irreversible inactivation. Max. inactivation: 45% (banana), 83% (carrot), 87% (orange and tomato). Increasing electric field strength and treatment time enhanced the enzyme inactivation	Espachs-Barroso et al. (2006)

(continued)

Table 10.4 Effect of PEF Processing on the Stability of Fruit and Vegetables Enzymes in Buffer Solution or in Fruit/Vegetable Juices (Continued)

Enzymes	Enzyme and buffer solution/medium	Processing condition	Enzyme stability[b]	References
	Pectinex 100 Lr (Novo Nordisk Ferment, Neumatt, Switzerland) in distilled water at 5% mass fraction; electrical conductivity: 11.48 S/m, pH 4.73	Square wave pulse; pulse width: 40 μs; electric field strength: 5–24 kV/cm; total treatment time: 16 ms; pulse frequency: 0.5 or 5 Hz	Irreversible inactivation Max. inactivation: approximately 95–98% treated at 5 kV/cm for 16 ms; 12 kV/cm for 8–12 ms; 20 kV/cm for 6–8 ms and 24 kV/cm for 4 ms Increases in electric field strength and treatment time enhanced the enzyme inactivation Model based on Bayesian framework used to describe enzyme inactivation	Giner et al. (2005a)
	Desalted pectinex 100 Lr (Novo Nordisk Ferment, Neumatt, Switzerland) diluted in bi-distilled water (1:5 v/v); pH 4.6	Exponential decay pulse; monopolar mode; electric field strength: 19, 25, 30, 33, 36 and 38 kV/cm; inlet temperature: 4°C; pulse numbers: 10 to 100	Irreversible inactivation Max. inactivation: approximately 86.8% treated at 38 kV/cm for 340 μs Increases in electric field strength, pulse number, treatment time and electrical density energy input enhanced the enzyme inactivation Fermi, Hülsheger or Weibull equations to describe enzyme inactivation kinetics	Giner et al. (2005b)

Grapefruit juice	Monopolar mode, electric field strength: 20, 30, and 40 kV/cm; preheating at 23, 35, and 50°C; 15 Hz; pulse width: 1 µs; treatment time: 25, 50, 75, and 100 µs	Irreversible inactivation Max. inactivation: 96.8% obtained using a combination of preheating to 50 °C, and a PEF treatment time of 100 µs at 40 kV/cm Increasing electric field strength and treatment time enhanced the enzyme inactivation First-order kinetic model to describe enzyme inactivation	Riener et al. (2009)
(Valencia) orange juice	Square-wave pulse; monopolar mode; inlet temperature < 40°C; pulse width: 1.4 µs, pulse frequency: 600 Hz; electric field pulse: 20–35 kV/cm; treatment time 59 µs	Irreversible inactivation Max. inactivation: 90% treated at 35 kV/cm for 59 µs	Yeom et al. (2000a)
(Valencia) orange juice	Six co-field tubular chambers (diameter 0.635 cm); electric field pulse: 35 kV/cm; treatment time 59 µs	Irreversible inactivation Max. inactivation: 88%	Yeom et al. (2000b)
(Valencia) orange juice	Square-wave pulse; bipolar mode; inlet temperature 10–50°C; pulse frequency: 700 Hz; electric field pulse: up to 35 kV/cm; treatment time up to 250 ms	Irreversible inactivation Max. inactivation: 90% treated at 25 kV/cm and 50°C for 250 ms Increasing electric field strength and temperature enhanced the enzyme inactivation	Yeom et al. (2002)

(continued)

　　　　　　　　　　　　　　　　　　　　　　　Indrawati Oey

Table 10.4 Effect of PEF Processing on the Stability of Fruit and Vegetables Enzymes in Buffer Solution or in Fruit/Vegetable Juices (Continued)

Enzymes	Enzyme and buffer solution/ medium	Processing condition	Enzyme stability[b]	References
	Orange juice	Square wave pulse; monopolar pulse; 1000 pulses; pulse width: 1 μs; pulse frequency: 1 or 2 Hz; electric field strength: up to 35 kV/cm	No inactivation but increase in activity probably due to cell permeabilization and release of intracellular pectinmethylesterase	Van Loey et al. (2002)
	(Navelina) orange juice	Square-wave pulse; bipolar mode; temperature < 40°C; pulse width: 4 μs, pulse frequency: 200 Hz; electric field pulse: 35 kV/cm; treatment time 1 ms; energy input of 5390 MJ/m³	Irreversible inactivation Max. inactivation: 81.6%	Elez-Martínez et al. (2006)

Orange juice	Square wave pulse; mono- or bipolar mode; electric field strength: 5, 15, 25, 30, and 35 kV/cm; treatment time: 100, 300, 600, 1000, and 1500 µs; pulse width: 4 µs; pulse frequency: 200 Hz.	Irreversible inactivation Max. inactivation: 80% treated at 35 kV/cm for 1500 µs with 4 µs bipolar pulses at 200 Hz without exceeding 37.5 °C Increasing electric field strength and treatment time enhanced the enzyme inactivation At 30 kV/cm, 600 µs and square bipolar pulses of 4 µs, increasing pulse frequency up to 200 Hz significantly enhanced the inactivation maximally up to 50%. At 30 kV/cm, 600 µs, 200 Hz and square bipolar pulses, increasing pulse width (0 to 10 µs) enhanced inactivation maximally up to 70% Several models (first-order fractional conversion, Fermi and Hülsheger) used to describe enzyme inactivation	Elez-Martinez et al. (2007)
(Valencia) orange juice-milk beverage	Square wave pulse; bipolar mode; six co-field chambers; pulse width: 2.5 µs; treatment time: 50 µs; electric field strength: 30 kV/cm	Irreversible inactivation Max. inactivation: 90%	Sampredo et al. (2009)

(continued)

Table 10.4 Effect of PEF Processing on the Stability of Fruit and Vegetables Enzymes in Buffer Solution or in Fruit/Vegetable Juices (Continued)

Enzymes	Enzyme and buffer solution/ medium	Processing condition	Enzyme stability[b]	References
	Mixed orange and carrot juice (80: 20)	Square wave pulse; bipolar mode; six co-field treatment chambers; pulse width: 2 µs; electric field strength: 25 kV/cm; treatment condition: 280 ms/112 pulses/767 Hz/ max. temperature 68°C (P1) and 330 ms/132 pulses/ 904 Hz/max. temperature: 70°C (P2)	Irreversible inactivation Max. inactivation : 75.6 % (P1) and 81% (P2)	Rivas et al. (2006)
	Strawberry juice	Square wave pulse; bipolar mode; 8 co-field treatment chambers; flow rate: 60 mL/men.; electric field strength 35 kV/cm; treatment time: 1700 µs; pulse width: 4 µs; pulse frequency: 100 Hz; temperature < 35°C	Irreversible inactivation Max. inactivation: 87 %	Aguiló-Aguayo et al. (2009)

	Squeezed tomato juice	40 pulses, 87 kV/cm, 0.5 Hz, pulse width: 2 µs; no details on pulse shape and mode	Irreversible inactivation Max. inactivation : 55%	Nguyen and Mittal (2007)
	Tomato juice	Square wave pulse; bipolar mode; 8 co-field treatment chambers; flow rate: 60 mL/men; electric field strength 35 kV/cm; treatment time: 1500 µs; pulse width: 4 µs; pulse frequency: 100 Hz; temperature < 35°C	Irreversible inactivation Max inactivation: approximately 82%	Aguiló-Aguayo et al. (2008a)
Peroxidase (POD)	Commercial purified horseradish POD in 10 mM potassium phosphate and NaCl; 10 to 13 Units/mL or 80 to 100 mg/mL	Square wave pulse; bipolar mode; electric field strength up to 41.8 kV/cm; treatment time up to 126 ms	Irreversible inactivation Max. inactivation : 18.1% treated at 34.9 kV/cm for 126 µs Increasing electric field strength and treatment time enhanced the enzyme inactivation	Yang et al. (2004)

(continued)

Table 10.4 Effect of PEF Processing on the Stability of Fruit and Vegetables Enzymes in Buffer Solution or in Fruit/Vegetable Juices (Continued)

Enzymes	Enzyme and buffer solution/ medium	Processing condition	Enzyme stability[b]	References
	Commercial purified horseradish POD, details on medium and enzyme concentration not completely described	Exponential decay pulse; coaxial and parallel-plane treatment chamber, electric field strength: 5–25 kV/cm; treatment time: 0.5–15-µs; pulse frequency: 1 Hz, pulse numbers : 207 to 1449	Irreversible inactivation Max. inactivation: 14.4% and 16.7% in parallel-plane chamber and coaxial cylinder chamber, respectively, treated at 207 pulses of 25 kV/cm Increasing electric field strength and treatment time enhanced the enzyme inactivation	Zhang et al. (2006)
	Apple juice	Square wave, pulse frequency: 15 Hz; pulse width: 1 µs; electric field strength : 20, 30 and 40 kV/cm; treatment time : 25, 50 and 100 µs; pre-treatment heating temperatures: 23, 35 and 50 °C	Irreversible inactivation Max. inactivation: 68 % treated with a combination of preheating to 50 °C and a PEF treatment time of 100 µs at 40 kV/cm Increasing electric field strength and treatment time enhanced the enzyme inactivation	Riener et al. (2008b)

Apple juice	Square-wave pulse; monopolar mode; pulse width: 1 μs; pulse frequency: 15 Hz; pulse numbers: 100; treatment time: 100 μs; electric field strength: 40 kV/cm	Irreversible inactivation. Max inactivation: approximately 50% treated for 100 μs at 40 kV/cm and 15 Hz	Noci et al. (2008)
Grape juice	Square wave pulse; bipolar mode; 8 co-linear treatment chambers; inlet temperature: 12 to 15°C; electric field strength: 25, 30 and 35 kV/cm; treatment time: 1–5 ms; pulse frequency: 200, 6000 and 1000 Hz	Irreversible inactivation. Max inactivation: approximately 50% treated for 5 ms at 35 kV/cm and 600 Hz. Increasing electric field strength and treatment time enhanced the enzyme inactivation. Polynomial model to describe enzyme inactivation kinetics	Marsellés-Fontanet and Martín-Belloso (2007)
(Navelina) orange juice	Square-wave pulse; bipolar mode; temperature < 40°C; pulse width: 4 μs, pulse frequency: 200 Hz; electric field pulse: 35 kV/cm; treatment time 1 ms; energy input of 5390 MJ/m3	Irreversible inactivation. Complete inactivation. Weibull distribution model used to describe enzyme inactivation	Elez-Martínez et al. (2006)

(continued)

Table 10.4 Effect of PEF Processing on the Stability of Fruit and Vegetables Enzymes in Buffer Solution or in Fruit/Vegetable Juices (Continued)

Enzymes	Enzyme and buffer solution/ medium	Processing condition	Enzyme stability[b]	References
	Ripe tomato juice	Monopolar or bipolar mode; constant electric field strength of 35 kV/cm, T < 35 °C, pulse frequency: 50 to 250 Hz, pulse width: 1 to 7 μs; treatment time: 1000 to 2000 μs	POD inactivation more effective in bipolar than monopolar mode Prolonging treatment time enhanced the POD inactivation A pulse frequency of 200 Hz was enough to reach a minimum value of residual POD activity Significant interaction between pulse frequency and treatment time for POD inactivation. Different combinations of both variables gave the same POD inactivation Bipolar mode had more pulse width (5.5 μs) effect on POD inactivation than monopolar mode A complete irreversible POD inactivation was observed at 35 kV/cm for 2000 μs using 7 μs-bipolar pulses at 200 Hz Second-order polynomial response function used to describe enzyme inactivation	Aguiló-Aguayo et al. (2008b)

Enzyme/Substrate	Medium/conditions	Treatment	Observations	Reference
	Tomato juice	Square wave pulse; bipolar mode; 8 co-field treatment chambers; flow rate: 60 mL/men.; electric field strength 35 kV/cm; treatment time: 1500 μs; pulse width: 4 μs; pulse frequency: 100 Hz; temperature < 35°C	Irreversible inactivation Max inactivation: approximately 97%	Aguiló-Aguayo et al. (2008a)
Polygalacturonase (PG)	Pectinex 100 L (Novo Nordisk Ferment, Neumatt, Switzerland) in distilled water with 5% mass fraction; electrical conductivity 13.17 mS/cm, pH 4.62	Exponential decay pulses of 40 and 160 ms; bipolar mode, electric field strength 5.18 to 19.39 kV/cm, pulse number ranged up to 400, T < 25°C	Irreversible inactivation Max. inactivation: 98% treated for 32.4 ms at 10.28 kV/cm Increasing electric field strength and treatment time enhanced the enzyme inactivation First-order kinetic model was applied to calculate inactivation rate constants (ranged from 32 to 590 ms^{-1})	Giner et al. (2003)

(continued)

Indrawati Oey

Table 10.4 Effect of PEF Processing on the Stability of Fruit and Vegetables Enzymes in Buffer Solution or in Fruit/Vegetable Juices (Continued)

Enzymes	Enzyme and buffer solution/ medium	Processing condition	Enzyme stability[b]	References
	Pectinex 100 Lr (Novo Nordisk Ferment, Neumatt, Switzerland) in distilled water at 2.5% mass fraction; electrical conductivity: 0.765 to 0.773 S/m, pH 4.61	Square wave pulse, monopolar mode; electric field strength: 15 to 38 kV/cm; treatment time: 300 to 1100 µs	Irreversible inactivation Max. inactivation: 76.5% treated at 38 kV/cm for 1100 µs Increasing electric field strength and treatment time enhanced the enzyme inactivation At 15–20 kV/cm for 300 µs, augmentation of the PG activity measured after PEF (110.9 %) was observed 2 consecutive irreversible first-order steps used to describe enzyme inactivation	Giner-Seguí et al. (2006)
	Squeezed tomato juice	40 pulses, 87 kV/cm, 0.5 Hz, pulse width: 2 µs; no details on pulse shape and mode	Irreversible inactivation Max. inactivation : 55%	Nguyen and Mittal (2007)

Strawberry juice	Square wave pulse; bipolar mode; 8 co-field treatment chambers; flow rate: 60 mL/men.; electric field strength 35 kV/cm; treatment time: 1700 µs; pulse width: 4 µs; pulse frequency: 100 Hz; temperature < 35°C	Irreversible inactivation Max. inactivation: 26 %	Aguiló-Aguayo et al. (2009)
Tomato juice	Square wave pulse; bipolar mode; 8 co-field treatment chambers; flow rate: 60 mL/men.; electric field strength 35 kV/cm; treatment time: 1500 µs; pulse width: 4 µs; pulse frequency: 100 Hz; temperature < 35°C	Irreversible inactivation Max inactivation: approximately 12%	Aguiló-Aguayo et al. (2008a)

(continued)

Table 10.4 Effect of PEF Processing on the Stability of Fruit and Vegetables Enzymes in Buffer Solution or in Fruit/Vegetable Juices (Continued)

Enzymes	Enzyme and buffer solution/ medium	Processing condition	Enzyme stability[b]	References
Polyphenoloxidase (PPO)	Commercial purified mushroom PPO in 10 mM potassium phosphate and NaCl; 150 to 200 mg/mL or 300 to 400 units/mL	Square wave pulse; bipolar mode; electric field strength up to 41.8 kV/cm; treatment time up to 126 ms	Irreversible inactivation. Max. inactivation: 38.2% treated at 33.6 kV/cm for 126 μs. Increasing electric field strength and treatment time enhanced the enzyme inactivation. Increasing electrical conductivity of the enzyme medium enhanced the enzyme inactivation (above 20 kV/cm)	Yang et al. (2004)
	Clear apple and pear extracts (conductivity: 227 and 242 mS/cm, respectively)	Exponential decay pulse; bipolar mode; electric field strength: up to 24.6 kV/cm; T <15°C; treatment time up to 6 ms; pulse width: 0.02 ms	Irreversible inactivation. Max. inactivation of 3.15% and 38.0% for PPO in apple extract at 24.6 kV/cm and pear extract treated at 22.3 kV/cm both for 6 ms total treatment time. Increasing electric field strength and treatment time enhanced the enzyme inactivation. First-order reaction inactivation	Giner et al. (2001)

Squeezed apple juice	Square wave pulse; monopolar mode; 1000 pulses; pulse width: 1–40 µs; electric field strength: 7–31 kV/cm ; pulse frequency: 1–10 Hz	Irreversible inactivation Max. inactivation of 32% after 1000 pulses of 40 µs at 7 kV/cm and 10 Hz, probably due to temperature increase	Van Loey et al. (2002)
Apple juice	Square wave, pulse frequency: 15 Hz; pulse width: 1 µs; electric field strength : 20, 30 and 40 kV/cm; treatment time : 25, 50 and 100 µs; pre-treatment heating temperatures: 23, 35 and 50°C	Irreversible inactivation Max. inactivation: 71% treated with a combination of preheating to 50 °C and a PEF treatment time of 100 µs at 40 kV/cm Increasing electric field strength and treatment time enhanced the enzyme inactivation	Riener et al. (2008b)
Apple juice	Square wave pulse; pulse width: 3 to 8 µs; electric field strength: 15, 25 and 35 kV/cm (laboratory scale) and 30 kV/cm (pilot plant scale); energy input: 8.5 and 65.5 kJ/kg (laboratory scale) and 100 to 130 kJ/kg (pilot plant scale); colinear electrode configuration; preheating temperature ranged from 20 to 60°C	Laboratory scale: complete irreversible inactivation after a combined preheating at 60°C and PEF treatment at 25 and 35 kV/cm Pilot plant scale: max. inactivation of 48% after PEF treatment at 30 kV/cm; 173 Hz; 100 kJ/kg combined with preheating at 40°C	Schilling et al. (2008)

(continued)

Table 10.4 Effect of PEF Processing on the Stability of Fruit and Vegetables Enzymes in Buffer Solution or in Fruit/Vegetable Juices (Continued)

Enzymes	Enzyme and buffer solution/medium	Processing condition	Enzyme stability[b]	References
	Apple juice	Square-wave pulse; monopolar mode; pulse width: 1 μs; pulse frequency: 15 Hz; pulse numbers: 100; treatment time: 100 μs; electric field strength: 40 kV/cm	Irreversible inactivation Max inactivation: approximately 50% treated for 100 μs at 40 kV/cm and 15 Hz	Noci et al. (2008)
	Grape juice	Square wave pulse; bipolar mode; 8 co-linear treatment chambers; inlet temperature: 12 to 15°C; electric field strength: 25, 30, and 35 kV/cm; treatment time: 1–5 ms; pulse frequency: 200, 6000, and 1000 Hz	Irreversible inactivation Complete inactivation after PEF treatment for 5 ms at 35 kV/cm and 600 Hz Increasing electric field strength and treatment time enhanced the enzyme inactivation Polynomial model to describe enzyme inactivation kinetics	Marsellés-Fontanet and Martín-Belloso (2007)

[a] Partially purification was done mostly using ammonium sulfate precipitation.
[b] Enzyme stability referring to the residual enzyme activity measured after PEF processing (post factum).
[c] Total treatment time for square wave pulse was calculated as a mean of pulse width multiplied by number of pulses.

10.5   Conclusions

Fruit and vegetables are rich in enzymes that coexist with the associated substrates and (bio)compounds. Since consumers demand (processed) food products with less additives, science-based and intelligent process control of endogenous food ingredients such as enzymes becomes a big challenge for food technologists to achieve a balance between food quality and safety.

Enzyme sensitivity towards HP and PEF processing is varied dependent on food matrix and process intensity. In general, a significant enzyme inactivation requires a more severe condition of HP and PEF processing rather than for inactivation of microbial vegetative cells. In the same food matrix, the stability rank of enzymes towards HP or PEF could be different from conventional thermal processing. Hence, each novel processing technology offers a unique possibility and opportunity to control enzyme-catalyzed reactions while destroying microorganisms. So far, effects of novel technologies have been studied more intensively on enzyme stability rather than on enzymatic reaction during processing. To be able to control enzyme catalyzed reaction in fruit- and vegetable- based food products during processing, effects of process parameters should be studied not only on the stability of enzymes, substrates, and reaction products, but also on the enzymatic reaction rate. Hereto, thorough and integrated researches on mechanistic and kinetic basis are still needed to have a better understanding of enzyme stability and activity during novel processing.

Abbreviations

HP: hydrostatic pressure
HPL: hydroperoxide lyase
LOX: lipoxygenase
MYR: myrosinase
PEF: high-intensity pulsed electric field
PG: polygalacturonase
PGA: polygalac turonic acid
PME: pectin methyl esterase
PPO: polyphenoloxidase

References

Aguiló-Aguayo, I., R. Soliva-Fortuny, and O. Martín-Belloso. 2008a. Comparative study on color, viscosity and related enzymes of tomato juice treated by high-intensity pulsed electric fields or heat. *European Food Research and Technology* 227: 599–606.
Aguiló-Aguayo, I., I. Odriozola-Serrano, L. J. Quintão-Teixeira, and O. Martín-Belloso. 2008b. Inactivation of tomato juice peroxidase by high-intensity pulsed electric fields as affected by process conditions. *Food Chemistry* 107: 949–955.

Aguiló-Aguayo, I., G. Oms-Oliu, R. Soliva-Fortuny, and O. Martín-Belloso. 2009. Changes in quality attributes throughout storage of strawberry juice processed by high-intensity pulsed electric fields or heat treatments. *LWT—Food Science and Technology* 42: 813–818.

Akyol, C., H. Alpas, and A. Bayındırlı. 2006. Inactivation of peroxidase and lipoxygenase in carrots, green beans, and green peas by combination of high hydrostatic pressure and mild heat treatment. *European Food Research and Technology* 224: 171–176.

Balny, C. and P. Masson. 1993. Effects of high pressure on proteins. *Food Reviews International* 9(4): 611–628.

Baron, A., J. M. Dénes, and C. Durier. 2006. High-pressure treatment of cloudy apple juice. *LWT—Food Science and Technology* 39: 1005–1013.

Basak, S., H. S. Ramaswamy, and B. K. Simpson. 2001. High pressure inactivation of pectin methyl esterase in orange juice using combination treatments. *Journal of Food Biochemistry* 25: 509–526.

Brandts, J. F., R. J. Oliveira, and C. Westort. 1970. Thermodynamics of protein denaturation: Effect of pressure on the denaturation of ribonuclease A. *Biochemistry* 9: 1038–1047.

Bull, M. K., K. Zerdin, E. Howe, et al. 2004. The effect of high pressure processing on the microbial, physical and chemical properties of Valencia and Navel orange juice. *Innovative Food Science and Emerging Technologies* 5: 135–149.

Castro, S. M., A. Van Loey, J. A. Saraiva, C. Smout, and M. Hendrickx. 2005. Process stability of *Capsicum annuum* pectin methylesterase in model systems, pepper puree and intact pepper tissue. *European Food Research and Technology* 221: 452–458.

Castro, S. M., A. Van Loey, J. A. Saraiva, C. Smout, and M. Hendrickx. 2006a. Inactivation of pepper (*Capsicum annuum*) pectin methylesterase by combined high-pressure and temperature treatments. *Journal of Food Engineering* 75: 50–58.

Castro, S. M., A. Van Loey, J. A. Saraiva, C. Smout, and M. Hendrickx. 2006b. Identification of pressure/temperature combinations for optimal pepper (*Capsicum annuum*) pectin methylesterase activity. *Enzyme and Microbial Technology* 38: 831–838.

Castro, S. M., J. A. Saraiva, J. A. Lopes-da-Silva, et al. 2008. Effect of thermal blanching and of high pressure treatments on sweet green and red bell pepper fruits (*Capsicum annuum* L.). *Food Chemistry* 107: 1436–1449.

Cheftel, J. C. 1991. Applications des hautes pressions en technologie alimentaire. *Industries Alimentairs et Agricoles* 108: 141–153.

Dalmadi, I., G. Rapeanu, A. Van Loey, C. Smout, and M. Hendrickx. 2006. Characterization and inactivation by thermal and pressure processing of strawberry (*Fragaria ananassa*) polyphenol oxidase: a kinetic study. *Journal of Food Biochemistry* 30: 56–76.

Da Poian, A. T., A. C. Oliveira, and J. L. Silva. 1995. Cold denaturation of an icosahedral virus. The role of entropy in virus assembly. *Biochemistry* 34: 2672–2677.

De Haan, S. W. H. 2007. Circuitry and pulse shapes in pulsed electric field treatment of food. In *Food Preservation by pulsed electric fields. From research to application*, eds. H. L. M. Lelieveld, S. Notermans, and S. W. H. de Haan, 43–69. Cambridge: Woodhead Publishing Limited.

Elez-Martínez, P., R. C. Soliva-Fortuny, and O. Martín-Belloso. 2006. Comparative study on shelf life of orange juice processed by high intensity pulsed electric fields or heat treatment. *European Food Research and Technology* 222: 321–329.

Elez-Martínez, P., M. Suárez-Recio, and O. Martín-Belloso. 2007. Modeling the reduction of pectin methyl esterase activity in orange juice by high intensity pulsed electric fields. *Journal of Food Engineering* 78: 184–193.

Espachs-Barroso, A., A. Van Loey, M. Hendrickx, and O. Martín-Belloso. 2006. Inactivation of plant pectin methylesterase by thermal or high intensity pulsed electric field treatments. *Innovative Food Science and Emerging Technologies* 7: 40–48.

Fachin, D., A. M. Van Loey, B. L. Nguyen, I. Verlent, Indrawati, and M. E. Hendrickx. 2002a. Comparative study of the inactivation kinetics of pectinmethylesterase in tomato juice and purified form. *Biotechnology Progress* 18: 739–744.

Fachin, D., A. Van Loey, Indrawati, L. Ludikhuyze, and M. Hendrickx. 2002b. Thermal and high-pressure inactivation of tomato polygalacturonase: a kinetic study. *Journal of Food Science* 67: 1610–1615.

Fachin, D., A. M. Van Loey, B. L. Nguyen, I. Verlent, Indrawati, and M. E. Hendrickx. 2003. Inactivation kinetics of polygalacturonase in tomato juice. *Innovative Food Science and Emerging Technologies* 4: 135–142.

Fang, L., B. Jiang, and T. Zhang. 2008. Effect of combined high pressure and thermal treatment on kiwifruit peroxidase. *Food Chemistry* 109: 802–807.

García, A. F., P. Butz, and B. Tauscher. 2002. Mechanism-based irreversible inactivation of horseradish peroxidase at 500 MPa. *Biotechnology Progress* 18: 1076–1081.

Giner, J., V. Gimeno, G. V. Barbosa-Cánovas, and O. Martín. 2001. Effects of pulsed electric field processing on apple and pear polyphenoloxidases. *Food Science and Technology International* 7: 339–345.

Giner, J., V. Gimeno, M. Palomes, G. V. Barbosa-Cánovas, and O. Martín. 2003. Lessening polygalacturonase activity in a commercial enzyme preparation by exposure to pulsed electric fields. *European Food Research and Technology* 217: 43–48.

Giner, J., E. Bailo, V. Gimeno, and O. Martín. 2005a. Models in a Bayesian framework for inactivation of pectinesterase in a commercial enzyme formulation by pulsed electric fields. *European Food Research and Technology* 221: 255–264.

Giner, J., P. Grouberman, V. Gimeno, and O. Martín. 2005b. Reduction of pectinesterase activity in a commercial enzyme preparation by pulsed electric fields: comparison of inactivation kinetic models. *Journal of the Science of Food and Agriculture* 85: 1613–1621.

Giner-Seguí, J., E. Bailo-Ballarín, S. Gorinstein, and O. Martín-Belloso. 2006. New kinetic approach to the evolution of polygalacturonase (EC 3.2.1.15) activity in a commercial enzyme preparation under pulsed electric fields. *Journal of Food Science* 71: 262–269.

Guerrero-Beltrán, J. A., G. V. Barbosa-Cánovas, G. Moraga-Ballesteros, M. J. Moraga-Ballesteros, and B. G. Swanson. 2006. Effect of pH and ascorbic acid on high hydrostatic pressure-processed mango puree. *Journal of Food Processing and Preservation* 30: 582–596.

Guiavarc'h, Y., O. Segovia, M. Hendrickx, and A. Van Loey. 2005. Purification, characterization, thermal and high-pressure inactivation of a pectin methylesterase from white grapefruit (*Citrus paradisi*). *Innovative Food Science and Emerging Technologies* 6: 363–371.

Hawley, S. A. 1971. Reversible pressure-temperature denaturation of chymotrypsinogen. *Biochemistry* 10(13): 2436–2442.

Hayashi, R., T. Kinsho, and H. Ueno. 1998. Combined applications of subzero temperature and high pressure on biological materials. In *High Pressure Food Science, Bioscience and Chemistry*, ed. N.S. Isaacs, 166–174. Cambridge: The Royal Society of Chemistry.

Heremans, K. 1982. High pressure effects on proteins and other biomolecules. *Annual Review on Biophysics and Bioengineering* 11: 1–21.

Heremans, K. 1993. The behaviour of proteins under pressure. In *High Pressure Chemistry, Biochemistry and Materials Science*, eds. R. Winter and J. Jonas, 443–469. The Netherlands: Kluwer Academic Publisher.

Hsu, K. C. 2008. Evaluation of processing qualities of tomato juice induced by thermal and pressure processing. *LWT—Food Science and Technology* 41: 450–459.

Indrawati, A. M. Van Loey, L. R. Ludikhuyze, and M. E. Hendrickx. 1999. Soybean lipoxygenase inactivation by pressure at subzero and elevated temperatures. *Journal of Agricultural and Food Chemistry* 47: 2468–2474.

Indrawati, A. Van Loey, L. Ludikhuyze, and M. Hendrickx. 2000a. Kinetics of pressure inactivation at subzero and elevated temperature of lipoxygenase in crude green beans (*Phaseolus vulgaris* L.) extract. *Biotechnology Progress* 16(1): 109–115.

Indrawati, L. R. Ludikhuyze, A. M. Van Loey, and M. E. Hendrickx. 2000b. Lipoxygenase inactivation in green beans (*Phaseolus vulgaris* L.) due to high pressure treatment at subzero and elevated temperatures. *Journal of Agricultural and Food Chemistry* 48: 1850–1859.

Indrawati, A. M. Van Loey, L. R. Ludikhuyze, and M. E. Hendrickx. 2001. Pressure temperature inactivation of lipoxygenase in green peas (*Pisum sativum*): a kinetic study. *Journal of Food Science* 66: 686–693.

Knorr, D., K. H. Engel, R. Vogel, B. Kochte-Clemens, and G. Eisenbrand. 2008. Statement on the treatment of food using a pulsed electric field. Opinion of the senate commission on food safety (SKLM) of the German Research Foundation (DFG). *Molecular Nutrition and Food Research* 52: 1539–1542.

Li, Y. Q., Q. Chen, X. H. Liu, and Z. X. Chen. 2008. Inactivation of soybean lipoxygenase in soymilk by pulsed electric fields. *Food Chemistry* 109: 408–414.

Ly-Nguyen, B., A. M. Van Loey, D. Fachin, et al. 2002a. Strawberry pectin methylesterase (PME): Purification, characterization, thermal and high-pressure inactivation. *Biotechnology Progress* 18: 1447–1450.

Ly-Nguyen, B., A. M. Van Loey, D. Fachin, I. Verlent, Indrawati, and M. E. Hendrickx. 2002b. Partial purification, characterization, and thermal and high-pressure inactivation of pectin methylesterase from carrots (*Daucus carrota* L.). *Journal of Agricultural and Food Chemistry* 50: 5437–5444.

Ly Nguyen, B., A. Van Loey, D. Fachin, I. Verlent, Indrawati, and M. Hendrickx. 2002c. Purification, characterization, thermal, and high-pressure inactivation of pectin methylesterase from bananas (cv *Cavendish*). *Biotechnology and Bioengineering* 78: 683–691.

Ly-Nguyen, B., A. M. Van Loey, C. Smout, et al. 2003. Mild-heat and high-pressure inactivation of carrot pectin methylesterase: a kinetic study. *Journal of Food Science* 68: 1377–1383.

Ludikhuyze, L., V. Ooms, C. Weemaes, and M. Hendrickx. 1999. Kinetic study of the irreversible thermal and pressure inactivation of myrosinase from broccoli (*Brassica oleraceae* L. cv *italica*). *Journal of Agricultural and Food Chemistry* 47: 1794–1800.

Ludikhuyze, L., L. Rodrigo, and M. Hendrickx. 2000. The activity of myrosinase from broccoli (*Brassica oleraceae* L. cv Italica): influence of intrinsic and extrinsic factors. *Journal of Food Protection* 63: 400–403.

Marsellés-Fontanet, Á., and O. Martín-Belloso. 2007. Optimization and validation of PEF processing conditions to inactivate oxidative enzymes of grape juice. *Journal of Food Engineering* 83: 452–462.

Masson, P. 1992. Pressure denaturation of proteins. In *High Pressure and Biotechnology*, eds. C. Balny, R. Hayashi, K. Heremans, and P. Masson, 89–99. France: Colloque INSERM 224, John Libbey Eurotext Ltd.

Meersman, F., E. Nies, and K. Heremans. 2008. On the thermal expansion of water and the phase behavior of macromolecules in aqueous solution. *Zeitschrift für Naturforschung* 63b: 785–790.

Min, S., S. K. Min, and Q. H. Zhang. 2003a. Inactivation kinetics of tomato juice lipoxygenase by pulsed electric fields. *Journal of Food Science* 68: 1995–2001.

Min, S., Z. T. Jin, and Q. H. Zhang. 2003b. Commercial scale pulsed electric field processing of tomato juice. *Journal of Agricultural and Food Chemistry* 51: 3338–3344.

Min, S., G. A. Evrendilek, and H. Q. Zhang. 2007. Pulsed electric fields: processing system, microbial and enzyme inhibition, and shelf life extension of foods. *IEEE Transactions on Plasma Science* 35: 59–73.

Mozhaev, V. V., K. Heremans, J. Frank, P. Masson, and C. Balny. 1994. Exploiting the effects of high hydrostatic pressure in biotechnological applications. *Trends in Biotechnology* 12: 493–500.

Mozhaev, V. V., R. Lange, E. V. Kudryashova, and C. Balny. 1996. Application of high hydrostatic pressure for increasing activity and stability of enzymes. *Biotechnology and Bioengineering* 52: 320–331.

Nguyen, P., and G. S. Mittal. 2007. Inactivation of naturally occurring microorganisms in tomato juice using pulsed electric field (PEF) with and without antimicrobials. *Chemical Engineering and Processing* 46: 360–365.

Nienaber, U., and T. H. Shellhammer. 2001. High-pressure processing of orange juice: kinetics of pectinmethylesterase inactivation. *Journal of Food Science* 66: 328–331.

Noci, F., J. Riener, M. Walkling-Ribeiro, D. A. Cronin, D. J. Morgan, and J. G. Lyng. 2008. Ultraviolet irradiation and pulsed electric fields (PEF) in a hurdle strategy for the preservation of fresh apple juice. *Journal of Food Engineering* 85: 141–146.

Nunes, C. S., S. M. Castro, J. A. Saraiva, M. A. Coimbra, M. E. Hendrickx, and A. M. Van Loey. 2006. Thermal and high-pressure stability of purified pectin methylesterase from plums (*Prunus domestica*). *Journal of Food Biochemistry* 30: 138–154.

Oey, I., M. Lille, A. Van Loey, and M. Hendrickx. 2008. Effect of high pressure processing on colour, texture and flavour of fruit- and vegetable-based food products: a review. *Trends in Food Science and Technology* 19: 320–328.

Peeters, L., D. Fachin, C. Smout, A. Van Loey, and M. E. Hendrickx. 2004. Influence of β-subunit on thermal and high-pressure process stability of tomato polygalacturonase. *Biotechnology and Bioengineering* 86: 543–549.

Perez, O. E., and A. M. R. Pilosof. 2004. Pulsed electric fields effects on the molecular structure and gelation of β-lactoglobulin concentrate and egg white. *Food Research International* 37: 102–110.

Phunchaisri, C., and A. Apichartsrangkoon. 2005. Effects of ultra-high pressure on biochemical and physical modification of lychee (*Litchi chinensis* Sonn.). *Food Chemistry* 93: 57–64.

Pilar, M., and A. Vercet. 2006. Effect of PEF on enzymes and food constituents. In *Pulsed Electric Fields Technology for the Food Industry: Fundamentals and Applications*, eds. J. Raso and V. Heinz, 131–151. New York: Springer.

Polydera, A. C., E. Galanou, N. G. Stoforos, and P. S. Taoukis. 2004. Inactivation kinetics of pectin methylesterase of greek Navel orange juice as a function of high hydrostatic pressure and temperature process conditions. *Journal of Food Engineering* 62: 291–298.

Plaza, L., T. Duvetter, S. Monfort, E. Clynen, L. Schoofs, A. M. Van Loey, and M. E. Hendrickx. 2007. Purification and thermal and high-pressure inactivation of pectinmethylesterase isoenzymes from tomatoes (*Lycopersicon esculentum*): A novel pressure labile isoenzyme. *Journal of Agricultural and Food Chemistry* 55: 9259–9265.

Privalov, P. L. 1990. Cold denaturation of proteins. *Critical Reviews in Biochemistry and Molecular Biology* 25(4): 281–305.

Qi, M., A. E. Anderson, D. Z. Chen, S. Sun, and K. J. Auborn. 2005. Indole-3-carbinol prevents PTEN loss in cervical cancer in vivo. *Molecular Medicine* 11: 59–63.

Rapeanu, G., A. Van Loey, C. Smout, and M. Hendrickx. 2005a. Effect of pH on thermal and/or pressure inactivation of Victoria grape (*Vitis vinifera sativa*) polyphenol oxidase: a kinetic study. *Journal of Food Science* 70: E301–E307.

Rapeanu, G., A. Van Loey, C. Smout, and M. Hendrickx. 2005b. Thermal and high-pressure inactivation kinetics of polyphenol oxidase in Victoria grape must. *Journal of Agricultural and Food Chemistry* 53: 2988–2994.

Rapeanu, G., A. Van Loey, C. Smout, and M. Hendrickx. 2006. Biochemical characterization and process stability of polyphenoloxidase extracted from Victoria grape (*Vitis vinifera* ssp. *Sativa*). *Food Chemistry* 94: 253–261.

Riahi, E., and H. S. Ramaswamy. 2003. High-pressure processing of apple juice: kinetics of pectin methyl esterase inactivation. *Biotechnology Progress* 19: 908–914.

Riener, J., F. Noci, D. A. Cronin, D. J. Morgan, and J. G. Lyng. 2008a. Combined effect of temperature and pulsed electric fields on soya milk lipoxygenase inactivation. *European Food Research and Technology* 227: 1461–1465.

Riener, J., F. Noci, D. A. Cronin, D. J. Morgan, and J. G. Lyng. 2008b. Combined effect of temperature and pulsed electric fields on apple juice peroxidase and polyphenoloxidase inactivation. *Food Chemistry* 109: 402–407.

Riener, J., F. Noci, D. A. Cronin, D. J. Morgan, and J. G. Lyng. 2009. Combined effect of temperature and pulsed electric fields on pectin methyl esterase inactivation in red grapefruit juice (*Citrus paradisi*). *European Food Research and Technology* 228:373–379.

Rivas, A., D. Rodrigo, A. Martínez, G. V. Barbosa-Cánovas, and M. Rodrigo. 2006. Effect of PEF and heat pasteurization on the physical–chemical characteristics of blended orange and carrot juice. *LWT—Food Science and Technology* 39: 1163–1170.

Rodrigo, D., R. Jolie, A. Van Loey, and M. Hendrickx. 2006a. Combined thermal and high pressure inactivation kinetics of tomato lipoxygenase. *European Food Research and Technology* 222: 636–642.

Rodrigo, D., C. Cortés, E. Clynen, L. Schoofs, A. Van Loey, and M. Hendrickx. 2006b. Thermal and high-pressure stability of purified polygalacturonase and pectinmethylesterase from four different tomato processing varieties. *Food Research International* 39: 440–448.

Rodrigo, D., R. Jolie, A. Van Loey, and M. Hendrickx. 2007. Thermal and high pressure stability of tomato lipoxygenase and hydroperoxide lyase. *Journal of Food Engineering* 79: 423–429.

Sampedro, F., D. Rodrigo, and M. Hendrickx. 2008. Inactivation kinetics of pectin methyl esterase under combined thermal–high pressure treatment in an orange juice–milk beverage. *Journal of Food Engineering* 86: 133–139.

Sampedro, F., D. J. Geveke, X. Fan, D. Rodrigo, and Q. H. Zhang. 2009. Shelf-life study of an orange juice–milk based beverage after pef and thermal processing. *Journal of Food Science* 74: 107–112.

Schilling, S., S. Schmid, H. Jaeger, et al. 2008. Comparative study of pulsed electric field and thermal processing of apple juice with particular consideration of juice quality and enzyme deactivation. *Journal of Agricultural and Food Chemistry* 56: 4545–4554.

Sila, D. N., C. Smout, Y. Satara, V. Truong, A. Van Loey, and M. Hendrickx. 2007. Combined thermal and high pressure effect on carrot pectinmethylesterase stability and catalytic activity. *Journal of Food Engineering* 78: 755–764.

Silva, J. L., and G. W. Weber. 1993. Pressure stability of proteins. *Annual Review of Physical Chemistry* 44: 89–113.

Silva, J. L., D. Foguel, A. T. Da Poian, and P. E. Prevelige. 1996. The use of hydrostatic pressure as a tool to study viruses and other macromolecular assemblages. *Current Opinion in Structural Biology* 6: 166–175.

Soysal, C., Z. Söylemez, and F. Bozoglu. 2004. Effect of high hydrostatic pressure and temperature on carrot peroxidase inactivation. *European Food Research and Technology* 218: 152–156.

Stoforos, N. G., S. Crelier, M. C. Robert, and P. S. Taoukis. 2002. Kinetics of tomato pectin methylesterase inactivation by temperature and high pressure. *Journal of Food Science* 67: 1026–1031.

Suzuki, K. 1960. Studies on the kinetics of protein denaturation under high pressure. *The Review of Physical Chemistry of Japan* 29: 49–55.

Van Buggenhout, S., I. Messagie, I. Van der Plancken, and M. Hendrickx. 2006. Influence of high-pressure–low-temperature treatments on fruit and vegetable quality related enzymes. *European Food Research and Technology* 223: 475–485.

Van Eylen, D., I. Oey, M. Hendrickx, and A. Van Loey. 2007. Kinetics of the stability of broccoli (*Brassica oleracea* cv. *Italica*) myrosinase and isothiocyanates in broccoli juice during pressure/temperature treatments. *Journal of Agricultural and Food Chemistry* 55: 2163–2170.

Van Eylen, D., I. Oey, M. Hendrickx, and A. Van Loey. 2008a. Behavior of mustard seed (*Sinapis alba* L.) myrosinase during temperature/pressure treatments: a case study on enzyme activity and stability. *European Food Research and Technology* 226: 545–553.

Van Eylen, D., I. Oey, M. Hendrickx, and A. Van Loey. 2008b. Effects of pressure/temperature treatments on stability and activity of endogenous broccoli (*Brassica oleracea* L. cv. Italica) myrosinase and on cell permeability. *Journal of Food Engineering* 89: 178–186.

Van Eylen, D., N. Bellostas, B. W. Strobel, et al. 2009. Influence of pressure/temperature treatments on glucosinolate conversion in broccoli (*Brassica oleraceae* L. cv Italica) heads. *Food Chemistry* 112: 646–653.

Van Loey, A., B. Verachtert, and M. Hendrickx. 2002. Effects of high electric field pulses on enzymes. *Trends in Food Science & Technology* 12: 94–102.

Verlent, I., A. Van Loey, C. Smout, T. Duvetter, B. Ly Nguyen, and M. E. Hendrickx. 2004a. Changes in purified tomato pectinmethylesterase activity during thermal and high pressure treatment. *Journal of Agricultural and Food Chemistry* 84: 1839–1847.

Verlent, I., A. Van Loey, C. Smout, T. Duvetter, and M. E. Hendrickx. 2004b. Purified tomato polygalacturonase activity during thermal and high-pressure treatment. *Biotechnology and Bioengineering* 86: 63–71.

Verlent, I., C. Smout, T. Duvetter, M. E. Hendrickx, and A. Van Loey. 2005. Effect of temperature and pressure on the activity of purified tomato polygalacturonase in the presence of pectins with different patterns of methyl esterification. *Innovative Food Science and Emerging Technologies* 6: 293–303.

Verlent, I., M. Hendrickx, L. Verbeyst, and A. Van Loey. 2007. Effect of temperature and pressure on the combined action of purified tomato pectinmethylesterase and polygalacturonase in presence of pectin. *Enzyme and Microbial Technology* 40: 1141–1146.

Wang, R., X. Zhou, and Z. Chen. 2008. High pressure inactivation of lipoxygenase in soy milk and crude soybean extract. *Food Chemistry* 106: 603–611.

Weber, G. 1995. Van't Hoff revisited: enthalpy of association of protein subunits. *Journal of Physical Chemistry* 99: 1052–1059.

Yang, R. J., S. Q. Li, and Q. H. Zhang. 2004. Effects of pulsed electric fields on the activity of enzymes in aqueous solution. *Journal of Food Science* 69: 241–248.

Yeom, H. W., Q. H. Zhang, and P. Dunne. 1999. Inactivation of papain by pulsed electric fields in a continuous system. *Food Chemistry* 67: 53–59

Yeom, H. W., C. B. Streaker, Q. H. Zhang, and D. B. Min. 2000a. Effects of pulsed electric fields on the activities of microorganisms and pectin methyl esterase in orange juice. *Journal of Food Science* 65: 1359–1363.

Yeom, H. W., C. B. Streaker, Q. H. Zhang, and D. B. Min. 2000b. Effects of pulsed electric fields on the quality of orange juice and comparison with heat pasteurization. *Journal of Agricultural and Food Chemistry* 48: 4597–4605.

Yeom, H. W., Q. H. Zhang, and G. W. Chism. 2002. Inactivation of pectin methyl esterase in orange juice by pulsed electric fields. *Journal of Food Science* 67: 2154–2159.

Zhang, Y. S., P. Talalay, C. G. Cho, and G. H. Posner. 1992. A major inducer of anticarcinogenic protective enzymes from broccoli—isolation and elucidation of structure. *Proceedings of the National Academy of Sciences of the United States of America* 89: 2399–2403.

Zhang, R., L. Cheng, L. Wang, and Z. Guan. 2006. Inactivation effects of pef on horseradish peroxidase (HRP) and pectinesterase (PE). *IEEE Transactions on Plasma Science* 34: 2630–2636.

Zhao, W., R. Yang, R. Lu, Y. Tang, and W. Zhang. 2007. Investigations of the mechanisms of pulsed electric fields on inactivation of enzyme: lysozyme. *Journal of Agricultural and Food Chemistry* 55: 9850–9858.

Zhao, W., and R. Yang. 2008. Comparative study of inactivation and conformational change of lysozyme induced by pulsed electric fields and heat. *European Food Research and Technology* 228: 47–54

Zipp, A., and W. Kauzmann. 1973. Pressure denaturation of metmyoglobin. *Biochemistry* 12(21): 4217–4228.

chapter eleven

Biosensors for fruit and vegetable processing

Danielle Cristhina Melo Ferreira, Lucilene Dornelles Mello,
Renata Kelly Mendes, and Lauro Tatsuo Kubota

Contents

11.1 Introduction

Quality and conformity according to the food and regulatory agencies are the key issues common to all industrial products, and these parameters present great economic importance. A good control system and the analysis of the composition and properties of raw materials and food products are essential for food quality and safety due to the growing interest of both consumers and food industry in these concepts. Quality attributes include color, flavor, texture (crispness, yield points, consistency, viscosity, etc.), and nutritional and microbial content. Safety depends on the content of hazardous microorganisms and toxic compounds, including toxins produced by microorganisms and naturally occurring toxic compounds, antinutritional compounds, and pesticides (Terry et al., 2005; Castillo et al., 2004).

The increase in consumption of high-quality products associated with the demands from society for more environmentally friendly production practices results in the necessity of continuous improvement of before- and post-harvest quality detection methods. Current analytical practices in the food industry are time consuming and require skilled labor. These analytical techniques require time-consuming separation, expensive instrumentation, and the use of chemicals with high purity. Most of these drawbacks can be overcome by applying enzymatic analysis (Newman et al., 1998; Prodromidis and Karayannis et al., 2002). In this sense, biosensors have been used for many years to provide process control data in food processing due to the characteristics of the produced devices to be of small size, accurate, and economically viable emerging technology for rapid diagnostics sector applied to agricultural products (Hall, 2002).

Biosensors are defined as analytical devices composed of a biospecific recognition system or a biologically derived material (cells, receptors, enzymes, antibodies, antigens, ions, proteins, oligonucleotides, etc.) used in combination with or integrated within a physicochemical transducer or transducing microsystems (optical, acoustic, electrochemical, thermal, piezoelectric), which converts the biological response into a measurable signal. These sensors usually generate a digital electronic signal that is proportional to the concentration of a specific substance or group of substances. Although the signal may in principle be continuous, devices can be configured to yield single measurements to meet specific market requirements (Tothill, 2001). Miniaturization, reduced cost, and the improved processing power of modern microelectronics have increased the analytical capabilities of such devices and given them access to a wider range of applications. Different biosensor formats have been developed for single target substance and for broad-spectrum monitoring (Newman and Turner, 1994).

Fundamental studies in biosensor development can be summarized in three aspects: molecular recognition elements, tools for biosensor construction, and transducers. The type of active biological component or the mode of signal transduction or the combination of these aspects distinguishes the biosensor design.

The choice of the biological material and the transducer depends on the properties of each sample of interest and the type of physical magnitude to be measured. Figure 11.1 shows some analytes possible to be analyzed by immobilization of the different biological components, separately, in several transducers. The immobilized biocomponent is the main part to determine the degree of selectivity or specificity of the biosensor (Zhang et al., 2000).

Among the various types of biosensors, the most common ones are where the monitoring of the analyte is based on the use of enzymes. The wide applicability of enzymatic biosensors is related to their benefits such as sensitivity, rapid responses, low cost, high specificity, detection of low concentrations of analyte, variable number of commercially available enzymes, and a variety of methodologies employed in the construction of these biological sensors (Choi et al., 2005; Albareda-Sirvent et al., 2001; Nunes et al., 1999; Mao et al., 2008).

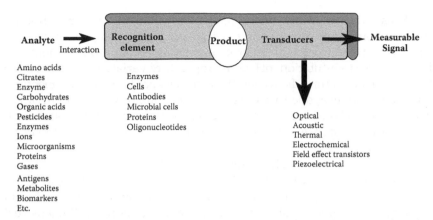

Figure 11.1 Biocomponents, recognition elements, and transducers employed in biosensor construction.

11.2   Biosensor components

11.2.1   Biosensor recognition elements

Sensors are normally classified either by the type of recognition or transduction element. A sensor can be classified as a biosensor if its recognition material is an enzyme or series of enzymes, whole microbial cells, antibodies, or receptors (Davis et al., 1995). Nowadays, other engineering biomaterials are used as molecular recognition elements such as peptides, nucleic acids, aptamers, and molecularly imprinted polymers (Wang et al., 1996; Feng et al., 2008; Kindschy and Alocilja, 2004). Among the biosensors, enzymatic ones are the most common sensors in the applications related to foods (Mello and Kobota, 2002).

Several enzymes such as cholinesterase, urease, and tyrosinase have been employed in the construction of biosensors in single or multi-enzyme arrangements using different transducers such as amperometric, optical, and conductimetric (Scheper et al., 1996; Thévenot et al., 2001; Zhang et al., 2000; Jin et al., 2004; Krawczyk et al., 2000; Chang et al., 2002). Among the enzymes commercially available, the oxidases are the most frequently used. This category of enzymes offers the advantage of stability, and in some situations does not require coenzymes or cofactors (Phadke, 1992; Wong et al., 2008). Since glucose is the most important component of carbohydrates, its monitoring and determination are well-known by the applications of glucose-based biosensors in food science area (Wong et al., 2008; Newman and Turner, 2005). Currently, most glucose biosensors utilize glucose oxidase as their recognition element that catalyzes the oxidation of glucose to gluconolactone.

Sensors based on enzymes as recognition elements are very attractive, due to a variety of measurable reaction products arising from catalytic processes. Biosensors with microbial cells as recognition elements are less sensitive to inhibition, pH, and temperature variations, and normally have a longer lifetime. However, these biosensors present low selectivity and slow response, due to a variety of metabolic processes occurring in a living cell (Davis et al., 1995; Phadke, 1992). Problems like selectivity and the slow response of microbial sensors can be overcome by the use of enzyme biosensors.

When compared with biosensors based on enzymes, the hybrid receptors such as DNA and RNA probes have shown promising application in food analysis for microorganism detection. Commercially, biosensing DNA probes exist for the detection of foodborne pathogens such as *Salmonella, Listeria, Escherichia coli,* and *Staphylococcus aureus* (Boer and Beumer, 1999).

11.2.2 Immobilization procedures

Immobilization of the biological element to its transducer is a key point in the resulting bioanalytical device. The major requirements in the choice of suitable enzyme deposition method include efficient and stable immobilization of the biological macromolecules on transducer surface, chemical inertness of the host structure, accessibility of the immobilized molecule, and maximum retention of its biological activity. In general, the selection of an appropriate immobilization method depends on the nature of the biological element, type of the transducer used, physicochemical properties of the substance analyzed, and operating conditions for the biosensor (Guilbault, 1982).

The main enzyme immobilization methods include the physical adsorption onto a solid surface, entrapment in a matrix (using gels or polymers), cross-linking with bifunctional or multifunctional reagents, covalent binding to a surface, self-assembled biomolecules, electrochemical polymerization, and sol-gel entrapment (Terry et al., 2005).

The physical adsorption method is the oldest and simplest immobilization method and is based on van der Waals attractive forces between the biocomponent and the transducer. The advantage of this method is its simplicity and the great variety of beads that could be used for immobilization. However, there is the possibility of the adsorbed biocomponent loss if pH, ionic strength, or temperature of the environment changes during measurements. Therefore, these biosensors are rarely used, except for a few cases and are suitable for single-use for non-repeatable measurement. Disposable biosensors can be mass-produced by mixing graphite power with an enzyme that has been dissolved in a binder solution and then screen printing the paste onto a planar working electrode (Fox, 1991; Bickerstaff, 1997).

Among the enzyme immobilization methods, the majority of works has been carried out via covalent coupling due to its high stability. It can be used to achieve the immobilization of a biocomponent to a membrane matrix or directly onto the surface of the transducer. Replaceable membranes have been much employed as platform for the immobilization of enzymes to avoid electrode fouling, and have shown good results in terms of increased shelf-life, sensitivity, and background of the resulting signal of the biosensors. In this approach, the enzyme in solution is immobilized in a substance permeable membrane to maintain close contact and prevent its leaching into the sample solution. In this case, the enzymes are immobilized by adsorption or covalent attachment using the cross-linking agents, and they are used mainly in electrochemical transducers. The support based on membranes include dialysis (Campuzano et al., 2007), polycarbonate (Basu et al., 2006), collagen, and biological membranes (Choi et al., 2005; Yang et al., 2006; Wu et al., 2004).

Natural membranes such as bamboo inner shell or eggshell membranes characterized by possessing suitable gas and water permeability for substrates and products were also tested for immobilization (Yang et al., 2006; Wu et al., 2004). Their biological properties are believed to provide a biocompatible microenvironment around the immobilized enzyme molecules that become more stable.

Immobilization protocols are based on the reaction between the same terminal functional groups of the protein (not essential for its catalytic activity) and reactive groups on the solid surface of the insoluble bed. Functional groups available in the enzymes or proteins mainly are originated from the side chain of the terminal amino acids. They include, for example, the amino groups from lysine, carboxyl groups from aspartate and glutamate, sulfhydryl groups from cysteine, and phenolic hydroxyl groups from tyrosine. These matrix as membranes with different active functional groups, and are able to immobilize biocomponents with great efficiency and facility. Bifunctional reagents (homo or hetero functional) have also been used in the immobilization of enzymes and proteins. The method is based on the macroscopic particle formation as a result of the formation of covalent binding between molecules of inert bed with functional reagents. Some of the most used homofunctional reagents include glutaraldehyde, carbodiimide, and others; while the heterofunctionals include trichloro triazine, 3-metoxi diphenyl methane-4,4' di isocyanate (Roos et al., 2004; Mello and Kobota, 2002). In these methods, the materials that have been used as matrix include metals (gold, silver, and platinum), material hydroxylic polymers (Campuzano et al., 2007), synthetic polymers (nylon) (Kirgöz et al., 2006), and silica and carbon as inorganic materials (Choi et al., 2005; Yang et al., 2006; Wu et al., 2004).

The noble metals (gold, silver, and platinum) are usually used as surfaces in both electrochemical and optical systems. They can be used either

as a pure surface or as an oxidized form. Enzymes can be bound to monolayers or the enzyme can be modified to contain thiol moieties, which
bound to the metal surface. Then, self-assembled monolayers are basically
interfacial layers between a metal surface and a solution. Both approaches
based on self-assembled monolayers have as a great advantage to provide control over the orientation and distribution of the immobilized
enzymes and as a consequence producing the highly ordered immobilization matrices and hence reproducible enzymes electrodes (Chaki and
Vijayamohanan, 2002).

Concerning the enzyme immobilization using self-organized films,
another trend is the use of bilayer lipid membranes that work as models
of biological membranes (Ruiz et al., 2007). These methods are based on
films composed of multiple layers or organized lipid-like molecules. These
structures are characterized to be insoluble and amphiphilic, having a
hydrophilic end and a hydrophobic end including phospholipids and fatty
acids, which can be deposited on surfaces using the Langmuir-Blodgett
method. This disposition is usually accomplished by slow insertion of the
support through the gas–liquid interface. Although the adsorption is usually weak in Langmuir-Blodgett films, an important characteristic of these
immobilized amphiphilic molecules is the stability on the surfaces due to
intermolecular forces. This characteristic associated to a high degree of
lateral mobility of membrane molecules, simulates the behavior in natural
membranes. Films prepared by the Langmuir-Blodgett method have been
deposited on both electrochemical and optical supports and used for the
detection of substances by direct interaction with these films (Scouten et al.,
1995).

The enzyme immobilization or entrapment, especially of oxireductases
class, by electropolimerization has also proven to be suited to the preparation of biosensors for fruit and vegetable quality control. In this case, the
enzyme is bulk-entrapped in a polymer matrix that remains on the electrode surface from the solution containing the dissolved monomer and
enzyme. The monomer is electrochemically oxidized at a polymerization
potential, giving rise to free radicals. These radicals are adsorbed onto the
electrode surface and subsequently undergo a wide variety of reactions
leading to the polymer network. The electropolimerization should preferably occur in an aqueous solution with a neutral pH to the biological
component be incorporated into the polymer film in a suitable form. The
process is governed by the electrode potential and by the reaction time,
which allow controlling the thickness of the resulting film. Conducting
polymers widely used are polypyrrole (Böyükbayram et al., 2006), poly-
aniline (Ruiz et al., 2007), and polyindole (Badea et al., 2003). This immobilization technique is widely used for amperometric biosensors and to a
less extent for other electrochemical transducers such as conductimetry
and potentiometry (Scouten et al., 1995).

The enzyme immobilization method applied in most cases for optical systems is the sol-gel entrapment. This approach is based on the growing of siloxane polymer chains around the biomolecule within an inorganic oxide network. In the sol-gel network, the porous nature makes possible that entrapped species remain accessible to interact with external chemical species or analytes (Gupta and Chaudhury, 2007).

11.2.3 Transducers

Several transducers have been successfully employed toward the development of biosensors that the most commonly used one is electrochemical transducers (Prodromidis and Karayannis, 2002; Chaubey and Malhotra, 2002). Optical transducers exploit properties such as simple light absorption, fluorescence/phosphorescence, bio/chemiluminescence, reflectance, Raman scattering, and refractive index. They offer advantages such as speed, safety, sensitivity, and robustness, as well as permitting *in situ* sensing and real time measurements (Table 11.1) (Vo-Dinh et al., 2001).

Table 11.1 Biosensor Transduction Systems

Transducers	
Electrochemical	Potentiometry
	Amperometry
	Voltammetry
	Electrochemical impedance spectroscopy
	ISFET (ion-sensitive field effect transistors)
	CHEMFET (chemical field-effect transistor)
Electrical	Surface conductivity scattering
	Conductivity
	Capacitance
Optical	Fluorescence
	Luminescence
	Reflection
	Absorption
	Surface plasmon resonance
	Evanescent waves
Piezoelectrical	QCM (quartz crystal microbalance)
Thermal	Calorimetry
	Thermistor
Magnetic	Paramagnetism

Piezoelectric transducers are another type of transducer frequently used in biosensors. They are often referred to as "mass-sensitive" techniques because of the mass or thickness measurements (Song et al., 2008). The use of piezoelectric devices in chemical and biochemical sensing was largely employed with quartz crystal microbalance (QCM) measurements. QCMs have continued to be the most popular piezoelectric transducers due to their easiness and ready availability. QCM is an ideal workhorse for studying the attachment of antibody sensors (Collings and Caruso, 1997).

As commented above, the most common transducers are electrochemical, monitoring electrochemical changes that occur when chemicals interact with a sensing surface. The electrical changes can be based on a change in the measured voltage, potential, current, or resistance. Electrochemical biosensors are the best suited sensors for various field applications (Prodromidis and Karayannis, 2002).

Amperometric biosensors are based on enzymes that either consume oxygen or produce hydrogen peroxide, or produce (indirectly) the reduced form of β-nicotinamide adenine dinucleotide (phosphate), NAD(P)H, during the course of the catalytic reaction with the substrate of interest:

$$\text{Substrate} + O_2 \xrightarrow{\text{enzyme}} \text{Product} + H_2O \tag{11.1}$$

$$\text{Substrate} + O_2 \xrightarrow{\text{enzyme}} \text{Product} + H_2O_2 \tag{11.2}$$

$$\text{Substrate} + NAD^+ \xrightarrow{\text{enzyme}} \text{Product} + NADH + H^+ \tag{11.3}$$

The most common method of monitoring O_2 is based on a Clark electrode. The Clark electrode was employed in the development of the first amperometric biosensor for glucose analysis using the glucose oxidase enzyme. It consists of a platinum cathode, which O_2 is reduced, and a reference electrode (usually silver/silver chloride), immersed in an electrolyte solution, usually potassium chloride and covered by a semi-permeable membrane, through which O_2 diffuses. The electrode operates at potentials of -0.6 to -0.9 V (vs. Ag/AgCl), and a current proportional to O_2 concentrations is produced, according to the following reactions.

$$\text{Ag Anode: } 4\,Ag + 4Cl^- \rightarrow 4\,AgCl + 4e^- \tag{11.4}$$

$$\text{Pt Cathode: } O_2 + 4H^+ + 4e^- \rightarrow 2\,H_2O \tag{11.5}$$

These biosensors are called *first-generation* biosensors (Freire et al., 2003) and the formation of the product or the consumption of the reagent can be monitored to measure the analyte concentration. The oxygen-based sensor offers the advantage of no electrochemical interference from other sample constituents. In spite of this, measurements based on oxygen have a level of limitations. The response is low and the dependence on dissolved oxygen can reduce the accuracy and reproducibility of the device. Moreover, because the high background of signal, the minimum detectable concentration is not very low. An alternative to overcome these drawbacks is the detection of H_2O_2 generated in the reaction (Prodromidis and Karayannis, 2002; Wagner and Guilbault, 1984).

Hydrogen peroxide generated also can be measured by amperometric method by oxidation at anode of a solid (platinum, glassy carbon) electrode, polarized at $+0,65V$ vs. Ag/AgCl:

$$H_2O_2 \xrightarrow{+0.65 \text{ V } vs\, Ag/AgCl} O_2 + 2H^+ + 2e^- \tag{11.6}$$

In the case of biosensors involving NADH, the mechanism of its oxidation has not been fully understood until now, but the scheme of reaction usually accepted in the literature is as follows:

$$\text{NADH} \underset{\text{slow}}{\xrightarrow{-e}} \text{NADH}^{\bullet+} \underset{+H^+}{\overset{-H^+}{\rightleftharpoons}} \underbrace{\text{NAD}^{\bullet} \underset{+e^-}{\overset{-e^-}{\rightleftharpoons}}}_{\text{(intermediate)}} \overset{(\text{fast})}{\text{NAD}^+} \tag{11.7}$$

Amperometric biosensors modified with mediators are referred to as *second-generation* biosensors. Mediators are redox substances used in cases of the reactions taking place on a bare electrode with a low kinetic rate transfer mechanism, to increase the applied potential over the thermodynamic system. The mediator substance when attached on the electrode surface can support the change transfer improving the rate of electron transfer. The active form of the mediator is regenerated electrochemically on the surface of the electrode and creates an electron shuttling. As a result, the applied potential can be decreased to the value of the standard potential of the mediator avoiding interferences and increasing the sensitivity (Prodromidis and Karayannis, 2002; Wagner and Guilbault, 1984).

The direct enzyme-electrode coupling biosensors based on direct electron transfer mechanism are called *third generation*. In this case, the electron is directly transferred from the electrode to the enzyme and then to the substrate molecule (or vice versa). In this mechanism the electron acts as a second substrate for the enzymatic reaction and results in generating a catalytic current. The substrate transformation (electrode process) is essentially a catalytic process (Ghindilis et al., 1997; Habermüller et al., 2000).

For food analysis, the majority of the electrochemical biosensors are based on the amperometric transducers in combination with the enzyme class of oxidases due to the substrates and biosensor sensitivity. Among the amperometric transducers, those that are based on monitoring of hydrogen peroxide present a higher sensitivity than those based on the detection of oxygen consumed (Mello and Kubota, 2002).

Other amperometric biosensors are used for indirect detection of microbial contamination in foodstuffs. Several microorganisms can be detected amperometrically by their enzyme-catalyzed electrooxidation/electroreduction or their involvement in a bioaffinity reaction (Boer and Beumer, 1999; Fitzpatrick et al., 2000). In these systems, an enzyme-linked amperometric immunosensor is utilized for the detection of bacteria by means of the antigen/antibody combination. In this case, a heat-killed bacteria, such *S. typhimurium*, is sandwiched between antibody-coated magnetic beads and an enzyme-conjugated antibody (Brooks et al., 1992). Other amperometric immunoassays include enzyme-channeling reactions and electrochemical regeneration of mediators within the membrane layer of an anion-exchange enzyme-antibody modified electrode (Rishpon and Ivnitski, 1997). Other biosensors sensing the microorganisms are based on a partially immersed immunosensor in a solution, resulting in the formation of a supermeniscus on the electrode surface. This supermeniscus plays a role in providing optimal hydrodynamic conditions for the current generation process (Hamid et al., 1998). All these immunoassays cited have a relatively short assay time.

11.3   Use of biosensors as analytical tools for fruit and vegetable processing

After Clark works in biosensor technology (Updike and Hicks, 1967), biosensors became a powerful alternative to conventional analytical techniques such as chromatography, spectrophotometry, and so on. They join the specificity and sensitivity of biological systems in a practical way (Castillo et al., 2004). Biosensor research and development have been directed mainly towards clinical application, exemplified by well-known glucose biosensor.

The determination of food components by using specific biosensors is the particular interest for food field. In environmental and food monitoring, the biosensor is still a technology under improvement. Despite promising research in the development of biosensors, there are not many reports of applications in the agricultural product monitoring due to the slow technology transfer of integrated biosensor systems to the marketplace.

   Biosensors based on enzymatic methods of analysis are closely related to chemistry and technology of fruit and vegetable processing. For example, beverage industries need rapid and affordable methods to determine species that have not previously been monitored or to replace existing but inefficient or expensive procedures. Some of the more recent and important reports about the prototypes in development applied to fruit and vegetable quality monitoring are presented in Tables 11.2 and 11.3. Some biosensors listed in the tables are used to determine more than one compound or used in combination for simultaneous measurements. The tables show the analytes, applications, and detection range of the biosensors with respective transducers. As can be seen, the electrochemical biosensors dominate the biosensors literature due to measure facility of the enzymatic reactions with their substrates. Amperometry is based on the measurement of the current resulting from the electrochemical oxidation or reduction of an electroactive species, and this resulting current is directly correlated to the bulk concentration of the electroactive species or its production or consumption rate within the adjacent biocatalytic layer (Thevenot et al., 1999). There are only a few optical detection methods employing enzymatic biosensors for vegetable and fruit processing. Low frequency of use is probably due to the low sensitivity of the spectrophotometric detection, the expensive equipment, and the complicated labeling protocols of the fluorescence techniques.

   Table 11.3 shows enzyme-based biosensors for the determination of carbohydrates being the main focus on glucose determination. The reason for the extensive activity in developing a glucose biosensor has been the analytical and biotechnological importance of glucose itself among the other substances, particularly for the beverage industry. In addition, it is the convenient work using the glucose oxidase enzyme, which offers nearly suitable properties related to activity, selectivity, stability, and commercial availability.

   Another important carbohydrate is sucrose, because it is a component of fruit beverages. The determination of this sugar is one of the most important routinely performed tests for quality control in the beverage industry. A promising study involving sucrose determination in different fruit juice samples was performed by using a multienzyme amperometric biosensor in a flow injection analysis (FIA) system (Majer-Baranyi et al., 2008). Glucose oxidase, invertase, and mutarotase were immobilized on a pig's membrane by using a glutaraldehyde for cross-linking. These three enzymes convert the sucrose to hydrogen peroxide, which is monitored via anodic oxidation in an amperometric detector. Because the glucose present in the sample interferes with the analysis, its amount was measured with the glucose sensor and the response was subtracted from the total response of the sucrose sensor. The correlation between the results of the standard reference method and biosensor analysis was excellent.

Table 11.2 Enzymatic Sensors Applied in Fruit and Vegetables Processing

Analyte	Application	Biocomponent	Transducer*	Detection range	Reference
Glucose, fructose, sucrose, galactose	Juice, milk	D-glucose dehydrogenase, D-fructose dehydrogenase, β-galactosidase invertase	Amp.	2–50 mmol L^{-1} (glucose), 3–25 mol L^{-1} (fructose), 12 mmol L^{-1} (sucrose), 1–3 mmol L^{-1} (galactose)	Maestre et al. 2005
Glucosinolates	Vegetables	Glucose oxidase, myrosinase	Amp.	0.005–1.6 mmol L^{-1}	Tsiafoulis et al.2003
Ethanol	Wine	Alcohol dehydrogenase	Optic	0.008–0.024% (v/v)	Páscoa et al. 2006
Malic acid and Lactic acid	Samples fermentation	Malic enzyme Lactate oxidase	Amp.	$1{\times}10^{-5}$–$4{\times}10^{-4}$ (L-malic acid) $5{\times}10^{-6}$–$1{\times}10^{-3}$ (L-lactic acid)	Esti et al. 2004
Acetic acid	Wine	Acetate kinase, pyruvate kinase, pyruvate oxidase	Amp.	0.05–20 mmol L^{-1}	Mizutani et al. 2003
D-malate	Juice	D-malate dehydrogenase	Optic	0.02–50 µmol L^{-1}	Mori and Shiraki 2008

Analyte	Sample	Enzyme	Transducer	Linear range	Reference
Pesticide	Vegetables, fruits	Acetylcholinesterase	Calorim.	1×10^{-7}–1×10^{-9} mol L^{-1} (carbofuran)	Nikolelis et al.2005
Ascorbic acid	Juice	Ascorbate oxidase	Calorim.	2.4–350 mmol L^{-1}	Vermeir et al. 2007
L-amino acids	Juice and milk samples	L-amino acid oxidase	Amp.	5–200 µmol L^{-1}	Wcislo et al. 2007
L-glutamate	Food samples	L-glutamate oxidase, L-glutamate dehydrogenase	Amp.	0.02–3 mg L^{-1}	Basu et al. 2006
L-glutamate	Tomato	L-glutamate oxidase	Amp.		Pauliukaite et al. 2006
Glucosinolates	Vegetables	Glucose oxidase, myrosinase	Optic	0–4 mmol L^{-1}	Choi et al.2005

Table 11.3 Enzymatic Biosensors for Carbohydrate Analysis in Fruit and Vegetable Processing

Analyte	Application	Biocomponent	Transducer	Detection range	Reference
Glucose	Juice	Glucose oxidase	Amp.	0.05–0.50 mmol L^{-1}	Pauliukaite et al. 2008
D-glucose and L-lactate	Fermented beverages	Glucose oxidase, lactate oxidase	Amp.	1–100 mmol L^{-1} (glucose) 1–50 mmol L^{-1} (L-lactate)	Sato and Okuma 2006
Glucose	Juice fruit and yoghurt drink	Glucose oxidase	Amp.	0.05–1.0 mmol L^{-1}	Ivekovic et al. 2004
Glucose and malic acid	Wine fermentation	Glucose oxidase, malic enzyme	Amp.	< 5 × 10^{-3} mol L^{-1}(glucose), 0–1 × 10^{-3} mol L^{-1} (malic acid)	Lupu et al. 2004
Glucose	Juice	Glucose oxidase	Amp.	0.05–4 mmol L^{-1}	Badea et al. 2003
Glucose	Juice	Glucose oxidase	Amp.	1 × 10^{-5} mol L^{-1}	Derwinska et al. 2003
Glucose, sucrose, ascorbic acid	Fruit	Glucose oxidase, mutarotase, ascorbate oxidase	Amp.	0–7 mmol L^{-1} (glucose, sucrose and ascorbic acid),	Jawaheer et al. 2003
Glycoalkaloid	Potato	Acetylcholinesterase, butyrylcholinesterase	ISFET	0.2–100 µmol L^{-1}	Soldatkin et al. 2005
Glycoalkaloid	Potato	Butyrylcholinesterase	ISFET	0.2–100 µmol L^{-1}	Arkhypova et al. 2003

The designed biosensors to determine the content of glucose are based on some of the following reactions:

$$\text{D-glucose} + O_2 \xrightleftharpoons{\text{glucose oxidase}} \text{D-gluconic acid} + H_2O_2 \qquad (11.8)$$

$$\text{D-glucose} + \text{ATP} \xrightleftharpoons{\text{hexokinase}} \text{D-glucose 6-phosphate} + \text{ADP} \qquad (11.9)$$

D-glucose 6-phosphate + NADP$^+$

$$\xrightleftharpoons{\substack{\text{Glucose-6-phosphate} \\ \text{dehydrogenase}}} \text{6-phosphoglucono-}\gamma\text{-lactone} + \text{NADPH} + H^+$$

$$(11.10)$$

Other analogous electrodes for other compounds include those which share a common dependence on O_2 as co-substrate which rely instead on nicotinamide adenine dinucleotide (NAD$^+$) or nicotinamide adenine dinucleotide phosphate (NADP$^+$) as cofactors. A good example is the monitoring of fermented products, which is usually based on the reaction of dehydrogenases or oxidases to produce oxidized products from substrate using the nicotinamide adenine dinucleotide (NAD$^+$) or oxygen as oxidizer. In most cases, the resulting electrodes are specific in their response and respond essentially to only one class of compound. However, it was observed that enzymes possessing broader substrate specificity have been employed, such as oxidases, peroxidases, and catalases (Prada et al., 2003; Mazzei et al., 2007; ElKaoutit et al., 2007).

Enzyme-based biosensors can also be used for alcoholic beverage production. For example, glycerol is an important product of alcohol fermentation and its level is related to wine quality. Other important markers of the fermentation process include the control of malic and lactic acid levels in the malolactic and alcoholic fermentation process. Biosensors based on two enzymes, either lactate oxidase or lactate dehydrogenase, can be used for monitoring wine quality (Shkotova et al., 2008; Csöregi et al., 2001). Enzymatic biosensors for these compounds usually follow the catalyst cycle of alcohol dehydrogenase (ADH):

$$\text{Ethanol} + \text{NAD}^+ \xrightarrow{\text{ADH}} \text{Acetaldehyde} + \text{NADH} + H^+ \qquad (11.11)$$

$$\text{Glycerol} + \text{NAD}^+ \xrightarrow{\text{GDH}} \text{Dihydroxyacetone} + \text{NADH} + H^+ \qquad (11.12)$$

$$\text{Malate} + \text{NADP}^+ \xrightarrow{\text{ME}} \text{pyruvate} + \text{NADPH} + CO_2 + H^+ \qquad (11.13)$$

$$\text{Lactate} + \text{NAD}^+ \xrightarrow{\text{LDH}} \text{pyruvate} + \text{NADH} \qquad (11.14)$$

where GDH is glycerol phosphate dehydrogenase, ME is malic enzyme, and LDH is lactate dehydrogenase.

Biosensors have been developed for the detection of phenolic compounds. Tyrosinase based biosensors are restricted to the monitoring of phenolic compounds with at least one free *ortho*-position. On the other hand, the laccase biosensor can detect free *para-* and *meta*-position, but its catalytic cycle is complicated and its major part is different from tyrosinase. Also, the formed products during the reaction of the laccase still are not well understood (Jarosz-Wilkolazka et al., 2004). Peroxidases exhibit low specificity for electrons donor as the phenolic compounds and can be used for the phenols detection with certain selectivity and sensitivity (Ruzgas et al., 1996).

The presence of D-amino acids is important for fruit juices, where they are considered molecular markers of bacterial activity. It has been shown that significant amounts of free D-alanine were only found in juices affected by bacterial contaminations (Gandolfi et al., 1994). In this context, a screen-printed biosensor was developed based on D-amino acid oxidase for amperometric determination of D-alanine in fruit juice samples (Wcislo et al., 2007). A carbon electrode was modified with the enzyme D-amino acid oxidase and the mediator Prussian Blue. The enzyme catalyzes the oxidation of the amino acid by oxygen to pyruvic acid and hydrogen peroxide. The electrode process of Prussian Blue reduction is a direct source of current necessary for amperometric detection. The biosensor was applied in fruit juice samples and the results were satisfactory when compared with the standard measurement methods.

Organic acids such as citrate, isocitrate, malate, oxalate, and tartarate are important for fruit and vegetable processing. Among these organic acids, isocitrate is an abundant acid in many fruits, especially apple, strawberry, and grape. Therefore, the determination of isocitrate is useful as an index of the ripeness of fruits as well as a quality control in the beverage and fermentation industries. An interesting study involving the isocitrate determination was carried out using a potentiometric biosensor in an FIA system (Kim and Kim, 2003). In this work, the enzyme isocitric dehydrogenase was immobilized on a glass substrate via covalent binding. By the enzyme reaction, α-ketoglutarate and CO_2 are produced from isocitric acid in the presence of $NADP^+$ as a cofactor. Under alkaline conditions, CO_2 is converted to CO_3^{2-}, which can be measured using an ion selective electrode coupled with an FIA system. The measurements were performed in fruit juices and the results were similar to ones carried out using gas chromatography. A similar work was carried out for the determination of citric acid in fruit juices by using a biosensor with citrate lyase and oxaloacetate decarboxylase (Kim, 2006). The biosensor was constructed using these enzymes generating carbonate ions under alkaline conditions in an FIA system. The carbonate ions could be monitored potentiometrically with a carbonate selective electrode. The biosensor system was applied to determine citrate concentration in different fruit juices under optimum conditions, and there were no significant differences in results obtained by the biosensor and gas chromatography.

Natural plant toxins and some compounds may be present inherently in fruits and vegetables, which may be products from microorganisms or compounds that protect the plant against threats such as bacteria, fungi, and insects. They are often secondary metabolites and in some cases show high toxicity. Some substances can be destroyed by post-harvest processing treatments such as thermal treatment, but some natural toxins remain unchanged. Table 11.4 shows the analysis of alkaloid compounds by cholinesterase–based enzymatic biosensors with potentiometric transduction. The design of these biosensors depends on an enzyme inhibition-based analysis for food safety. The analyzed compounds are the most common class of glycoalkaloids present in potatoes (solanine) and tomatoes (tomatina). Dzyadevych et al. (2004) have developed an enzymatic sensor with an immobilized enzyme butyryl cholinesterase (BuChE) as a biorecognition element and the biosensor responses remaining stable almost three months. The measurement is based on the following basic reaction:

$$(CH_3)_3 - \overset{+}{N} - (CH_2)_2 - O - \underset{\underset{O}{\|}}{C} - (C_3H_7) \xrightarrow[H_2O]{BuChE} (C_3H_7)CO$$

$$+ HO(CH_2)_2 \overset{+}{N}(CH_3)_3 + H^+$$

$$(11.15)$$

This reaction results in proton generation, inducing a pH change in the sensor membrane, allowing potentiometric monitoring. Glycoalkaloids inhibit cholinesterases, as has been shown in previous *in vitro* and *in vivo* studies. The level of enzyme inhibition due to the action of glycolakaloids can be evaluated by a comparison of the biosensor responses before and after contact with a glycoalkaloid solution.

Another important application is the analysis of sulfite residues in fermented beverages that contribute to the quality of the beverages. Sulfite is used as an additive in food and beverages to prevent oxidation and bacterial growth and to control enzymatic reactions during production and storage. In a reported work, a biosensor was developed using sulfite oxidase from plant tissue that catalyses the final reaction in oxidative degradation of sulfur-containing amino acids cysteine and methionine (Isaac et al., 2006).

$$SO_3^{2-} + O_2 + H_2O \xrightarrow{\text{Sulfite oxidase}} SO_4^{2-} + H_2O_2 \qquad (11.16)$$

The principle of the measurements is based on the determination of the decrease of oxygen concentration due to the activity of enzyme present in the bioactive material.

Table 11.4 Biosensors Applied in the Analysis of Toxic Chemicals

Analyte	Application	Biocomponent	Transducer	Detection range	Reference
Natural toxin	Potato	Butyrylcholinesterase	Potent.	0.2–100 μmol L^{-1} (Glycoalkaloid)	Soldatkin et al. 2005
Natural toxin	Potato	Acetylcholinesterase, butyrylcholinesterase	Potent.	0.2–100 μmol L^1 (Glycoalkaloid)	Arkhypova et al. 2003
Natural toxin	Potato	Butyrylcholinesterase	Potent.	0.03–5 μmol L^{-1} (Glycoalkaloid)	Korpan et al. 2006
Azide	Fruit juice	Catalase	Amp. (Clark-type electrode)	25–300 μmol L^{-1}	Sezgintürk et al. 2005

One of the applications of the enzymatic biosensors is related to the measurement of pyruvate concentration in onion juice. Pyruvate concentration is used as the quality assurance indicator of pungency or flavor intensity. An amperometric disposable pyruvate biosensor has been developed using a pyruvate oxidase based on the following enzyme reaction:

$$\text{Pyruvate} + \text{HPO}_4^{2-} + \text{O}_2 \xrightarrow{\text{Pyruvate oxidase}} \text{acetylphosphate} + \text{H}_2\text{O}_2 + \text{CO}_2$$

(11.17)

This research has demonstrated the possibility of replacing the standard colorimetric assay used ubiquitously for determining pyruvate concentration with a faster method using a mediated amperometric biosensor (Abayomi et al., 2006).

Another key point for the use of biosensors is related to the detection of pesticides that they are used in agricultural production of fresh fruits and vegetables. Pesticides become an important source of contamination of soil and aquatic environments, representing a serious problem for public health. Among the pesticides, organophosphorus and carbamate compounds are widely used due to their lower environmental persistence than organochlorades and their high activity for insect control. Despite these, the residues of these pesticides in post-harvested foods represent toxicity to humans and other mammals. Their mode of action in vertebrates and insects is based on the inhibition of the activity of acetylcholinesterase (AChE) enzyme in the hydrolysis of the neurotransmitter acetylcholine, which is responsible for the transmission of nervous impulses (Rosenberry, 1975). The inhibition of the activity of AChE results in an excess of acetylcholine in the synapses, taking to a continuous impulse transmission and inducing symptoms ranging from increased salivation and headache to convulsion and suppressed breathing, which can result in death (Polanka et al., 2008). The measure of the anticholinesterase activity of these products can be used as a screening test for the evaluation of the presence of pesticide residues in food matrices. The enzyme inhibition-based biosensor development relies on a quantitative measurement of the enzyme activity before and after exposure to a target analyte (Amine et al., 2006). Biosensors based on AChE inhibition have been restricted to only a limited number of compounds because the AChE sensitivity to other substances does not meet the standards of sensitivity required. As shown in Table 11.5 the inhibition can be evaluated using the screen-printed electrodes as transducers for amperometric measurements. These disposable electrode materials are of great interest because they provide the possibility of construction of simple portable devices for fast screening purposes and in-field/on-site monitoring (Tudorache and Bala, 2007). Rapid contamination monitoring is essential for food quality assurance, and biosensors appear to be a potential tool for commercial products in this field.

Table 11.5 Biosensors Applied in the Analysis of Pesticides: All Using Screen-Printed Electrode with Amperometric Transduction

Analyte	Application	Biocomponent	Detection range	Reference
Pesticide	Fruit	Acetylcholinesterase		Ramírez et al. 2008
Pesticide	Infant food	Acetylcholinesterase		Waibel et al. 2006
Pesticide	Fruit, vegetables	Acetylcholinesterase		Grosmanova et al. 2005
Pesticide	Wheat	Choline oxidase	0.05–40 µg mL^{-1} (dichlorvos)	Longobardi et al.2005
Pesticide	Wine fermentation	Acetylcholinesterase	7–26 ppb (parathion-methyl)	Suprun et al.2005
Pesticide	Grapes	Acetylcholinesterase	10–75 µgL^{-1} (Paraoxon)	Boni et al.2004
Pesticide	Infant food	Acetylcholinesterase	> 20 µg Kg^{-1} (parathion)	Schulze et al.2004
Pesticide	Grape juice	Acetylcholinesterase	0.1–1.5 mmol L^{-1}	Ivanov et al. 2003

In spite of the great number of publications on biosensors applied in food analysis, only a few biosensors are commercially available. The major limitations are the limited lifetime of the biological component, the dependence on free-diffusing mediators, and the requirement of coenzymes. (Castillo et al., 2004).

11.4 Conclusions

It is apparent that biosensors are now increasingly becoming a part of the mainstream real-time analytical measurement tools. This chapter presents some of the developed works that are related to the use of enzyme-based biosensors as analytical tools in fruit and vegetable processing, and their applications in the development of analytical methods for quality control. In general, the use of enzymatic biosensors for fruit and vegetable analysis can provide a route to a specific, sensitive, rapid, and in some cases inexpensive method for monitoring a range of target analytes. It was observed that the electrochemical transducers are employed in most of the biosensors. However, as has been well observed, enzymatic biosensor technology even in this area still needs improvement.

Nowadays, enzymatic biosensors can be used in many processes. Biosensors and biosensor-related techniques have several obstacles commercial use to their such as the diversity of compounds and the complexity of matrices in samples, the variability in data quality requirements, and the broad range of possible monitoring. In this sense, research has been performed to get enzymes with better stability and find a way of obtaining enzymes at a low cost (Mao et al., 2008). The biomimetic approaches (Martín-Esteban, 2004; Ye and Haupt, 2004) as well as abiotic biosensors have attracted great attention and they will be a good alternative to expand the application in the food area.

In the food sector, one of the most important problems is the time-consuming and laborious process of food quality-control analysis. Innovative devices and techniques are being developed that can facilitate the preparation of food samples and their precise and inexpensive analysis. From this point of view, the development of nanosensors to detect microorganisms and contaminants is a particularly promising application of nanotechnology (Sozer and Kokini, 2009).

Abbreviations

AChE: acetylcholinesterase
Amp: amperometric
BuChE: butyryl cholinesterase
CHEMFET: chemical field-effect transistor
FIA: flow injection analysis

GDH:	glycerol phosphate dehydrogenase
ISFET:	ion-sensitive field effect transistors
LDH:	lactate dehydrogenase
ME:	malic enzyme
Poten.:	potentiometric
QCM:	quartz crystal microbalance

References

Abayomi, L. A., L. A. Terry, S. F. White, and P. J. Warner. 2006. Development of a disposable pyruvate biosensor to determine pungency in onions (*Allium cepa* L.). *Biosensors and Bioelectronics* 21: 2176–2179.

Albareda-Sirvent, M., A. Merkoçi, and S. Alegret. 2001. Pesticide determination in tap water and juice samples using disposable amperometric biosensors made using thick-film technology. *Analytica Chimica Acta* 442: 35–44.

Amine, A., H. Mohammadi, I. Bourais, and G. Palleschi. 2006. Enzyme inhibition-based biosensors for food safety and environmental monitoring. *Biosensors and Bioelectronics* 21: 1405–1423.

Arkhypova, V. N., S. V. Dzyadevych, A. P. Soldatkin, A. V. El'skaya, C. Martelet, and N. Jaffrezic-Renault. 2003. Development and optimisation of biosensors based on pH-sensitive field effect transistors and cholinesterases for sensitive detection of solanaceous glycoalkaloids. *Biosensors and Bioelectronics* 18: 1047–1053.

Badea, M., A. Curulli, and G. Palleschi. 2003. Oxidase enzyme immobilisation through electropolymerised films to assemble biosensors for batch and flow injection analysis. *Biosensors and Bioelectronics* 18: 689–698.

Basu, A. K., P. Chattopadhyay, U. Roychudhuri, and R. Chakraborty. 2006. A biosensor based on co-immobilized L-glutamate oxidase and L-glutamate dehydrogenase for analysis of monosodium glutamate in food. *Biosensors and Bioelectronics* 21: 1968–1972.

Bickerstaff, G. 1997. *Immobilization of Enzymes and Cells: Methods in Biotechnology*. New Jersey: Humana Press.

Boer, E., and R. R. Beumer. 1999. Methodology for detection and typing of foodborne microorganisms. *International Journal of Food Microbiology* 50: 119–130.

Boni, A., C. Cremisini, E. Magaro, M. Tosi, W. Vastarella, and R. Pilloton. 2004. Optimized biosensors based on purified enzymes and engineered yeasts: Detection of inhibitors of cholinesterases on grapes. *Analytical Letters* 37(8): 1683–1699.

Böyükbayram, A. E., S. Kiralp, L. Toppare, and Y. Yağci. 2006. Preparation of biosensors by immobilization of polyphenol oxidase in conducting copolymers and their use in determination of phenolic compounds in red wine. *Bioelectrochemistry* 69: 164–171.

Brooks, J. L., B. Mirhabibollahi, and R. G. Kroll. 1992. Experimental enzyme-linked amperometric immunosensors for the detection of Salmonella in foods. *Journal of Applied Bacteriology* 73: 189–196.

Campuzano, S., M. Gamella, B. Serra, A. J. Reviejo, and J. M. Pingarrón. 2007. Integrated electrochemical gluconic acid biosensor based on self-assembled monolayer-modified gold electrodes. Application to the analysis of gluconic acid in musts and wines. *Journal of Agricultural and Food Chemistry* 55: 2109–2114.

Castillo, J., S. Gaspar, S. Leth, M. Niculescu, A. Mortari, I. Bontidean, V. Soukharev, S. A. Dorneanu, A. D. Ryabov, and E. Csöregi. 2004. Biosensors for life quality: Design, development and applications. *Sensors and Actuators B* 102(2): 179–194.

Chaki, N. K. and K. Vijayamohanan. 2002. Self-assembled monolayers as a tunable platform for biosensor applications. *Biosensors and Bioelectronics* 17(1–2): 1–12.

Chang, S. S. C., K. Rawson, and C. J. Mcneil. 2002. Disposable tyrosinase-peroxidase bi-enzyme sensor for amperometric detection of phenols. *Biosensors and Bioelectronics* 17: 1015–1023.

Chaubey, A., and B. D. Malhotra. 2002. Mediated biosensor. *Biosensors and Bioelectronics* 17: 441–456.

Choi, M. M. F., M. M. K. Liang, and A. W. M. Lee. 2005. A biosensing method with enzyme-immobilized eggshell membranes for determination of total glucosinolates in vegetables. *Enzyme and Microbial Technology* 36: 91–99.

Collings, A. F., and F. Caruso. 1997. Biosensors: recent advances. *Reports on Progress in Physics* 60: 1397–1445.

Csöregi, E., S. Gáspár, M. Niculescu, B. Mattiasson, and W. Schuhmann. 2001. *Focus on Biotechnology: Amperometric Enzyme-Based Biosensors for Application in Food and Beverage Industry*, Vol. 7, 105–129. The Netherlands: Kluwer Academic Publishers.

Davis, J., D. H. Vaughan, and M. F. Cardosi. 1995. Elements of biosensors construction. *Enzyme and Microbial Technology* 17: 1030–1035.

Derwinska, K., K. Miecznikowski, R. Koncki, P. J. Kulesza, S. Glab, and M. A. Malik. 2003. Application of Prussian blue based composite film with functionalized organic polymer to construction of enzymatic glucose biosensor. *Electroanalysis* 15: 1843–1849.

Dzyadevych, S. V., V. N. Arkhypova, A. P. Soldatkin, A. V. El'skaya, C. Martelet, and N. Jaffrezic-Renault. 2004. Enzyme biosensor for tomatine detection in tomatoes. *Analytical Letters* 37(8): 1611–1624.

ElKaoutit, M., I. Naranja Rodriguez, K. R. Temsamani, M. D. de la Vega, and J. L. Hidalgo-Hidalgo de Cisneros. 2007. Dual Laccase-tyrosinase based Sonogel-carbon biosensor for monitoring polyphenols in beers. *Journal of Agricultural and Food Chemistry* 55: 8011–8018.

Esti, M., G. Volpe, L. Micheli, E. Delibato, D. Campagnone, D. Moscone, and G. Palleschi. 2004. Electrochemical biosensor for monitoring malolactic fermentation in red wine using two strains of *Oenococcus oeni*. *Analytica Chimica Acta* 513: 357–364.

Feng, K., C. Sun, Y. Kang, J. Chen, J.-H. Jiang, G.-L. Shen, and R.-Q. Yu. 2008. Label-free electrochemical detection of nanomolar adenosine based on target-induced aptamer displacement. *Electrochemistry Communications* 10(4): 531–535.

Fitzpatrick, J., L. Fanning, S. Hearty, P. Leonard, B. M. Manning, J. Q. Quinn, and R. O'Kennedy. 2000. Applications and recent developments in the use of antibodies for analysis. *Analytical Letters* 33: 2563–2609.

Fox, P. F. 1991. *Food Enzymology*. Amsterdam, Netherlands: Elsevier Applied Science.

Freire, S. R., C. A. Pessoa, and L. T. Kubota. 2003. Self-assembled monolayers applications for the development of electrochemical sensors. *Química Nova* 26: 381–389.

Gandolfi, I., G. Palla, R. Marchelli, A. Dossena, S. Puelli, and C. Salvadori. 1994. D-alanine in fruit juices: a molecular marker of bacterial activity, heat treatments and shelf-life. *Journal of Food Science* 59: 152–154.

Ghindilis, A. L., P. Atanasov, and E. Wilkins. 1997. Enzyme-catalyzed direct electron transfer: Fundamentals and analytical applications. *Electroanalysis* 9: 661–674.

Grosmanova, Z., J. Krejci, J. Tynek, P. Cuhra, and S. Barsova. 2005. Comparison of biosensoric and chromatographic methods for the detection of pesticides. *International Journal of Environmental Analytical Chemistry* 85: 885–893.

Guilbault, G. G. 1982. Immobilized enzymes as analytical reagents. *Applied Biochemistry and Biotechnology* 7: 85–98.

Gupta, R. and N. K. Chaudhury. 2007. Entrapment of biomolecules in sol–gel matrix for applications in biosensors: Problems and future prospects. *Biosensors and Bioelectronics* 22: 2387–2399.

Habermüller, K., M. Mosbach, and W. Schuhmann. 2000. Electron-transfer mechanisms in amperometric biosensors. *Fresenius' Journal of Analytical Chemistry* 366: 560–568.

Hall, R. H. 2002. Biosensor technologies for detecting microbiological foodborne hazards. *Microbes and Infection* 4: 425–432.

Hamid, I. A., D. Ivnitski, P. Atanasov, and E. Wilkins. 1998. Fast amperometric assay for *E. coli* O157:H7 using partially immersed immunoelectrodes. *Electroanalysis* 10: 758–763.

Isaac, A., C. Livingstone, A. J. Wain, R. G. Compton, and J. Davis. 2006. Electroanalytical methods for the determination of sulfite in food and beverages. *TrAC: Trends in Analytical Chemistry* 25(6): 589–598.

Ivanov, A., G. Evtugyn, H. Budnikov, F. Ricci, D. Moscone, and G. Palleschi. 2003. Cholinesterase sensors based on screen-printed electrodes for detection of organophosphorus and carbamic pesticides. *Analytical and Bioanalytical Chemistry* 377: 624–631.

Ivekovic, D., S. Milardovic, and B. S. Grabaric. 2004. Palladium hexacyanoferrato hydrogel as a novel and simple enzyme immobilization matrix for amperometric biosensors. *Biosensors and Bioelectronics* 20: 872–878.

Jarosz-Wilkolazka, A., T. Ruzgas, and L. Gorton. 2004. Use of laccase-modified electrode for amperometric detection of plant flavonoids. *Enzyme and Microbial Technology* 35(2–3): 238–241.

Jawaheer, S., S. F. White, S. D. D. V. Rughooputh, and D. C. Cullen. 2003. Development of a common biosensor format for an enzyme based on biosensor array to monitor fruit quality. *Biosensors and Bioelectronics* 18: 1429–1437.

Jin, S., Z. Xu, J. Chen, X. Liang, Y. Wu, and X. Qian. 2004. Determination of organophosphate and carbamate pesticide based on enzyme inhibition using a pH-sensitive fluorescence probe. *Analytica Chimica Acta* 523: 117–123.

Kim, M. 2006. Determining citrate in fruit juices using a biosensor with citrate lyase and oxaloacetate decarboxylase in a flow injection analysis system. *Food Chemistry* 99: 851–857.

Kim, M. and M. J. Kim. 2003. Isocitrate analysis using a potentiometric biosensor with immobilized enzyme in a FIA system. *Food Research International* 36: 223–230.

Kindschy, L. M. and E. C. Alocilja. 2004. A review of molecularly imprinted polymers for biosensor development for food and agricultural applications. *Transactions of the American Society of Agricultural Engineers* 47(4): 1375–1382.

Kirgöz, U. A., D. Odaci, S. Timur, A. Merkoçi, S. Alegret, N. Beşün, and A. Telefoncu. 2006. A biosensor based on graphite epoxy composite electrode for aspartame and ethanol detection. *Analytica Chimica Acta* 570: 165–169.

Korpan, Y. I., F. M. Raushel, E. A. Nazarenko, A. P. Soldatkin, N. J. Renault, and C. Martelet. 2006. Sensitivity and specificity improvement of an ion sensitive field effect transistors-based biosensor for potato glycoalkaloids detection. *Journal of Agricultural and Food Chemistry* 54: 707–712.

Krawczyk, T. K., M. Moszczynska, and M. Trojanowicz. 2000. Inhibitive determination of mercury and other metal ions by potentiometric urea biosensor. *Biosensors and Bioelectronics* 15: 681–691.

Longobardi, F., M. Solfrizzo, D. Campagnone, M. D. Carlo, and A. Visconti. 2005. Use of electrochemical biosensor and gas chromatography for determination of dichlorvos in wheat. *Journal of Agricultural and Food Chemistry* 53: 9389–9394.

Lupu, A., D. Compagnone, and G. Palleschi. 2004. Screen-printed enzyme electrodes for the detection of marker analytes during winemaking. *Analytica Chimica Acta* 513: 67–72.

Maestre, E., I. Katakis, A. Narváez, and E. Domínguez. 2005. A multianalyte flow electrochemical cell: application to the simultaneous determination of carbohydrates based on bioelectrocatalytic detection. *Biosensors and Bioelectronics* 21: 774–781.

Majer-Baranyi, K., N. Adányi, and M. Váradi. 2008. Investigation of a multienzyme based amperometric biosensor for determination of sucrose in fruit juices. *European Food Research and Technology* 228(1): 139–144.

Mao, X.-L., J. Wu, and Y.-B. Ying. 2008. Application of electrochemical biosensors in fermentation. *Chinese Journal of Analytical Chemistry* 36(12): 1749–1755.

Martín-Esteban, A. 2004. Molecular imprinting technology: a simple way of synthesizing biomimetic polymeric receptors. *Analytical and Bioanalytical Chemistry* 378: 1875.

Mazzei, F., F. Botrè, and G. Favero. 2007. Peroxidase based biosensors for the selective determination of D, L-lactic acid and L-malic acid in wines. *Microchemical Journal* 87: 81–86.

Mello, L. D., and L. T. Kubota. 2002. Review of the use of biosensors as analytical tools in the food and drink industries. *Food Chemistry* 77: 237–256.

Mizutani, F., Y. Hirata, S. Yabuki, and S. Iijima. 2003. Flow injection analysis of acetic acid in food samples by using trienzyme/poly (dimethylsiloxane)-bilayer membrane-based electrode as the detector. *Sensors and Actuators, B: Chemical* 91: 195–198.

Mori, H. and S. Shiraki. 2008. Determination of D-malate using immobilized D-malate dehydrogenase in a flow system and its application to analyze the D-malate content of beverages. *Journal of Health Science* 54: 72–75.

Newman, J. D. and A. P. Turner. 1994. Biosensors the analyst's dream. *Chemistry & Industry* 16: 374–378.

Newman, J. D. and A. P. F. Turner. 2005. Home blood glucose biosensors: a commercial perspective. *Biosensors and Bioelectronics* 20: 2435–2453.

Newman, J. D., L. J. Tigwell, and P. J. Warner. 1998. Biotechnology strategies in healthcare: A transatlantic perspective. *Financial Times Report*, June.

Nikolelis, D. P., M. G. Simantiraki, C. G. Siontorou, and K. Toth. 2005. Flow injection analysis of carbofuran in foods using air stable lipid film based on acetylcholinesterase. *Analytica Chimica Acta* 537: 169–177.

Nunes, G. S., D. Barceló, B. S. Grabaric, J. M. Diaz-Cruz, and M. L. Ribeiro. 1999. Evaluation of a highly sensitive amperometric biosensor with low cholinesterase charge immobilized on a chemically modified carbon paste electrode for trace determination of carbamates in fruit, vegetable and water samples. *Analytica Chimica Acta* 399: 37–49.

Páscoa, R. M. N. J., S. S. M. P. Vidigal, I. V. Tóth, and A. O. S. S. Rangel. 2006. Sequential injection system for the enzymatic determination of ethanol in wine. *Journal of Agricultural and Food Chemistry* 54: 19–23.

Pauliukaite, R., M. Schoenleber, P. Vadgama, and C. M. A. Brett. 2008. Development of electrochemical biosensors based on sol-gel enzyme encapsulation and protective polymer membranes. *Analytical and Bioanalytical Chemistry* 390: 1121–1131.

Pauliukaite, R., G. Zhylyak, D. Citterio, and U. E. S. Keller. 2006. L-glutamate biosensor for estimation of the taste of tomato specimens. *Analytical and Bioanalytical Chemistry* 386: 220–227.

Phadke, R. S. 1992. Biosensors and enzyme immobilized electrodes. *Biosystems* 27: 203–206.

Polanka, M., D. Jun, H. Kalasz, and K. Kuca. 2008. Cholinesterase biosensor construction—A review. *Protein & Peptide Letters* 15(8): 795–798.

Prada, A. G. V., N. Penã, M. L. Mena, A. J. Reviejo, and J. M. Pingarrón. 2003. Graphite-Teflon composite bienzyme amperometric biosensors for monitoring of alcohols. *Biosensors and Bioelectronics* 18: 1279–1288.

Prodromidis, M. I., and M. I. Karayannis. 2002. Enzyme based amperometric biosensor for food analysis. *Electroanalysis* 14: 241–261.

Ramírez, G. V., D. Fornier, M. T. R. Silva, and J. L. Marty. 2008. Sensitive amperometric biosensor for diclorvos quantification: application to detection of residues on apple skin. *Talanta* 74: 741–746.

Rishpon, J., and D. Ivnitski. 1997. An amperometric enzyme-channeling immunosensor. *Biosensors and Bioelectronics* 12: 195–204.

Roos, A. D., C. Grassin, M. Herweijer, K. M. Kragh, C. H. Poulsen, J. B. Soe, J. F. Sorensen, and J. Wilms, 2004. Industrial enzymes: Enzymes in food application. In: *Enzymes in Industry*, Aehle, W. (ed.). Weinheim, Germany: Wiley-VCH Verlag GmbH & Co. KGaA.

Rosenberry, T. L. 1975. Acetylcholinesterase. *Advances in Enzymology and Related Areas of Molecular Biology* 43: 103–218.

Ruiz, J. G., A. A. Lomillo, and F. J. Muñoz. 2007. Screen-printed biosensors for glucose determination in grape fruit. *Biosensors and Bioelectronics* 22: 1517–1521.

Ruzgas, T., E. Csöregi, J. Emnéus, L. Gorton, and G. Marko-Varga. 1996. Peroxidase-modified electrodes: Fundamentals and application. *Analytica Chimica Acta* 330(2–3): 123–138.

Sato, N., and H. Okuma. 2006. Amperometric simultaneous sensing system for D-glucose and L-lactate based on enzyme-modified bilayer electrodes. *Analytica Chimica Acta* 565: 250–254.

Scheper, T. H., J. M. Hilmer, F. Lammers, C. Müller, and M. Reinecke. 1996. Biosensor in bioprocess monitoring. *Journal of Chromatography A* 725: 3–12.

Schulze, H., R. D. Schmid, and T. T. Bachmann. 2004. Activation of phosphorothionate pesticides based on a cytochrome P450 BM-3 (CYP102-A1) mutant for expanded neurotoxin detection in food using acetylcholinesterase biosensors. *Analytical Chemistry* 76: 1720–1725.

Scouten, W. H., J. H. T. Luong, and R. S. Brown. 1995. Enzyme or protein immobilization techniques for applications in biosensor design. *Trends in Biotechnology* 13(5): 178–185.

Sezgintürk, M. K., T. Göktuğ, and E. Dinçkaya. 2005. A biosensor based on catalase for determination of highly toxic chemical azide in fruit juices. *Biosensors and Bioelectronics* 21: 684–688.

Shkotova, L. V., T. B. Goriushkina, C. Tran-Minh, J.-M. Chovelon, A. P. Soldatkin, and S. V. Dzyadevych. 2008. Amperometric biosensor for lactate analysis in wine and must during fermentation. *Materials Science and Engineering C* 28: 943–948.

Soldatkin, A. P., V. N. Arkhypova, S. V. Dzyadevych, A. V. El'skaya, J. M. Gravoueille, N. J. Renault, and C. Martelet. 2005. Analysis of the potato glycoalkalloids by using of enzyme biosensor based on pH-ISFETs. *Talanta* 65: 28–33.

Song, S., L. Wang, J. Li, C. Fan, and J. Zhao. 2008. Aptamer-based biosensors. *TrAC—Trends in Analytical Chemistry* 27(2): 108–117.

Sozer, N. and J. L. Kokini. 2009. Nanotechnology and its applications in the food sector. *Trends in Biotechnology* 27(2): 82–89.

Suprun, E., G. Evtugyn, H. Budnikov, F. Ricci, D. Moscone, and G. Palleschi. 2005. Acetylscholinesterase sensor based on screen-printed carbon electrode modified with Prussian blue. *Analytical and Bioanalytical Chemistry* 383: 597–604.

Terry, L. A., S. F. White, and L. J. Tigwell. 2005. The application of biosensors to fresh produce and the wider food industry. *Journal of Agricultural and Food Chemistry* 53: 1309–1316.

Thevenot, D. R., K. Toth, R. A. Durst, and G. S. Wilson. 1999. Electrochemical biosensors: Recommended definitions and classification. IUPAC, *Pure and Applied Chemistry* 71(12): 2333–2348.

Thevenot, D. R., K. Toth, R. A. Durst, and G. S. Wilson. 2001. Electrochemical biosensors: recommended definitions and classification. *Biosensors and Bioelectronics* 16: 121–131.

Tothill, I. E. 2001. Biosensors development and potential applications in the agricultural diagnosis sector. *Computers and Electronics in Agriculture* 30: 205–218.

Tsiafoulis, C. G., M. I. Prodomidis, and M. I. Karayannis. 2003. Development of a flow amperometric enzymatic method for the determination of total glucosinolates in real samples. *Analytical Chemistry* 75: 927–934.

Tudorache, M., and C. Bala. 2007. Biosensors based on screen-printing technology and their applications in environmental and food analysis. *Analytical and Bioanalytical Chemistry* 388: 565–578.

Updike, S. J., and G. P. Hicks. 1967. Enzyme electrode. *Nature* 214: 986–988.

Vermeir, S., B. M. Nicolaï, P. Verboven, P. Van Gerwen, B. Baeten, L. Hoflack, V. Vulsteke, and J. Lammertyn. 2007. Microplate differential calorimetric biosensor for ascorbic acid analysis in food and pharmaceuticals. *Analytical Chemistry* 79: 6119–6127.

Vo-Dinh, T., B. M. Cullum, and D. L. Stokes. 2001. Nanosensors and biochips: frontiers in biomolecular diagnostics. *Sensors and Actuators, B: Chemical* 74: 2–11.

Wagner, G., and G. G. Guilbault. 1984. *Food Biosensor Analysis*. New York: Marcell Dekker.

Waibel, M., H. Schulze, N. Huber, and T. T. Bachmann. 2006. Screen-printed bienzymatic sensor based on sol-gel immobilized *Nippostrongylus brasiliensis* acetylcholinesterase and a cytochrome P450 BM-3 (CYP102-A1) mutant. *Biosensors and Bioelectronics* 21: 1132–1140.

Wang, J., E. Palecek, P. E. Nielsen, G. Rivas, X. Cai, H. Shiraishi, N. Dontha, D. Luo, and P. A. M. Farias. 1996. Peptide nucleic acid probes for sequence-specific DNA biosensors. *Journal of the American Chemical Society* 118: 7667–7670.

Wcislo, M., D. Campagnone, and M. Trojanowicz. 2007. Enantioselective screen-printed amperometric biosensor of the determination of D-amino acids. *Bioelectrochemistry* 71: 91–98.

Wong, C. M., K. H. Wong, and X. D. Chen. 2008. Glucose oxidase: natural occurrence, function, properties and industrial applications. *Applied Microbiology and Biotechnology* 78: 927–938.

Wu, B., G. Zhang, S. Shuang, and M. M. F. Choi. 2004. Biosensor for determination of glucose with glucose oxidase immobilized on an eggshell membrane. *Talanta* 64: 546–553.

Yang, X., Z. Zhou, D. Xiao, and M. M. F. Choi. 2006. A fluorescent glucose biosensor based on immobilized glucose oxidase on bamboo inner shell membrane. *Biosensors and Bioelectronics* 21: 1613–1620.

Ye, L. and K. Haupt. 2004. Molecularly imprinted polymers as antibody and receptor mimics for assays, sensors and drug discovery. *Analytical and Bioanalytical Chemistry* 378: 1887–1897.

Zhang, S., G. Wright, and Y. Yang. 2000. Materials and techniques for electrochemical biosensor design and construction. *Biosensors and Bioelectronics* 15: 273–277.

chapter twelve

Enzymes in fruit and vegetable processing
Future Trends in Enzyme Discovery, Design, Production, and Application

Marco A. van den Berg, Johannes A. Roubos,
and Lucie Pařenicová

Contents

12.1   Introduction

The world annual production of fruits and vegetables is almost 1400 million tons (fruits, 42%; vegetables, 58%). Fresh fruits and vegetables are of course eaten directly, but a large part of the yearly harvest is processed towards a variety of daily consumer goods such as fruit juice, wine, tomato sauce, canned fruits, and vegetables. For example, the global production of citrus fruits in 2007 was around 115 million tons, including 64 mtons of oranges (FAO Foodstat, http://faostat.fao.org/), of which 2/3 is eaten fresh and 1/3 is processed. This ratio is similar for apples (yearly 64 mtons), bananas (81 mtons), and grapes (66 mtons); together the top four fruits produced in large quantities.

Enzymes play an important role in the processing of fruits and vegetables. In the United States alone, approximately 53% of the fruits on the market are processed: ~6% is canned, ~42% is juiced, ~2% is frozen, and ~4% is dried (U.S. Department of Agriculture, http://www.ers.usda.gov/Data/FoodConsumption/FoodAvailSpreadsheets.htm/). Due to the fast growth of the world population during the last two millennia, the number and type of food products have increased (i.e., more stable and convenience food products). Juices and other types of processed fruits and vegetables were developed to satisfy that need. The manufacturing of juices involves extraction of the liquid fraction and subsequent preservation for prolonged storage, resulting in either a clear or cloudy product. In the 1930s the use of enzymes to facilitate the filtering of extracted juices was introduced, with Pectinol K (of Röhm & Haas) as one of the first products launched for the production of clear apple juice. Enzymes often lead to high cost reduction since less mechanical energy is required for processing, whereas juice yield might increase several percentages due to specific cell wall degradation by enzymes. Over the past decades, the use of enzymes for fruit and vegetable processing has grown into a mature industry with annual sales over $50 million. The range of applied enzymes (Table 12.1) has grown to increase yield during manufacturing but also to develop new products, applications, and health ingredients.

Enzyme applications in fruit and vegetable processing are well accepted and studied, but there are still many opportunities for further improvement and new applications. Increased knowledge on the structural and kinetic behavior of enzymes, enzymes from alternative sources (biodiversity), and improved production processes will decrease overall cost-prices and pave the way for future industrial applications.

Table 12.1 Main Enzyme Classes Used in Industrial Fruit
and Vegetable Processing

Enzyme Class	Major application areas
Pectinase	Maceration
	Juices and wine extraction and clarification
	Peeling
Neutral cellulase	Vegetable juices
Amyloglucosidase	Hydrolysis of starch
Alpha-amylase	Hydrolysis of starch
Glucose oxidase	Juice processing
Ferulic acid esterase	Maceration

12.2   Discovery of enzymes for fruit processing

Although enzymes are applied in minor quantities during the various fruit processing applications, they are an essential part of the industry. The enzymes are produced by classical fermentations using various microbes (i.e., bacilli, filamentous fungi, and yeasts). Historically, the enzymes and the corresponding production hosts were discovered from analyzing fruit and vegetable processing lines (see for example Etchells et al., 1958). This has developed into an efficient industrial manufacturing of a wide range of enzymes, wherein the yields were boosted via classical mutagenesis and process optimization. With the development of molecular biology in the 1970s and genomics in the 1990s, discovery of new enzymes entered a new era. Today researchers can access online databases to facilitate the search, identification, and development of new enzymes.

12.2.1   Genome sequencing

Fungi and bacilli are broadly exploited for production of homologous plant degrading enzymes (such as carbohydrolases, proteases, and lipases) and their genomes encode many previously unknown enzyme activities. Specifically, fungal enzymes perform particularly well in industrial applications. For example, during fruit processing the enzymes need to function in acid pH environments, which is ideal for the acidic fungal pectinases but less suitable for most basic bacterial and plant pectinases (see for example Duvetter et al., 2006).

One of the first genome mining projects for food processing enzymes was done with the filamentous fungus *Aspergillus niger*, which is able to secrete large amounts of a wide variety of enzymes and metabolites. In nature, the enzymes are needed to release nutrients from complex biopolymers while metabolite excretion gives the fungus a competitive advantage. These natural characteristics are exploited by industry in both solid state and submerged fermentations for the production of polysaccharide-degrading enzymes (particularly amylases, pectinases, and xylanases) and organic acids (mainly citric acid). *A. niger* has a long history of safe use (Van Dijck et al., 2003; Schuster et al., 2002; Van Dijck, 2008); therefore it is an ideal host for the producing of a range of food-grade enzymes.

The genome sequence of CBS 513.88, an *A. niger* strain used for industrial enzyme-production, was published in 2007 (Pel et al., 2007; Cullen, 2007). CBS 513.88 is the ancestor of currently used industrial enzyme production strains. The strain was derived from *A. niger* NRRL 3122, a classically improved strain selected for increased glucoamylase A production. The genome data were used for a systematic identification of *A. niger* enzyme-coding genes. Strong function predictions were made for 6,506 of the 14,165 open reading frames identified, which confirmed that aspergilli contain a wide spectrum of enzymes for polysaccharide, protein, and lipid degradation. For example, 88 putative pectinase encoding genes were discovered, of which ~2/3 were novel (Martens-Uzunova et al., 2006; Pel et al., 2007; Table 12.2). The identification of this wide range of new genes enabled the targeted development of new enzymes for food processing applications, facilitated by a fast and controlled development of dedicated production strains (see paragraph 12.4). The availability of more genome sequences of species well capable of degrading plant materials like *Trichoderma reesei* (Martinez et al., 2008) will further boost the discovery of new enzymes.

12.2.2   Omics-facilitated enzyme discovery

The genome sequencing efforts initiated a number of new genome-based investigations: transcriptomics, proteomics, metabolomics, and fluxomics. Basically, all these tools are to facilitate the application of the genome sequences for (1) new enzyme discovery and (2) strain and process improvement. DNA micro arrays can be used to measure the transcription of genes that play an important role under the testing conditions. These give a detailed snapshot of cell physiology and indicate which genes are encoding the active enzymes. On average 6000–8000 genes show detectable transcript levels (see Pel et al., 2007). Next, proteome analysis of intracellular and extracellular samples is applied to create a (quantitative) list of protein levels (Jacobs et al., 2009). To characterize a pectinase mixture (Figure 12.1), data obtained from multiple fermentation regimes (varying temperature, pH, feed, medium composition, etc.) can be used to understand the inducing factors and subsequently used to influence the composition of the mixture in the right direction. Furthermore, not all genome-encoded pectinase genes are expressed (Table 12.2). By selective cloning and overexpression (see paragraph 12.4) it is now possible to test and evaluate new enzymes rapidly.

Table 12.2 Genes Encoding Pectinase Degrading Enzymes in the *A. niger* Genome, Transcription, and Expression in a Classical Pectinase-Producing Strain

Enzyme	Gx (#)	Tx (#)	Px (#)	CAZy Classification
Endo-polygalacturonase	8	8	5	Glycoside Hydrolase Family 28
Exo-polygalacturonase	6	5	1	Glycoside Hydrolase Family 28
Pectin lyase	5	5	1	Polysaccharide Lyase Family 1
Pectate lyase	1	1	1	Polysaccharide Lyase Family 1
Pectin methyl esterase	3	2	2	Carbohydrate Esterase Family 8
Rhamnogalacturonase	6	3	1	Glycoside Hydrolase Family 28
Rhamnogalacturonolyase	2	2	1	Polysaccharide Lyase Family 4
Rhamnogalacturan acetyl esterase	1	1	0	Carbohydrate Esterase Family 12
Pectin acetyl esterase	4	4	3	Carbohydrate Esterase Family 12
Ferulic acid esterase	8	4	0	Carbohydrate Esterase Family 1
Arabinase	6	3	1	Glycoside Hydrolase Family 43
α-Arabinofuranosidase	6	4	2	Glycoside Hydrolase Family 3, 51, 54
α-Galactosidase	6	5	0	Glycoside Hydrolase Family 27, 36
ß-Galactosidase	8	4	3	Glycoside Hydrolase Family 35
Galactanase	2	1	1	Glycoside Hydrolase Family 53
α-Rhamnosidase	8	4	0	Glycoside Hydrolase Family 78
α-Fucosidase	1	1	0	Glycoside Hydrolase Family 29
α-Xylosidase	1	1	0	Glycoside Hydrolase Family 31
ß-Xylosidase	3	2	0	Glycoside Hydrolase Family 43
α-Glucoronidase	1	0	0	Glycoside Hydrolase Family 67
ß-Glucoronidase	2	1	0	Glycoside Hydrolase Family 2

Note: Gx (#), number in genome; Tx (#), number visible in transcriptome; Px (#) number visible in proteome; CAZy classification according to CAZy database—Carbohydrate-Active Enzymes database (http://www.cazy.org/).

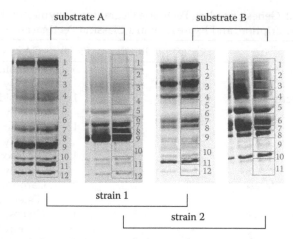

Figure 12.1 Search for cell wall-degrading enzymes by comparative proteomics. Total lane digestion of supernatant and subsequent analysis by LC-MS/MS allows for detection of >50 enzymes.

12.3   Design of enzymes for fruit and vegetable processing

12.3.1   Structure-function relation

Pectins have the most complex structure from all known polysaccharides and the commercial pectinases applied for pectin degradation are a mixture of various enzymes. Detailed knowledge on substrate-enzyme interactions and enzyme kinetics facilitate further improvement and applications, but also the identification of putative new enzymes. For example, the active sites and critical amino acid residues of enzymes like pectin lyase (Sanchez-Torres et al., 2003) and endopolygalacturonase I (van Pouderoyen et al., 2003) have been described. These findings will facilitate the design of optimized enzymes, which then can be produced in large quantities in suitable hosts like Aspergilli (Archer, 2000).

Improved knowledge on plant-borne inhibitors of pectinases, like the proteinous pectinmethylesterase and polygalacturonase inhibitors (PMEI and PGIP, respectively, Giovane et al., 2004; Di Matteo et al., 2006), might help in overcoming two major issues in fruit and vegetable processing: loss of firmness in canned products and cloud-loss in pulp-containing juices. Plants have their own set of pectin-degrading enzymes, and these inhibitors, when present, could inhibit the softening of the products during storage. Otherwise, it has been shown that addition of PMEI to non-pasteurized orange juice prevented loss of cloudiness during storage (Castaldo et al., 2006). The structural interactions and relevant amino acid residues for binding are known, which will help in further fine-tuning the application of these inhibitors during fruit processing.

12.3.2   Tailor-made enzymes

Structural knowledge is used to improve the activity of enzymes towards their substrates. A recent example is available for pectin methylesterases, catalyzing the removal of methyl esters from the homogalacturonan backbone domain of pectin. The degree of methyl esterification determines if homogalacturonan is susceptible to cleavage by pectin lyase and polygalacturonase. Øbro et al. (2009) screened a library of 99 variant enzymes in which seven amino acids were altered by various different substitutions to identify the most critical amino acids and used the knowledge for optimization of the enzyme (i.e., pH spectrum, themolability, and thermostability). Recent developments like codon optimization of enzyme-encoding genes and synthetic biology will be used to design the most optimal enzyme-coding genes.

12.3.3   High-throughput screening for improved functionality

Optimally, screening for improved classical enzyme producers should be done under the actual application conditions rather than in a well-defined biochemical assay. However, it is not an easy task to develop such a complex screening assay for pectinases. Ideally, one would have a small depolymerization (depectinization) test, but this would lead to gel formation in a microtiter plate and thus prevent any further analysis. In several cases, a classical plate screening assay using pectin as a substrate is still used as an initial screening, like the ruthenium red assay (Taylor et al., 1988) or a CuSO₄ overlay (Figure 12.2), allowing fast and efficient screening of millions of mutants. Another approach can be applied when screening for

A

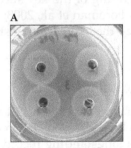

B

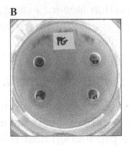

Figure 12.2 An example of plate screening assay for endopolygalacturonase–supernatant analysis: (A) polygalacturonic acid as substrate, McIlvain buffer pH = 6, CuSO₄ overlay; (B) polygalacturonic acid as substrate, 50 mM NaAc pH = 4.2, ruthenium red overlay.

specific pectinolytic activities—for instance, for endopolygalacturonases, pectin or pectate lyases, pectin methyl esterases, and others. Testing expression libraries of variant enzymes induced the development of medium to high-throughput methods. For the screening of the variant pectin methylesterases, Øbro et al. (2009) developed a microarray-based approach. Each mutant was incubated with a highly methyl-esterified lime pectin substrate and the samples were analyzed with an antibody that preferentially binds to homogalacturonans with a high degree of methyl esterification, allowing the rapid and correct identification of mutants. Another method allowing identification of a number of pectinolytic activities is the highly sensitive bicinchoninic acid (BCA) reducing value assay further adjusted by Meeuwsen et al. (2000) for screening of producing cells.

12.4   Industrial production of enzymes

The 2007 overview of Association of Manufacturers and Formulators of Enzyme Products (www.amfep.org) lists fungal species like *A. niger* and *A. oryzae* as main producers of industrial enzymes. Although these fungi can produce homologous proteins in dozens of grams per liter of fermentation broth, the production of heterologous proteins remains difficult. The current status and main aspects, such as proteolysis, secretion stress, mRNA processing, and so on, are well summarized by Lubertozzi and Keasling (2009). *A. niger* is broadly exploited for production of homologous (such as carbohydrolases, proteases, and lipases) and heterologous (such as lipases) enzymes.

Bacilli are another class of known good producers of homologous (like proteases and carbohydrolases) and a few heterologous (e.g., amylases) enzymes. Classical enzyme products, such as pectinases (e.g., Rapidase®), are being produced by diversity of prokaryotes and eukaryotes, for which production titers were optimized via strain mutagenesis and rational selection. Currently, industrial enzyme producers are using the available genome sequences for rapid understanding of the key success factors in enzyme production (see for examples Foreman et al., 2003; Guilemette et al., 2007; Jacobs et al., 2009) to optimize the expression hosts for homologous and heterologous enzymes.

12.4.1   Developing high-producing cell lines

To achieve cost-effective production of enzymes that fulfill all food safety requirements, several aspects have to be addressed. The most important are (1) use of a production host with a longstanding record of safe use in the biotech industry, (2) an expression vector that ensures high and stable expression of the enzyme under production conditions, and (3) a reproducible fermentation and downstream processing protocol.

To develop an efficient producing cell line, DSM started from a wild type *A. niger* isolate NRRL3122, which has been improved for decades, first by classical means for the production of glucoamylase (and acid amylase) and secondly by targeted genetic engineering, leading to its current production strains. The strain lineage has the name GAM (abbreviation of the enzyme name glucoamylase). The genetic analysis of the latest isolate of the GAM lineage, *A. niger* DS03043, showed that part of the improvement of glucoamylase production is due to the increased gene copy number (seven *gla*A genes), an event that is commonly observed in production strains that have undergone strain improvement by classical mutation and selection techniques (van Dijck et al., 2003). This strain was subsequently genetically modified to obtain a glucoamylase empty strain by deletion of the seven *gla*A loci in such a way that the empty loci could be individually detected (see van Dijck et al., 2003). This empty strain was additionally modified by inactivating the major extracellular aspartic protease *pep*A that led to a decrease of proteolysis and improved production capability. These strains are used to generate production strains for various enzymes and were approved as self-cloned by Dutch authorities (van Dijck et al., 2003).

As mentioned above, the second aspect for developing a robust and safe production system is an expression vector that ensures a high and stable expression of the gene of interest. In the case of DSM's *A. niger* PluGbug™, the *gla*A gene components—the *gla*A promoter and the *gla*A terminator—and the empty amplified *gla*A loci were exploited for this purpose (Figure 12.3). The gene of interest, either PCR amplified or from synthetic origin, is cloned behind the strong *gla*A promoter. After removing the *E. coli* part of the plasmid, it is transformed together with the *amd*S selection marker gene to *A. niger*. Using the *amd*S gene encoding acetamidase, the selection of transformants is done without antibiotics, thus ensuring the absence of any antibiotic marker in the production strain. As the expression vector contains the *gla*A 3' and 3" fragments, the expression cassette is targeted to one of the seven empty *gla*A loci. These loci are strongly expressed genomic loci. Subsequently the *amd*S marker is removed by forced recombination leading to a production strain containing solely copies of the gene of interest (Selten et al., 1995, 1998). After removing the selection marker, transformants are selected that usually contain more than five copies of the gene in one *gla*A locus. The further increase of the copy number of the gene of interest up to twenty and more occurs spontaneously via the process called gene conversion. Strains can be selected in which up to all seven *gla*A loci are filled with multiple copies of the gene of interest (see for details van Dijck et al., 2003). The final production strains are genetically checked and approved for use on large scale.

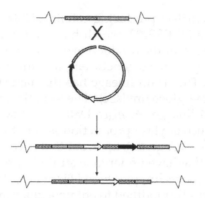

Figure 12.3 Example of the marker-gene free insertion of an expression unit. The expression unit, a linear piece of DNA, is integrated into one of the seven *glaA* loci by homology of the 3' and 3" regions. By varying the conditions of transformation multiple copies of the gene of interest arranged in tandem can be integrated in a single *glaA* locus. By selection on agar plates containing acetamide as the sole carbon source the transformants are selected. By counter-selection on agar plates containing fluoro-acetate variants can be selected from these transformants which have lost the *amdS* marker but which have retained (multiple copies of) the gene of interest. Legend: 3' *glaA* region, heavy dots; 3" *glaA* region, light dots; *amdS* gene, black arrow; *glaA* promoter, gray region; gene-of-interest, white arrow. (van Dijck, P. W. M., G. C. M. Selten, and R. A. Hempenius. 2003. On the safety of a new generation of DSM *Aspergillus niger* production strains. *Regulatory Toxicology Pharmacology* 38: 27–35. With permission.)

12.4.2   Codon optimization

An important aspect in the production of enzymes is the yield on the supplied feedstocks. Recent developments show that while maintaining the amino acid composition of the enzyme intact, it is possible to have significant improvements in yield by optimizing the codon usage (Rocha & Danchin, 2004). The codon usage and consequently the presence of corresponding tRNAs can differ significantly, even between closely related species. Optimization is often essential for obtaining good expression levels of proteins of heterologous origin in an expression host. The first applications in fungal products have been reported (Tokuako et al., 2008; Roubos et al., 2006; Roubos & van Peij, 2008) and further examples will follow.

12.4.3   Enzyme production and purification

Industrial enzymes are produced in highly robust and reproducible fermentations up to 200.000 l. Substrates range from defined ingredients to undefined ingredients, i.e., by-products from the food-industry as molasses, whey, cellulose, soybean, fish meal, yeast extract, etc. Depending on the actual product, many culture conditions are applied as some enzymes are degraded by certain proteases expressed by the host or are very labile at certain pH values. The exact protocols often remain the company's knowhow. Partial standardization can be obtained by using the same host for various products; in that case the difference between the different production strains is basically "only" the gene of interest (the genetic background of the strain and the expression vectors remain the same). This allows for a fast scale-up of the production process once a new production strain is developed. Using the identified pectinases from the available *A. niger* genome information and the standardized cell line generation technology described above, DSM screened the pectinolytic enzymes in *A. niger* to develop the Rapidase Smart self-cloned product. Compared to the classical pectinases, Rapidase Smart leads to a better product quality: slightly higher yield (+1–2%), no over-maceration, decreased stickiness of pomace, no increase in galacturonic acid or cellobiose in the juice (Figure 12.4), and no undesired side activities. Moreover, detailed life cycle assessment showed a significant decrease in the carbon footprint of the whole apple juice process.

Further fine-tuning of production strains and processes by industry is currently steered by transcriptomic and proteomic studies (Foreman et al., 2003; Jacobs et al., 2009). Leads are efficiently followed-up in mutants with improved gene targeting due to disruption of the non-homologous end-joining (NHEJ) repair pathway (Meyer et al., 2007), resulting in host strains that show lower degradation of heterologous enzymes (Jacobs et al., 2009).

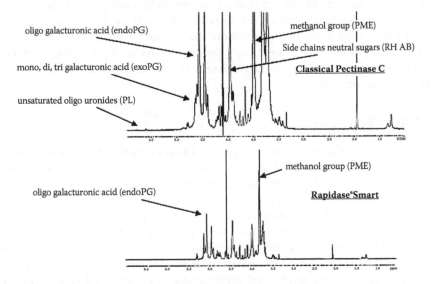

Figure 12.4 NMR analysis of reaction products after pectinase treatments of apple pectin. Classical pectinase compared with Rapidase Smart shows a much clearer product profile when compared to a classical pectinase.

Application of these approaches will lead to a further improvement in yield and purity of products.

Several well-known separation techniques are applied at an industrial scale, depending on the local infrastructure, the production host, and the sensitivity of the enzyme towards certain techniques. Most commonly applied is plate filtration (with or without a filter aid like dicalyte), but nowadays cross-flow microfiltration is used more and more, as well as centrifugation. The further purification steps depend strongly on the final quality needs: ultrafiltration and, if needed, chromatography.

12.5 Enzyme applications trends in fruit and vegetable processing

In recent years many new developments have been observed in the application of enzymes in fruit and vegetable processing. Several examples are summarized below.

12.5.1 Citrus peeling

The first step in the preparation of citrus juices is the peeling of the fruits. This is a mechanical process requiring energy. Pectinases are used to soften the peel by disruption of the albedo and thereby facilitate a significant reduction in energy costs. Current industrial practice is starting with pectinase treatment of whole fruit, followed by a vacuum infusion treatment with a pectinase solution like Rapidase® Intense (DSM) and Peelzyme (Novozymes) containing pectinesterase and polygalacturonase; thereafter the peel can be removed easily (1–2 hours) and the enzyme solution can be recycled.

12.5.2 Whole fruits

Processing of whole fruit or fruit parts requires several precautions to safeguard the firmness of the fruits. Pectins consist of very complex structures giving strength to fruits but are sensitive to mechanical pressure (shear), heating (chemical hydrolysis), pasteurization, storage (polymer dehydration), and osmotic pressure. Moreover, most pectinase preparations consist of multiple enzymes leading to weakening of the pectin polymers. For example, demethylation by pectinmethylesterase (PME) exposes the homogalacturonan backbone, which will be further degraded by enzymes like polygalacturonase (PG) and rhamnogalacturonase (RG), causing physical weakening of fruits. The FirmFruit® concept is based on the use of a PG- and RG-free PME (Rapidase® FP Super) in combination with calcium, which binds to the freed pectic acid *in situ* to form insoluble calcium pectates.

　　　Other developments are the inhibition of PG by plant-born PG inhibitors thereby preventing PG activity in the fungal enzyme mixtures (DiMatteo et al., 2006).

12.5.3　Immobilized enzymes

Traditionally, enzymes in the fruit and vegetable processing industry are applied as liquids or powders, while enzymes in pharma are often immobilized. Although it adds an additional step (and thus costs) in the preparation of the enzyme, improved characteristics like a lower pH optimum and increased half-life (through recycling) can turn this into an attractive opportunity. First examples are shown for polygalacturonase and tannase (Saxena et al., 2008; Sharma et al., 2008).

12.5.4　Preventing haze formation

There are several ways to clarify extracted juices. In fruit juices, this is traditionally done with bentonite, silica gel, or gelatine followed by filtration. Although these methods do work, they remove a considerable amount of the antioxidant phenolics, which can form haze-causing interactions with the proteins present. Landbo et al. (2006) showed that addition of gallic acid in combination with various proteases also reduced the haze formation drastically, but retained much more of the beneficial phenolics. The best-performing enzyme was Enzeco Fungal Acid Protease (Enzyme Development Corp) produced from *A. niger*, which reduces only 12% of the phenolics, while the traditional methods reduce up to 30%.

12.5.5　Preventing cloud loss

To produce cloudy juices, the process from extraction to pasteurization must be fast to minimize the effect of plant endogenous enzymes like pectinases and oxidases. Besides active inhibition of the plant enzymes by virtue of their inhibitors (see paragraph 12.3.1), α-amylases and glucoamylases can be added to reduce the starch content, as dissolved starch retrograde during cooled storage and can form precipitates.

12.5.6　Wine mouthfeel

Mannoproteins and other yeast compounds play a role in wine mouthfeel. Rapidase® Glucalees, an enzyme formulation based on a blend of pectinase from *Aspergillus niger* and β-(1,3)-D-glucanases from *Trichoderma harzianum*, is applied to increase the release of mannoproteins from the wine yeast.

12.5.7 Synergistic applications in red wine making

One of the latest developments is the combined application of enzymes and yeasts to improve the taste development and color stability of wines. For red wines color is crucial and depends on extraction yield, adsorption by lees, and stabilization. The combined application of a pectinase (Rapidase® MaxiFruit) and active dry yeast (Fermicru® XL) increase the color (stability) in red wines by optimal extraction and enzymatic conversion of grape phenolic constituents. Phenolic acids react with tartaric acid to form cinnamoyl-tartaric esters, which are hydrolyzed by the cinnamyl esterase of Rapidase MaxiFruit. The yeast cinnamate-decarboxylase then forms vinyl phenols, which react with the freed anthocyanins to form odorless, stable but color-rich pigments, the pyrano-anthocyanins (Figure 12.5).

12.5.8 Synergistic applications in white wine making

In contrast to red wine, white wines require limited polyphenol extraction. For this purpose Rapidase Expression, an enzyme naturally low in

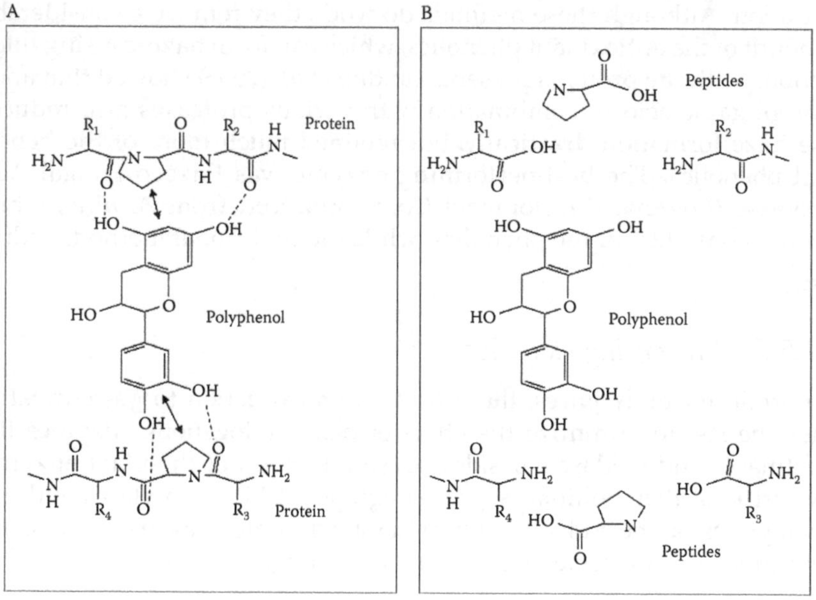

Figure 12.5 (A) Hypothesized chemistry between proteins and polyphenols of haze-causing hydrogen (dashed lines) and hydrophobic bonding (double headed arrows), respectively, in beer, wines, and fruit juices. (B) Proteases destroying the proteins prevent haze formation. (Adapted from Landbo, A. R. et al. 2006. Protease-assisted clarification of black currant juice: Synergy with other clarifying agents and effects on the phenol content. *Journal of Agricultural and Food Chemistry* 54: 6554–6563. With permission.)

cinnamyl-esterase and free of cellulases is applied. If used in combination with specific yeast strains like Collection Cépage® Sauvignon or Fermicru® 4F9, it is now possible to control the release and conversion of the thiol-precursors in the grapes into various aromas, leading to a 50–100% yield increase for 4-mercapto-4-methyl-pentan-2-one (4MMP), 3-mercaptohexanol (3MH), and 3-mercaptohexyl acetate (A3MH). The precursors could also be liberated via a higher maceration temperature, but that is not beneficial, as the extraction is not selective and the volatile thiols will evaporate.

12.5.9   Use of waste streams

Waste streams in the fruit and vegetable processing industry are significant and contain a range of valuable compounds, like carbohydrates, vitamins, and antioxidants. For example, 15–20 million ton of apple pomace is generated every year. The application of pectinases during maceration is crucial for economic extraction of these compounds (Bhushan et al., 2008). Fruits like berries contain a number of polyphenols and extraction by all sorts of commercial pectinases provide an increase in both antimicrobial and antioxidant activity of the obtained products (Puupponen-Pimiä et al., 2008). However, glycosidases can be present as side activity in pectinase solutions (Pricelius et al., 2009) and degrade these freed anthocyanins. Therefore, a fine-tuning of the applied enzyme mixture is needed to make optimal use of this potential in fruits.

Also, waste streams like apple pomace can be used as substrates for fermentation processes (reviewed by Vendruscolo et al., 2008). The potential of these applications will increase with the improvement of industrial cellulases needed for the pre-processing of the materials (see for example Heinzelman et al., 2009).

While acidophilic pectinases are used for extraction and clarification of fruit juices under acid conditions, alkaline pectinases can be applied under more neutral or basic conditions in the pre-treatment of waste streams from the food industry containing pectinaceous compounds (reviewed by Hoondal et al., 2002).

12.6   Conclusions

The fruit and vegetable industry is using advanced technologies to meet consumer demands for consistency and healthier products. Approximately 50% of the annual world production is processed (i.e., juiced, canned, frozen, or dried) using increasingly customized products containing enzymes. Enzymes, enzyme production, and enzyme applications are fine-tuned by using the latest technologies like directed evolution, codon optimization, and specialty products (like clever enzyme mixtures), respectively. This will lead to further improvements in taste, shelf life, and nutritional value of the derived products. Moreover, after adaptation of the processing conditions to these new enzymes and/or enzyme mixtures, the sustainability of the fruit and vegetable industry will be improved as the yields will be higher, while the waste streams and the net energy consumption will reduce.

Acknowledgments

We would like to thank our colleagues David Guerrand, Celine Fauveau, Patrice Pellerin, and JanMetske van der Laan for suggestions and comments.

Abbreviations

BCA	bicinchoninic acid
CBS	Centraal Bureau voor Schimmelcultures
GAM	glucoamylase
NRRL	Northern Regional Research Laboratories
PME	pectinmethylesterase
PMEI	pectinmethylesterase inhibitor
PG	polygalacturonase
PGIP	polygalacturonase inhibitor
PL	Pectin lyase
RG	rhamnogalacturonase
4MMP	4-mercapto-4-methyl-pentan-2-one
3MH	3-mercaptohexanol
A3MH	3-mercaptohexyl acetate
CAZy	Carbohydrate-Active Enzymes database

References

Archer, D. B. 2000. Filamentous fungi as microbial cell factories for food use. *Current Opinion in Biotechnology* 11: 478–483.

Bhushan, S., K. Kalia, M. Sharma, B. Singh, and P. S. Ahuja. 2008. Processing of apple pomace for bioactive molecules. *Critical Reviews in Biotechnology* 28: 285–296.

Castaldo, D., A. Lovoi, I. Quagliuolo, I. Servillo, C. Balestrieri, and A. Giovane. 2006. Orange juices and concentrates stabilization by a proteic inhibitor of pectin methylesterase. *Journal of Food Science* 56: 1632–1634.

Cullen, D. 2007. The genome of an industrial workhorse. *Nature Biotechnology* 25: 189–190.

Di Matteo, A., D. Bonivento, D. Tsernoglou, L. Federici, and F. Cervone. 2006. Polygalacturonase-inhibiting protein (PGIP) in plant defence: a structural view. *Phytochemistry* 67: 528–533.

Duvetter, T., I. Fraeye, D. N. Sila, I. Verlent, C. Smout, M. Hendrickx, and A. Van Loey. 2006. Mode of de-esterification of alkaline and acidic pectin methyl esterases at different pH conditions. *Journal of Agricultural and Food Chemistry* 54: 7825–7831.

Etchells, J. L., T. A. Bell, R. I. Monroe, P. M. Masley, and A. L. Demain. 1958. Populations and softening enzyme activity of filamentous fungi on flowers, ovaries, and fruit of pickling cucumbers. *Applied and Environmental Microbiology* 6: 427–440.

Foreman, P. K., D. Brown, L. Dankmeyer, R. Dean, S. Diener, N. S. Dunn-Coleman, et al. 2003. Transcriptional regulation of biomass-degrading enzymes in the filamentous fungus *Trichoderma reesei*. *Journal of Biological Chemistry* 278: 31988–31997.

Giovane, A., L. Servillo, C. Balestrieri, A. Raiola, R. D'Avino, M. Tamburrini, M. A. Ciardiello, and L. Camardella. 2004. Pectin methylesterase inhibitor. *Biochimica et Biophysica Acta-Proteins and Proteomics* 1696: 245–252.

Guillemette, T., N. N. M. E. van Peij, T. Goosen, K. Lanthaler, G. D. Robson, C. A. M. J. J. van den Hondel, H. Stam, and D. B. Archer. 2007. Genomic analysis of the secretion stress response in the enzyme-producing cell factory *Aspergillus niger*. *BMC Genomics* 8: article no. 158.

Heinzelman, P., C. D. Snow, I. Wu, C. Nguyen, A. Villalobos, S. Govindarajan, J. Minshull, and F. H. Arnold. 2009. A family of thermostable fungal cellulases created by structure-guided recombination. *Proceedings of the National Academy of Sciences of the United States of America* 106: 5610–5605.

Hoondal, G., R. Tiwari, R. Tewari, N. Dahiya, and Q. Beg. 2002. Microbial alkaline pectinases and their industrial applications: a review. *Applied Microbiology and Biotechnology* 59: 409–418.

Jacobs, D. I., M. M. A. Olsthoorn, I. Maillet, M. Akerotd, S. Breestraat, S. Donkers, et al. 2009. Effective lead selection for improved protein production in *Aspergillus niger* based on integrated genomics. *Fungal Genetics and Biology* 46(Supplement 1): S141–S152.

Landbo, A. R., M. Pinelo, A. F. Vikbjerg, M. B. Let, and A. S. Meyer. 2006. Protease-assisted clarification of black currant juice: synergy with other clarifying agents and effects on the phenol content. *Journal of Agricultural and Food Chemistry* 54: 6554–6563.

Lubertozzi, D., and J. D. Keasling. 2009. Developing *Aspergillus* as a host for heterologous expression. *Biotechnology Advances* 27: 53–75.

Martens-Uzunova, E. S., J. S. Zandleven, J. A. E Benen, H. Awad, H. J. Kools, G. Beldman, A. G. J. Voragen, J. A. Van Den Berg, and P. J. Schaap. 2006. A new group of exo-acting family 28 glycoside hydrolases of *Aspergillus niger* that are involved in pectin degradation. *Biochemical Journal* 400: 43–52.

Martinez, D., R. M. Berka, B. Henrissat, M. Saloheimo, M. Arvas, S. E. Baker, et al. 2008. Genome sequencing and analysis of the biomass-degrading fungus *Trichoderma reesei* (syn. *Hypocrea jecorina*). *Nature Biotechnology* 26: 553–560.

Meeuwsen, P. J., J. P. Vincken, G. Beldman, and A. G. Voragen. 2000. A universal assay for screening expression libraries for carbohydrases. *Journal of Bioscience and Bioengineering* 89: 107–109.

Meyer, V., M. Arentshorst, A. El-Ghezal, A. C. Drews, R. Kooistra, C. A. M. J. J. van den Hondel, and A. F. J. Ram. 2007. Highly efficient gene targeting in the *Aspergillus niger* kusA mutant. *Journal of Biotechnology* 128: 770–775.

Øbro, J., I. Sørensen, P. Derkx, C. T. Madsen, M. Drews, M. Willer, J. D. Mikkelsen, G. William, and T. Willats. 2009. High-throughput screening of *Erwinia chrysanthemi* pectin methylesterase variants using carbohydrate microarrays. *Proteomics* 9: 1861–1868.

Pel, H. J., J. H. de Winde, D. B. Archer, P. S. Dyer, G. Hofmann, P. J. Schaap, et al. 2007. Genome sequencing and analysis of the versatile cell factory *Aspergillus niger* CBS 513.88. *Nature Biotechnology* 25: 221–231.

Pricelius, S., M. Murkovic, P. Souter, and G. M. Guebitz. 2009. Substrate specificities of glycosidases from Aspergillus species pectinase preparations and elderberry anthocyanins. *Journal of Agricultural and Food Chemistry* 57: 1006–1012.

Puupponen-Pimiä, R., L. Nohynek, S. Ammann, K.-M. Oksman-Caldentey, and J. Buchert. 2008. Enzyme-assisted processing increases antimicrobial and antioxidant activity of bilberry. *Journal of Agricultural and Food Chemistry* 56: 681–688.

Rocha, E. P. C., and A. Danchin. 2004. An analysis of determinants of amino acids substitution rates in bacterial proteins. *Molecular Biology and Evolution* 21: 108–116.

Roubos, J. A., and N. N. M. E. Van Peij. 2008. A method for achieving improved polypeptide expression. *International Patent Application* WO/2008/000632.

Roubos, J. A., S. P. Donkers, H. Stam, and N. N. M. E. Van Peij. 2006. Method for producing a compound of interest in a filamentous fungal cell. *International Patent Application* WO/2006/077258.

Sánchez-Torres, P., J. Visser, and J. A. E. Benen. 2003. Identification of amino acid residues critical for catalysis and stability in *Aspergillus niger* family 1 pectin lyase A. *Biochemical Journal* 370: 331–337.

Saxena, S., S. Shukla, A. Thakur, and R. Gupta. 2008. Immobilization of polygalacturonase from *Aspergillus niger* onto activated polyethylene and its application in apple juice clarification. *Acta Microbiologica et Immunologica Hungarica* 55: 33–51.

Schuster, E., N. Dunn-Coleman, C. Frisvad, and P. W. M. van Dijck. 2002. On the safety of *Aspergillus niger*—a review. *Applied Microbiology and Biotechnology* 59: 426–435.

Selten, G. C. M., R. F. Van Gorcom, and B. M. Swinkels. 1995. Selection marker gene free recombinant strains, a method for obtaining them and the use of these strains. *International Patent Application* EP0635574.

Selten, G. C. M., B. M. Swinkels, and R. A. L. Bovenberg. 1998. Gene conversion as a tool for the construction of recombinant industrial filamentous fungi. *International Patent Application* WO/98/46772.

Sharma, S., L. Agarwal, and R. K. Saxena. 2008. Purification, immobilization and characterization of tannase from *Penicillium variable*. *Bioresource Technology* 99: 2544–2551.

Taylor, R. J., and G. A. Secor. 1988. An improved diffusion assay for quantifying the polygalacturonase content in *Erwinia* culture filtrates. *Phytopathology* 78: 1101–1103.

Tokuoka, M., M. Tanaka, K. Ono, S. Takagi, T. Shintani, and K. Gomi. 2008. Codon optimization increases steady-state mRNA levels in *Aspergillus oryzae* heterologous gene expression. *Applied and Environmental Microbiology* 74: 6538–6546.

van Dijck, P. W. M. 2008. The importance of Aspergilli and regulatory aspects of Aspergillus nomenclature. In: *Aspergillus in the Genomic Era*, J. Varga and R. A. Samson, eds., pp. 249–257. The Netherlands: Wageningen Academic Publishers.

van Dijck, P. W. M., G. C. M. Selten, and R. A. Hempenius. 2003. On the safety of a new generation of DSM *Aspergillus niger* production strains. *Regulatory Toxicology Pharmacology* 38: 27–35.

Van Pouderoyen, G., H. J. Snijder, J. A. Benen, and B. W. Dijkstra, 2003. Structural insights into the processivity of endopolygalacturonase I from *Aspergillus niger*. *FEBS Letters* 554: 462–466.

Vendruscolo, F., P. M. Albuquerque, F. Streit, E. Esposito, and J. L. Ninow. 2008. Apple pomace: a versatile substrate for biotechnological applications. *Critical Reviews in Biotechnology* 28: 1–12.

Index

"f" indicates material in figures. "t" indicates material in tables.

A

α-amino acid, 1
α-amylase, 14, 188, 343t, 353
α-arabinofuranosidase, 345t
α-farnesene, 23, 25, 58
α-farnesene synthase, 58
α-fucosidase, 345t
α-galactosidase, 345t
α-glucoronidase, 345t
α-helix, 1, 284, 288t
α-humulene, 60
α-keto acids, 58
α-ketobutylrate, 57
α-ketoglutarate, 328
α-phellandrene, 60–61
α-pinene, 60
α-rhamnosidase, 345t
α-rhamnoside, 2
α-terpinene, 61
α-terpineol, 225f
α-xylosidase, 345t
AAT, 48–52, 56, 60
Abrasion peeling, 27
Abscisic acid, 57–58, 93
ACC, 50, 52
ACC oxidase, 52, 55, 57, 95, 102
ACC synthase, 52, 55
Aceric acid, 78
Acetaldehyde, 55–56, 327
Acetamidase, 349
Acetate, 47–49, 51, 56, 57, 59–60, 355
Acetate kinase, 324t
Acetic acid, 234–235, 324t
Acetylase, 84
Acetylcholine, 331

Acetylcholinesterase, 325t, 326t, 330t, 331, 332t
Acetylmuramylhydrolase, 233
Acid dips, 151
Acid urease, 233
Actinidia deliciosa, 73; *See also* Kiwi fruit
Actinomycetes, 162
Activation energy, 10, 11, 132
Activation volume, 12
Activators, 7
Adenosine diphosphate (ADP), 62, 327
Adenosine triphosphate (ATP), 61–62, 327
ADH, 3t, 52–56, 58, 60, 324t, 327
Adhesion
    AGPs and, 80
    calcium and, 78, 89
    cell size and, 73
    cellulose and, 147
    HG and, 78
    HGA and, 89
    in orange, 149, 152
    pectin and, 147
    PL and, 86t
    texture and, 72, 73, 97
    in tomato, 89, 97
    turgor pressure and, 97
ADP, 62, 327
AFases, 84, 90, 345t
Affinity chromatography, 254t–258t, 276t–277t
Aftertaste, 61
Aglucone, 283
Aglycone, 226, 227
Akatsuki peach, 92
Alanine, 48, 50, 328
Albedo, 145–153, 148f, 154t, 155, 158–160

myrosinase and, 283
in pectin matrix, 78
in potato, 32–33
tests for, 324t
Glucose dehydrogenase, 324t
Glucose isomerase, 3t
Glucose oxidase, 3t, 183, 315, 320, 324t–326t,
    343t
Glucosidase, 222, 226–227
Glucosinolate, 32, 274t, 283, 324t–325t
Glucuronic acid, 78
Glucuronoarabinoxylan, 80
Glucuronoxylan, 79
Glutamate, 317, 325t
Glutamate dehydrogenase, 325t
Glutamate oxidase, 325t
Glutamine, 10, 50
Glutaraldehyde, 317, 323
Glutathione, 21–22, 25
Glycanase, 84
Glycerol, 327
Glycerol phosphate dehydrogenase,
    327–328
Glycine, 1, 80
Glycoalkaloid, 326t, 329
Glycoconjugate, 225
Glycohydrolase, 163–165
Glycoprotein, 75–76
Glycosidase, 23, 84, 162, 216–217, 226–227,
    355
Glycoside, 225–226, 235
Glycoside hydrolase, 345t
Gold, 317–318
Golden Delicious apple
    aroma of, 52
    expansins in, 92
    juice extraction from, 183
    LOX in, 52
    MdAAT2 gene cloned in, 49
    PME in, 265t, 276t
    texture of, 72f
Granny Smith apple, 23, 72f, 88
Grano de Oro orange, 151f
Grapefruit
    color of, 148
    composition of, 203–204
    Duncan, 152, 206t
    enzymatic peeling of, 147–155, 148f,
        154t, 156f, 162–168, 164t
    epicuticular wax in, 167
    Macfad, 152
    Marsh Seedless, 152
    Marsh White, 203

pressure, temperature, and, 255t
red, 150, 152, 203
texture of, 167
white, 203, 255t
Grapefruit juice
    attributes of, 152
    composition of, 205–207, 206t
    concentration of, 201
    CT for, 209
    D-value for, 205, 206t
    enzymatic peeling and, 147
    PEF and, 291t
    soluble solids in, 200
    taste of, 210
    z-value for, 205, 206t
Grape juice
    aroma and, 225
    biosensors for, 332t
    cellulases and, 183
    composition of, 231
    extraction of, 181
    in-line process for, 178f
    PEF and, 297t, 304t
    pH of, 219
    pressure, temperature, and, 259t
    pulp and, 182t
Grapes
    aroma of, 220, 225–226
    biosensors for, 332t
    Botrytis cinerea and, 231
    Cabernet S., 225f
    cellulose in, 216
    color of, 34, 182t, 220
    composition of
        isocitric acid, 328
        maceration and, 221f
        pectin, 221f
        sugars, 217f, 221f
        wine and, 216, 219–222, 226, 231, 235
    flavor of, 225
    in-line process for, 178f
    Monastrell, 216, 222, 225f
    must models, 34
    pressure, temperature, and, 259t–260t,
        270t
    production of, 342
    pulp of, 182t, 216, 219, 225
    red, 182t, 183
    rot in, 231
    seeds of, 220
    skin of
        aroma and, 220, 225
        color and, 220

Milton Keynes UK
Ingram Content Group UK Ltd.
UKHW021824071024
449327UK00021B/1423

9 780367 384128